THE NORTON HISTORY OF
TECHNOLOGY

DONALD CARDWELL was Professor of the History of
Science and Technology at the University of Manchester
Institute of Science and Technology (UMIST) from 1974 to
1984. Between 1954 and 1956 he was on the staff of the
British Association Committee on Science and Industry. He
lectured in the History and Philosophy of Science at Leeds
University, prior to moving to UMIST in 1963. He took a
major part in the movement to establish a science museum
on the UMIST campus. This has now developed into the
Manchester Museum of Science and Industry. He is an
Honorary Member and President (1991–93) of the
Manchester Literary and Philosophical Society. He has been
awarded the Dickinson Medal of the Newcomen Society
(London) and the Dexter Prize and Leonardo da Vinci
Medal of the Society for the History of Technology (USA).

His previous books include: *The Organization of Science in
England* (Heinemann, 2nd edn, 1980); *Steam Power in the
Eighteenth Century* (Sheed and Ward, 1963); *From Watt to
Clausius* (Heinemann and Cornell Univeristy Press, 1971;
reprinted, University of Iowa Press, 1989); *Technology Science
and History (Heinemann, 1972)*, published in the USA as
Turning Points in Western Technology (Neale Watson
Associates), as an Italian translation by Il Mulino, Bologna,
and as a Japanese translation by Kawada Shobo Shinsa;
and *James Joule: A Biography* (Manchester University Press,
1989). He has also edited two books: *John Dalton and the
Progress of Science* (Manchester University Press, 1968); and
Artisan to Graduate (Manchester University Press, 1974).

NORTON HISTORY OF SCIENCE
GENERAL EDITOR: ROY PORTER

already published:

Environmental Sciences PETER J. BOWLER
Chemistry W. H. BROCK
Technology DONALD CARDWELL
Astronomy and Cosmology JOHN NORTH

forthcoming:

Mathematics IVOR GRATTAN-GUINNESS
Physics L. PEARCE WILLIAMS
Biology ROBERT OLBY
Medicine ROY PORTER
Science in Society
LEWIS PYENSON & SUSAN SHEETS-PYENSON
Human Sciences ROGER SMITH

NORTON HISTORY OF SCIENCE
(Editor: Roy Porter)

THE NORTON HISTORY OF
TECHNOLOGY

Donald Cardwell

W·W·NORTON & COMPANY
New York   London

PREFACE TO
THE NORTON HISTORY OF SCIENCE

Academic study of the history of science has advanced dramatically, in depth and sophistication, during the last generation. More people than ever are taking courses in the history of science at all levels, from the specialized degree to the introductory survey; and, with science playing an ever more crucial part in our lives, its history commands an influential place in the media and in the public eye.

Over the past two decades particularly, scholars have developed major new interpretations of science's history. The great bulk of such work, however, has been published in detailed research monographs and learned periodicals, and has remained hard of access, hard to interpret. Pressures of specialization have meant that few survey works have been written that have synthesized detailed research and brought out its wider significance.

It is to rectify this situation that the Norton History of Science series has been set up. Each of these wide-ranging volumes examines the history, from its roots to the present, of a particular field of science. Targeted at students and the general educated reader, their aim is to communicate, in simple and direct language intelligible to non-specialists, well-digested and vivid accounts of scientific theory and practice as viewed by the best modern scholarship. The most eminent scholars in the discipline, academics well known for their skills as communicators, have been commissioned.

The volumes in this series survey the field and offer powerful overviews. They are intended to be interpretive, though not primarily polemical. They do not pretend to a timeless, definitive quality or suppress differences of viewpoint, but are meant to be books of and for their time; their authors offer their own interpretations of contested issues as part of a wider, unified story and a coherent outlook.

Carefully avoiding a dreary recitation of facts, each volume develops a sufficient framewrok of basic information to ensure that the beginner finds his or her feet and to enable student readers to use such books as their prime course-book. They rely upon chronology as an organizing framework, while stressing the importance of themes, and avoiding the narrowness of anachronistic 'tunnel history'. They incorporate the best up-to-the-minute research, but within a larger framework of analysis and without the need for a clutter of footnotes – though an attractive feature of the volumes is their substantial bibliographical essays. Authors have been given space to amplify their arguments and to make the personalities and problems come alive. Each volume is self-contained, though authors have collaborated with each other and a certain degree of cross-referencing is indicated. Each volume covers the whole chronological span of the science in question. The prime focus is upon western science, but other scientific traditions are discussed where relevant.

This series, it is hoped, will become the key synthesis of the history of science for the next generation, interpreting the history of science for scientists, historians and the general public living in a uniquely science-orientated epoch.

ROY PORTER
Series Editor

In Memory of My Friends

GORDON HESLING
Consultant Paediatrician (1920–1989)

WILFRED FARRAR
Historian of Chemistry (1920–1977)

CONTENTS

PREFACE

The history of technology begins in what has been called the eotechnic age with the simplest of tools and crafts. Indeed, the history of *Homo sapiens* begins with the first tools that archaeologists have found and it continues through the various cultures and civilizations up to the present age of the computer and space travel. This is the longest, the most general and, it can be claimed, the most basic of all forms of secular history. It is, or at least it can easily be made, the least tainted with local, national or racial bias. Not only is the range in time enormous but the multiplicity of inventions is enormous, ranging from the simplest – a drawing pin, for example – to the most complicated. Any attempt to present a complete history is therefore out of the question. Selection is essential. Fortunately there are two reasonable guidelines: how important was a particular invention or series of inventions and to what extent did it stimulate other inventions or innovations. These two often, but by no means always, coincide. It is possible to have an invention that is important but that does not stimulate other inventions.

One of the features of technology that has steadily become more important is its close association with science. This offers a third guideline. There is, and there always has been, a two-way relationship, with science gaining as much from technology as technology gains from science. In spite of this, historians of science and historians of technology seem to form more or less mutually exclusive groups, even though a satisfactory history either of science or of technology cannot be written without due allowance being made for the role of the other subject. It is hoped that the work that follows will contribute a little to the acceptance of this evident fact. Some historians of science have scorned the history of technology: we recall a late historian of science who characterized the difference between the history of science and the history of technology as the difference between thinkers and tinkers. In fact, science owes a great deal to technology and the history of science does not make sense unless account is taken of technology.

This book is based on lectures and seminars at the University of Manchester Institute of Science and Technology over the past 25 years. I hope it may be of use in similar courses in other institutions. It is, in the nature of the case, an interim work for it deals with what is, in some respects, a new subject: the history-of-science-and-technology. In fact, it can be regarded as a contribution towards the creation of an integrated and manageable history of technology The major limitation of the book is that it is substantially restricted to the physical technologies. I feel that to do any justice to the biological technologies would require at least one additional and parallel volume. At the same time, chemical technology has only a minor part in the story; the reason for this being that excellent histories of chemistry and chemical technology are available. Another limitation is what may be regarded as an undue emphasis on British experience. I would plead, in mitigation, that it is indisputable that the great movement known as the Industrial Revolution began in Britain in the eighteenth century and that it would be unrealistic to expect anyone writing from the heartland of that revolution to write about it as if it had happened on the other side of the earth.

It remains only for me to thank my friends and past and present colleagues for many stimulating discussions, comments and constructive criticisms. I am particularly grateful to Mr Michael Bailey, Dr K. R. Barlow, Dr Michael Duffy, Dr Kathleen Farrar and her husband, the late Dr Wilfred Farrar, Mr R. S. Fitzgerald, Dr Patrick Greene and the members of staff of the Manchester Museum of Science and Industry, Rev. Dr Richard Hills, Mr J. O. Marsh, Dr Arnold Pacey, Dr John Pickstone and Dr Alan Williams together with other members of what may be called the Manchester school, to the librarians and their staffs in Manchester and elsewhere and, last but not least, to many postgraduate and undergraduate students. The responsibility for mistakes is exclusively mine.

ILLUSTRATIONS

FIGURES

PART I

Clockwork
and Christianity

Introduction

'Every science', wrote James Clerk Maxwell[1], 'has some
instrument of precision, which may be taken as a material
type of that science which it has advanced, by enabling
observers to express their results as measured quantities.
In astronomy we have the divided circle, in chemistry the
balance, in heat the thermometer while *the whole system of
civilised life may be fitly symbolised by a foot rule, a set of weights
and a clock'* [author's italics]. Since a foot rule, a set of
weights and a clock are instruments of technology (as well
as being the products of technology) Maxwell's statement
constitutes as comprehensive an acknowledgement of the
central role of technology in human affairs as the literature
affords. It is the purpose of this book to illustrate and
confirm this claim; and, at the same time, to throw some
light on the nature of technology and the circumstances
under which technological progress takes place.

An impartial reader must wonder why the history of
technology does not occupy a prominent place on library
shelves or in school, college and university syllabuses. Its
importance is clear. Whatever our attitude and understanding it is undeniable that technology is now a major
determinant of most human activities, expectations and
even beliefs. It has virtually unlimited potential for good;
wrongly used it could, as we are so often reminded, bring
disaster to us all. Measured against the span of human
existence on earth the historical evidence for civil societies
shows that they emerged relatively only very recently. This
implies that politics, literature, philosophy, etc. – the staple
items of conventional history – are recent inventions; in

contrast human technics can be traced back to the very first appearance of human beings. In other words, the record of humanity on earth begins with the first archaeological evidence of human technics. The history of technics is therefore the most basic, the most comprehensive history of all. On the other hand practically every familiar object of everyday life includes or implies a highly elaborate technics. A video recorder is an obvious example; but even the simplest utensil or tool that has changed little in outward form is now usually composed of entirely new substances and manufactured by complex machinery. Could it be that the subject matter is too complicated and the material too multifarious to permit of a reasonably concise history of technology? But further thought suggests that this is not so much the cause as the consequence of the undeveloped state of the subject. Every branch of history necessitates a series of discriminatory judgements; what is held to be unimportant is ignored, or relegated to enthusiasts or antiquarians.

To attempt definitions at the outset would be to confuse the issues. Nevertheless some classification is required. We have on the one hand crafts, technics and inventions; on the other hand technology, applied science and inventions. And, common to both sides we have innovation, which we interpret as the action needed to put an invention into practice. Generally speaking, we say that inventions related to or springing from technics and crafts do not involve systematic knowledge and are, in a sense, empirical; inventions deriving from technology or applied science involve systematic or scientific knowledge. It follows that many more of the latter have been made since the Scientific Revolution in the seventeenth century. But – and here is the caveat – there is a great deal more to effective innovation than these simple definitions suggest. This, it is hoped, will become clear in the course of the work.

What, then, are, or should be, the common talking points of the history of technology? Before I suggest a tentative answer to this question let us summarize the representative works on the subject. First, there are encyclopedic studies, such as the seven volumes of the Oxford *History of Technology*, each chapter of which is written by a different author and deals with a specific technology. The benefit of specialist knowledge and skills is bought at the cost of somewhat inconsistent selection procedures and frequent failure to examine the links between technologies. These difficulties have been avoided in the cases of works by individual scholars such as Forbes, Klemm and Usher, who have restricted themselves to specific topics. Habbakuk, Landes, Musson, Robinson, Rosenberg and other economic historians have made penetrating studies of the circumstances that have favoured particular innovations and the economic and social consequences of those innovations. Usually, and understandably, economic historians have been rather less interested in the technology than in the economic and social factors associated with it. This, however, has served – and serves – to counterbalance the enthusiasms of the multitude of popularizers of the history of technology: the writers of lavishly illustrated books on locomotives and trains, on vintage cars and aircraft and on warfare. Finally, at the amateur and practical ends of the history of technology, there are the industrial archaeologists. They are concerned to record as fully as possible and, in important cases, to preserve the relics of industrialization. They recognized that many unique relics were either in decay or were in danger of being swept away and they argued that these should be preserved for posterity. This is eminently worth while but it does not take us to the heart of historical technology; it does not help to explain the processes whereby technology advances. But that, it would be countered, is not what industrial archaeology is all about.

While not in any way criticizing these approaches to the history of technology, I want to suggest that other and complementary approaches may be no less valid, even though they have not been developed to anything like the same extent. In particular it is possible to envisage a history of technology that is related to the history of science and to the history of ideas generally. But, it must be emphasized, technology so regarded is not to be thought of as the dependent variable, drawing its ideas from and parasitic upon science; rather it is an equal partner contributing at least as much to a common stock as it draws out. This is the approach that I follow in this book and it gives us criteria for deciding what is important and what may be left out.

This does not imply that only the most scientific branches of technology deserve consideration. As I remarked, another of my purposes is to examine the methods of technology and the modes of technological advance; by no means all of these are 'science-related'. Once technological innovation has become commonplace and, in effect, self-sustaining (like science), a rationale becomes possible that enables us to exercise discriminatory judgements. It is at this stage that technology develops from, or grows out of, technics or the assemblage of crafts and skills that for so long met the needs of society and, occasionally, led to new inventions. Technics, of course, still continues; as important, or more important, than ever. But my touchstones will be the universality and the consequences of a particular innovation together with the originality and insight displayed by the inventor. To be brief, I hope to identify some of the main themes in the secular progress of technology. These could constitute the framework of a systematic history of technology.

There is nothing revolutionary about my approach. As the word 'technology' implies a relationship between science, defined for the moment as systematic knowledge independent of any specific purpose, and the industrial arts, it is, on the contrary, a natural one to choose. Why then, in spite of the works of A. Wolf, A. C. Crombie and

Forbes and Dijksterhuis, has it not become the norm? There are several reasons. In the first place there has been a continuing lack of communication between the various historians of technology and historians of science (we shall consider this in a moment); there has been the specialization of studies that tends to inflict division rather than encourage unification, with the added disincentive that those who elect to study history may well have deliberately rejected science, and vice versa; finally, there is the still relatively undeveloped state of the history of science.

In spite of the fact that the Industrial Revolution that began in eighteenth-century England must, in principle, have had a profound effect on science, as on virtually all other human activities, many historians and philosophers of science have tended, until fairly recently, implicitly to take the view that science is a purely intellectual activity, concerned with abstract thought and disinterested experimentation and having no direct connection with utilitarian aims and achievements. This attitude has been due, in part, to the methodological interests of many of the most vocal philosophers of science; in part it has been due to a determination to protect science from the damaging restrictions associated with totalitarian political systems. To some extent, at least, it certainly reflects the values and the prejudices of older societies with well established class levels. The Rev. William Whewell (1794–1866), of Trinity College, Cambridge, wrote an exhaustive treatise on the history of science – the first British work on the subject – and another on the philosophy of science. He made practically no mention of technology. Samuel Smiles, a younger, near-contemporary, wrote popular, but not academically negligible, books on the history of technology, or rather of engineering. He made little mention of science. Between these two authors there was no traffic, no correspondence; they lived and wrote in different worlds. Their respective followers have perpetuated the breach. This was most unfortunate. Indeed, in England [sic] the study of technology has never enjoyed anything approaching the prestige of the study of the classics, mathematics, science and modern literature in the universities. And much the same is true of other European nations. This may well be the reason why the history of technology has been more effectively pursued in the USA than in Britain or some other European countries.

At this point we should look a little more closely at the agencies that formed the different academic syllabuses and the disciplines they represent. It was largely due to the reforming zeal of Revolutionary and Napoleonic France that the sciences were first organized into their present, tolerably coherent, disciplines. But there is no question that during the nineteenth century the German universities acquired – and deserved – enormous prestige as the world's leading schools for science teaching and research. The ideal of Wilhelm von Humboldt (1767–1835), perhaps the main architect of the success of the German university system in the nineteenth century, was that a university should advance pure learning and that students should acquire a love of disinterested learning – research – by carrying it out for themselves under a master who was an acknowledged scholar[2]. Practical or vocational studies, such as technology, that were supposed to require no more than the memorizing of facts, must be excluded. This was an educational creed that went back to Plato and Aristotle and was congenial to the governing elements of all European nations. But von Humboldt's ideal could never be fully realized. As the century wore on, specialization and the educational requirements – the need for more and more school teachers, lawyers, doctors, civil servants and administrators – of a rapidly developing nation state entailed that the Humboldtian programme was progressively diluted. Nevertheless the university ideal of disinterested learning and research remained and is strongly upheld today. In this way the notion of 'pure science' was born. But in practice this 'pure science' was defined administratively; it was the science pursued in universities and not in technical colleges. The model of pure science was imported into America, Britain and other countries by the many students who, having studied at German universities, returned home understandably enthusiastic about German science, research and education. The German technical colleges (*Technische Hochschulen*) and later technical universities could emphasize the importance of free research but they could hardly stress 'pure' learning. They, almost certainly, had far less influence on foreign opinion.

This is not, in any way, a criticism of the admirable system of higher education in Germany. The point is this: a large part of the history and philosophy of science, at least until recently, has been formed in the light of German university practice, a practice substantially followed in the rest of the civilized world. In other words, the history of the history of science reveals the effects of external bureaucratic agencies. To some extent, then, the exclusion of technology from the history of science is a consequence of the exclusion of technology from the German universities. To this it must be added that the autonomy of science – or of technology – is not denied and it is quite clear that extensive areas of the history of science – sidereal astronomy from the eighteenth century onward, for example – are related to technology only marginally and in respect of scientific instruments.

Academic fashions change even if sometimes rather slowly. During the 1920s and 1930s there was keen interest in such questions as whether electrons exist and, if so, in what sense? What bearing does the uncertainty principle have on the problem of free will? What meaning can be ascribed to the statement that space-time is curved? These, surely interesting, questions are no longer fashionable. In any case events since the 1940s have faced us with different and often urgent problems. How was it possible to organize scientists and technologists so that the infant technology of electronics could play an increasingly decisive part in the Second World War; and how could an atomic bomb be made, finally to end the War? What are the prospects for organized science and technology? What are the moral·responsibilities of scientists and technologists[3]? What can be done to limit or end the various forms of pollution? Do we face an ultimate energy crisis? What about global warming? Society, we can see, is now deeply and irreversibly involved in the affairs of technology and science; and we realize that science is now socially controlled as never before – and to an extent that would surely have horrified Lavoisier, Ampère, Faraday, Joule, Liebig, Clausius and the other leaders of the scientific world two hundred or even one hundred years ago. To the intelligent but non-scientific man or woman the triumphs of modern science are, in fact, usually, triumphs of technology. This cannot now be changed: science, technology and political affairs are closely interwoven, the warp and weft being national defence, education, institutionalized research, the needs of science-based industries and a host of other sensitive issues. The problems here are innumerable, material for countless extended debates and learned treatises. But the impact of technology and science on society and – no less important – the impact of society on science and technology are themes that lie outside the scope of this work.

What follows, then, is an attempt to identify the main critical or turning points in the history of technology and to elucidate the principles that were involved. I am not concerned with the fine details of individual inventions; these are adequately described in various specialized studies, to which references are made. Furthermore, limitations of space, of my own knowledge and of available information have combined to make it impossible for me to discuss such major turning points as the inventions of writing, of the water-wheel, of the calendar, of the boat or canoe, of the spear and the spade and indeed of many other key developments in the early records of humankind. I can hardly do more than briefly mention the achievements of the Chinese, Indian and Islamic civilizations. I have concentrated on the evolution of technics and technology since the beginning of the middle ages in Europe, in itself a big enough theme. My justification for this is twofold. In the first place the evidence suggests that some time in the early middle ages a remarkable change took place in humanity's perception of and attitude to Nature. This was related to and accompanied by an important series of basic inventions; inventions, that is, whose ultimate importance far transcended their immediate utility. They included the printing press, the weight-driven clock, the spinning wheel, the cannon, the blast furnace. These inventions, combined with the change in perception of nature and the revival of learning (in itself indicative of an innovative attitude), made possible the Scientific Revolution, as it has been aptly called, of the seventeenth century.

Although examples can be found in the ancient world, technology, as distinct from technics – even the advanced technics of, for example, the printing press – effectively began in France and England in the eighteenth century. It was clearly apparent in communications, the development of power sources, mechanization and the early chemical industry. One of its features was a willingness to use systematic methods of research and the latest findings of science; but, conspicuously, technology developed methods of its own. The rapid development of technology began at about the time that the Industrial Revolution took off in England. This was to have profound intellectual as well as social and economic consequences for Britain and the world.

It is certainly impossible to understand the development of modern technology without some knowledge of its scientific components, in particular the energy doctrines of Carnot, Joule, Kelvin and Clausius and the field theory of Faraday, Maxwell and their successors. But, as I indicate, it is difficult to see how these scientific components could have evolved in the absence of contemporaneous technology. The second scientific revolution, that of the mid nineteenth century, was accompanied by a new industrial revolution that steadily built up momentum and is well under way at the present time. In the light of the above outline I have divided this work into three parts: the first dealing with the development of technics up to the seventeenth century and the rise of modern science, the second dealing with the Industrial Revolution and the third carrying the story from the mid nineteenth century up to the present.

A note of explanation and apology is appropriate. I recognize that I may be charged with hero-worship, or at least of emphasizing the importance of individuals at the expense of the social system. But this is surely a matter of taste as well as of judgement. A historical narrative is none the worse, and in fact is improved, if the work and brief lives of outstanding figures are included as well as the details of the associated inventions. In addition, such figures are useful in that their work can be used to explain some of the main features of technological advance, which they helped to determine. I acknowledge, too, that I may be accused of 'triumphalism'. But by any reasonable standards technology *has* advanced, triumphantly. And to have attempted to make an assessment of what some would regard as the ill-effects of technology would have meant writing a totally different book. It is for the reader to judge the extent to which I have elucidated an acceptable framework for the history of technology, a framework in which command over energy lies close to the heart of the matter. And the reader must assess the extent to which I have made my case that a satisfactory history of science must be one that takes full account of the relations between science and technology, just as the history of technology must, *pace* Smiles and his disciples, recognize the importance of science in the advance of technology.

A Survey of Early Technics

By far the greater part of the story of humankind is a record of technics as revealed by archaeology. Humans, it has been said, are tool-makers and not merely tool-using creatures; this distinguishes them from the most intelligent of the other primates. The first record of humankind is that of their tools. Something is known about primitive tools, weapons and domestic utensils. We know rather less about primitive art. From burial remains and, to some extent, from primitive art, we can infer a little about religious beliefs. The absence of alphabets and written languages deprives us of knowledge of primitive law, music, poetry and literature, marriage and other social customs. For many millennia during the Paleolithic, or old stone age, and during the shorter Neolithic, or new stone age, progress must have been slow and for long periods, no doubt, barely recognizable to the modern student. Nevertheless the fact that tools were made implies three important things: first, that a degree of division of labour existed (stone-age factories for the manufacture of stone tools are known); second, that progressively higher skills were, if slowly, achieved; and, last, that there must have been some system of learning or of apprenticeship. Quite evidently the scope for further invention was severely hampered so long as humankind was restricted to tools of stone and wood. The universal experience is that, the more resources a community possesses, the more inventions it will make and adopt. A breakthrough in one direction will release a flood of inventions, many of which could not be inferred from the original breakthrough.

It appears most likely that distinctively human

settlements and activities first appeared in the African Rift Valley from which they spread east, north and west. What is clear is that a number of distinct and separate civilizations arose during the last ten thousand years or so and reached high standards of social organization and technical competence. The question then arises: why did these civilizations remain arrested at the stage they reached; why did they not go on to achieve progressively higher standards of economic organization and materially innovative competence?

The majority of what we may call the basic inventions were made before anything like a historical record was made. The use of fire, hunting and fishing equipment and simple weapons, spinning and weaving together with bleaching, dyeing, painting, colouring, pottery and glazing, house building, agriculture and the domestication of animals, coracles and other primitive boats, the making of domestic implements, water supply and irrigation, all these and other major inventions, such as an alphabet and writing, were made before history began. The names of the inventors – and they must have been many – have never been recorded. It is, in fact, likely that there were more inventors than inventions, for certain inventions were, in all probability, made several times by different people in different cultures who were ignorant of each other. Simultaneous and independent invention, a well authenticated feature of the modern world, doubtless occurred many times in remote antiquity.

The first great civilizations, the first major towns and cities, were perhaps those of the Minoan civilization, supposedly destroyed as a result of a volcanic eruption, on the island of Crete and the civilizations of Iraq (Babylon) and Egypt that depended on the great rivers, Nile, Tigris and Euphrates. Forms of writing began to appear in these parts of the world nearly five thousand years ago and arithmetic about a thousand years later. Writing and arithmetic were communicated, in the case of Babylon, on clay tablets and, in the case of Egypt, on papyri. Scores of thousands of dried clay tablets have been recovered and are kept in the great museums of the world, but preserving and interpreting them demand great skills.

Four thousand years ago Hammurabi (or Khammurabi), ruler of a small territory in Babylon, expanded his kingdom into a large state and laid down a comprehensive system of laws – the code of Hammurabi. It may be that in this effort to establish balance in society, to ensure that differences are settled by rational procedures, applicable to all people, we have the origins not only of civilized society but also of science. Damages caused must be compensated in civil actions; crimes have to be appropriately punished following trials at law. In this, argued Hans Kelsen[4], we have the origins of the scientific principle of causality. Cause, he pointed out, precedes effect as crime precedes punishment. It is not difficult to imagine that, in Nature, the same sequences apply. Natural phenomena are governed by quasi-legal, perhaps divine, sanction. If the sun failed to rise it would be punished, just as the temple guards would be punished if they failed to turn up for duty. How far is Nature governable by people? Here, a common impulse may be relevant: children and men (less frequently women) often attempt to smash (or punish) a toy or a tool that fails them. Xerxes, in frustration, is reputed to have had the sea flogged. According to Kelsen, there was a court in ancient Athens for the trial of tools that had injured their users. Similar examples can be found in historical and anthropological records. But long experience shows that Nature cannot be commanded in such ways.

Babylonian clay tablets, carrying cuneiform scripts inscribed by means of styluses, have revealed a remarkable skill in arithmetic and in predictive astronomy. From generation to generation the motions of the heavenly bodies were recorded by dedicated astronomers and numerical methods of prediction developed. No evidence, however, has yet been found to indicate that there was any astronomical or cosmological theory behind these observations and predictions. They were merely recorded and, deliberately or accidentally, made available for other civilizations to use as they wished.

Arithmetic and astronomy were not the only technics in which the Babylonians showed great skill[5]. The development of settled agriculture necessitated the apportionment and then the buying and selling of land. This, in turn, required the art of the surveyor and the making of charts, diagrams or maps showing the areas of land involved in transactions. Some clay tablets carry beautifully drawn diagrams showing the division of land and calculations of the dimensions involved. In Egypt resurveying the land after the Nile floods was a matter of some urgency. In societies as advanced as these, social arrangements would also require a calendar so that complex negotiations or discussions as well as religious ceremonials could be arranged well in advance. And this, in turn, implied another series of social inventions. But these, although of enormous range and importance, lie outside the scope of this work.

The discovery and exploitation of metals was, together with the inventions of numerical symbols, arithmetic and the alphabet, the most important of all ancient innovations. The discovery of copper came first and then came the invention of methods of hardening it – thus extending the range of its uses – by adding tin to make bronze – hence the 'bronze age' – and zinc to make brass. Gold, the soft metal that resists corrosion, or rust, was discovered and became a basis for many currencies. Lastly came the more intractable metal, iron. If a red-hot lump of iron ore was repeatedly hammered and reheated in a fire with charcoal (another invention) the ancient smiths found that they had an extremely strong and hard metal – wrought iron. They did not, of course, know anything about the chemical processes involved; understanding came only thousands of years later. They were also to find that by skilfully varying the heating and the amount of charcoal they could get a type of iron that was both strong and light. This was steel, but methods of systematic, large-scale production were quite unknown. These metals, which could readily be used to make a wide range of tools, previously formed of wood or stone, led also to the inventions of nails, screws, nuts and bolts so that we may say that metallurgy was, with applied mathematics, the first of the truly *strategic* technologies[6]: it fertilized and made possible advances in an enormous range of technics. These latter inventions could not possibly have been inferred from the original techniques of substituting metal for stone or wood to make tools. Metal working, in other words, made entirely new – previously inconceivable – tools and implements possible. The cardinal importance of metal working and the prestige of skilled metal workers are, as Samuel Smiles remarked, confirmed by the common surname of Smith. Furthermore, the word 'smith' is of Nordic, or Germanic, origin, which offers a clue as to the location of the early centres of metal working in western Europe.

Metal working was commonly carried out in mountainous regions, where the ores were to be found and where wood was often plentiful. An important clue here is the German word *Berg*, which translated means mountain. But *Bergakademie* means mining college, *Bergarbeit*, mining, *Bergbaukunst*, mining science; historically, the mining in question was non-ferrous metal mining in such areas as the Harz mountains. But the areas in which metal mining and working first began, at least in the west, were in the mountain regions of northern Iraq, Lebanon and Syria – Damascus steel was famous in medieval Europe – Anatolia and the Balkans.

At about the time when the ancient civilizations of the Middle East and the Mediterranean were reaching their peak a remarkable civilization was developing in China that was to endure with remarkable social continuity up to the last century or so. Whitehead[7], in striking if imprecise words, described China as representing the greatest volume of civilization known to history. Although some of his claims may fairly be disputed the researches of Dr Needham and his associates have indicated the extent to which the rest of the world is indebted to the Chinese for many key inventions. A number of these are mentioned in the chapters to follow. The Chinese were perhaps the first to make what we may call frivolous inventions: the kite, the revolving bookcase, firecrackers, etc. This is no criticism of Chinese inventiveness; rather the contrary for it confirms the versatility of Chinese inventors. The modern world enjoys a multitude of frivolous inventions.

Although there can be no doubt that inventiveness flourished in ancient India it is most regrettable that we have no history of science and civilization in India comparable to Dr Needham's work. This means that historians of western technology lack a full understanding of the extents to which the west is indebted to India for transmitting Chinese inventions as well as for transmitting home-made inventions. Equally unsatisfactory is our knowledge of the fate of western inventions that were transmitted to the east.

The establishment of cities and increasing skill in metal working marked the progress in technics, a progress that had some way to run in the ancient world before the rise of Greco-Roman civilization. The question raised by this progress, century after century, passing from one nation, or culture, to another, is as important as it is impossible to answer; at least in the present stage of our knowledge. What is it that causes the secular progress of technics, what is it that drives it forward? In some circumstances fear of enemies may be the driving force, in others economic considerations may be paramount. Perhaps some religions favoured the progress of technics while others almost certainly inhibited it. Yet again the inventive talent of individuals – invention for its own sake – may be sufficient, granted the willingness of contemporaries to accept his or her inventions. This last, however, begs the question and we are forced to the (unsatisfactory) conclusion that a variety of agencies, acting together, must determine technical progress, granted that social or environmental circumstances are not inimical. Overall, humanity's technics have advanced steadily since the earliest days. The ultimate driving force of this progress, however, remains obscure and hidden in our nature.

Gears from the Greeks

The Greek city states of the ancient world (*c.* 600 BC onward) occupied a geographically favourable position. The surrounding seas gave their peninsular and island cities a degree of security against invasion. At the same time the sea gave them – seafarers and merchants, then as now – opportunities to trade with, and profit from (in more ways than one), the great civilizations that lay round them to the south and east. They observed the different codes of law in practice and they concluded that there existed natural laws, or laws that all peoples obey. In contrast conventional laws were those that applied only in particular nations. They learned mathematics from their neighbours before making their own contributions, the most famous of which was Euclid's books of geometry. In this work geometry was systematized; in other words the art of the surveyor – practical land or earth measurement (whence the word 'geometry') – was transformed into an abstract science. Although not commonly acknowledged as such, this was the earliest instance of technics forming the basis for a scientific advance.

Greek astronomers profited from – that is, they copied – the astronomical observations made by the Babylonians but went on to frame elaborate astronomical theories. These were conspicuously rational[1]: their astronomy had no room for gods and goddesses or other spirits. They believed, reasonably enough, that all the heavenly bodies revolved daily around the earth, which was at the centre of the universe. Apart from the multitude of fixed stars, many of them arranged in constellations, there were seven planets that had additional, independent, if much slower, motions of their own. These planets – sun, moon, Mercury, Venus, Mars, Jupiter and Saturn – took one year, 28 days, 88 days, 226 days, two years, eleven years and thirty years respectively to travel round against the background of the fixed stars. It was easy to imagine that the sun and the moon were fixed to transparent spheres of some fifth element, unknown on earth. But the problem with the five minor planets was that their motions were erratic. In their journeys round the heavens they sometimes stopped, moved backwards for a time and then moved on again, in so-called retrograde motion. Such was Greek geometrical skill that they were able to postulate systems of homocentric spheres that could reproduce the retrograde motion of the planets. But it proved too difficult to use these models predictively. For the purpose of planetary forecasting astronomers had to use the computing method of epicycles and deferents. This sophisticated technique was developed by three men: Apollonius of Perga (*fl. c.* 200 BC), Hipparchus (*fl. c.* 150 BC) and Claudius Ptolemy (*fl. c.* AD 150), the last of whom perfected it. The use of epicycles and deferents for planetary prediction and the geocentric system as a picture of the universe remained the corner-stones of astronomy until the sixteenth century of our era. It must be added that one of the main uses of planetary astronomy was the casting of reliable horoscopes. Astrology was, after all, a reasonable science. The sun dominates life on earth; the moon has many, more subtle, influences: for example, it rules the tides. Surely, then, on commonsense grounds, the minor planets must have their own distinctive influences on the lives of men and women?

The Dominance of Greece

The Greeks made notable discoveries in astronomy in addition to devising methods of predicting the movements of the planets. Hipparchus discovered the precession of the equinoxes, a discovery that could only have been made by taking account of observations extending over the lifetimes of several astronomers. He drew up a catalogue of the positions of about 850 stars together with their relative (apparent) brightness. Aristotle gave sound reasons for believing that the (stationary) earth was spherical in form. One man, Aristarchus of Samos, is credited with the suggestion that the earth itself was a planet that orbited round the sun. However, the available evidence was far too thin to support such an outlandish theory. One thing is quite clear: although most of the ancient Greek astronomers were practising astrologers, they made astronomy into a science, with all the institutional features of science, such as the cumulative acquisition of knowledge from generation to generation. And the Greeks could have pointed to respectable advances in technics as well.

Greek astronomy, like any progressive science, was accompanied by the invention and manufacture of scientific instruments. They built orreries, or mechanisms to reproduce the motions of the planets, and they developed sighting instruments such as armillaries and quadrants. However, the crowning achievement of Greek astronomers and technicians was not revealed until 1900. In that year a group of sponge fishermen discovered the wreck of an old ship close to the shore of the tiny island of Anti-Kythera, between Crete and the mainland (Peloponnisos). The most remarkable item of salvage turned out to be the remains of a small instrument dating from about 87 BC. So corroded and damaged was it after two millennia under the sea that it was a long time before an acceptably accurate account of its purpose and operation could be given. The careful researches of the late Professor Price have shown that it was a calendrical computer with remarkably complex trains of bronze gear wheels[2]. At the heart of the instrument is a sophisticated differential that enabled the motions of the sun and the moon to be exhibited consistent with the phases of the moon. Professor Price conjectured that the small machine was worked by hand, although its exact purpose is unknown. It could have been intended for display, or even amusement, and therefore an instance of what we called frivolous technology. Whatever its purpose there can be no question of the immense practical skill and clear understanding of the kinematics of gearing that it reveals. These skills and the associated knowledge passed to the astronomical instrument makers of Islam and from them to medieval Europe. But it was not until the sixteenth century that a differential train was built into a European clock. For the present it is enough to point out that the Anti-Kythera machine belongs to the same family of mechanisms as James Watt's sun-and-planet gear, Morin's *compteur* and the automobile differential.

So much attention has been paid to Greek literature, mathematics and philosophy that Greek technics has been largely ignored; some have denied that the Greeks made any inventions at all. In fact, only since 1945 has Greek technics been seriously studied. In principle it has always been obvious that structures like the Acropolis could not have been erected by people using no more than their bare hands and abstract thought. The same inference applies to the construction of the massive Lion Gate at ancient Mycenae. In fact we know very little about Greek building technics, apart from the fact that it must have been relatively advanced. And, formally, we know no more about Greek ships and shipbuilding than what can be gleaned from pictures on vases. However, in recent years an imaginative project to build a replica trireme has thrown new light on these remarkable warships and the skills and knowledge that went into their design and construction.

Western civilization is rooted in the Greek achievement. If the Greeks learned much from other nations the question arises as to what effects they, in their turn, had on their neighbours. In the case of the Egyptians there was surely a creative diversity between the two civilizations. In Alexandria the Greek genius, which was undeniably philosophical in temper, and the Egyptian genius which was more obviously practical and technological, met and generated such typical works as Euclid's geometry, Ptolemy's astronomy and geography, Eratosthenes' estimate of the circumference of the earth, Heron's steam-driven toys, the first ideas on chemistry, the Archimedean screw (an Egyptian invention) and the foundation of the Great Library. Evidently there was enough common ground to allow a dialogue (to use the modern cliché) between Greek and Egyptian that led to a fruitful mix of science, scholarship and practical achievement. Perhaps the best exemplar of this cultural cross-fertilization was Archimedes (*c.* 287–212 BC), a native of the Greek colony in Sicily, who met the learned men of Alexandria. Archimedes, by applying mathematics to discover the law of the lever and the principles of hydrostatics, could fairly be described as the first true physicist. He could also be said to have had the aptitudes and interests of an engineer. As the world knows, he was reputedly killed by a soldier following the capture of Syracuse by the Roman army.

What might have been achieved had there been one or two more Alexandrias, in communication and competition with one another, in the great Roman empire that later included both Greece and Egypt? Something like modern science might have evolved many years before it did. But this is only speculation, put forward to make the point that it is of the essence of technological, or scientific, progress that there should be more than one centre of excellence for the whole enterprise to progress. Cross-fertilization on the individual and on the collective scale are essential. Political changes and military disasters were to lead to the break-up of the Roman empire and the eventual eclipse of Alexandria.

Technics in Antiquity

The Romans made few notable contributions to science or to technics. Their civil engineering techniques were not original; they were those of the Babylonians, the Greeks and the Egyptians. They made no notable contributions to early chemistry and metallurgy; nor were they active in mathematics, astronomy and instrument making. Their genius lay in administration and law. The empire they created was, it is claimed, tolerant of the diverse religions and philosophies held or practised by the various subject nations. If their tolerance, legal skills and military power made the empire possible the price was that there could hardly be, under these circumstances, a common culture apart from that of law. Even in this field allowance was made for diverse peoples under Roman rule; hence the *jus gentium*[3], or law of the nations, that resulted from a winnowing out of the codes that were common to all the constituent nations or tribes. It is from Roman law that the legal systems of the European nations, the Americas and many eastern states derive.

The inferences to be drawn – tentatively – from the experiences of the ancient world are, first, that for science and technics to continue to prosper more than one centre of excellence is required and, second, that the centres of excellence should share a common culture, a common philosophy. To this we can add that the common philosophy should not be so comprehensive or authoritative that it acts as a constraint on originality.

The late Bertrand Gille[4] offered sound explanations for the ultimate technological failures of the ancient world. He pointed out that the Mediterranean area and the Middle East lack iron ore and coal and are deficient in water power; soils tend to be thin and poor while forests are relatively few and trees are stunted. Some of these resources are deficient in other areas where ancient civilizations developed, notably in those parts of the

Americas where pre-Columbian civilizations flourished. But it must be added that, unlike water power, an abundance of iron was not an essential element of the industrial movement that began in medieval northern Italy and southern Germany and that acquired such irresistible force in eighteenth-century England. Brass and bronze can meet most of the requirements that iron serves, including armour and weapons. And, as the cases of the Central and South American cultures show, two other drawbacks can be significant. Ease of communication, we know, stimulates invention and innovation. The civilizations of the Aztec, the Maya and the Inca were geographically isolated, by sea, by jungle and by mountains. And isolation may well have prevented the advancement of the peoples of sub-Saharan Africa and of Australia and Tasmania. Rigorously imposed isolation was certainly a factor in the long stagnation of the magnificent Chinese civilization. Moreover, we may suspect that technics, science and a developing economy could never prosper in cultures dominated, as was the Aztec, by a religion that made a practice of regular human sacrifices.

The metaphysical and religious beliefs of different peoples are generally believed to influence the development of their scientific ideas. Thus the Chinese, who could observe few regularities in Nature, implicitly denied immutable laws of Nature. They did not, therefore, explain natural phenomena in terms of scientific law but rather as the undetermined outcome of the conflict between opposing powers. In a similar way they did not base their jurisprudence on the (Greek) principles of absolute law, as the Romans and, following them, the later Europeans did, but rather on the effects of juridical decisions on society as a whole. The question of guilt or innocence was subordinate to the question of the effects of any decision on social harmony. It is clear that, whatever their influence on Chinese science, these beliefs did not for many centuries hinder the impressive development of Chinese technics and inventiveness. In the long run, however, the moralistic and aesthetic elements in Chinese juridical law may have hindered economic development and hence technics by virtue of a bias in favour of spiritual harmony.

The Question of Slavery

The problem with all these popular, and much discussed, theories is simple. If a set of beliefs – religious, social, philosophical – is shown, plausibly enough, to be uncongenial to science and technology the assumed causal relationship – beliefs inhibit science – is not necessarily correct. It may be that, for quite different reasons, inability to develop science and technology allowed anti-scientific beliefs free rein. Equally, of course, the growth of science and technology might well be assumed to have encouraged beliefs sympathetic to science. The latter theory was commonly and confidently held by nineteenth-century rationalists.

Support for the view that regards science as the spearhead of human understanding is provided by the example of the Greeks. Their religion included an array of gods, goddesses and semi-divine individuals, all with human – and some all too human – attributes, and yet they were eminently successful in developing theories of Nature according to which phenomena are governed by rational and immutable laws. If, after a brilliant start, the impetus of science in Greek city states and colonies faltered, this could be ascribed to foreign conquest. Symbolically, the murder of Archimedes by a Roman soldier was something more than the death of a great mathematician; it marked the end of part of a most successful scientific enterprise. The part that lived on was centred on Alexandria. Of course other reasons can be found for the eventual stifling of science in the ancient world. For example, Greek and Roman society was characterized by the institution of slavery and this might be thought to have been the cause of its ultimate decline. Gille is critical of the thesis that slavery was a positive disincentive to technical progress. He puts forward good reasons to support his view: how, for example, could one account for the undeniable advance of science and technics in the ancient world during the period when slavery flourished? Nevertheless, it is impossible to accept Gille's criticism, at least in full.

Slavery seems to have characterized most, if not all, the ancient and older civilizations – Greek, Roman, Arabian, pre-Columbian American, etc. If we ignore the ethical objections, the brutal institution of slavery can be regarded as marking a stage in the advance of a civilization and as a response to technical problems and challenges. Slavery, in other words, is a wide-ranging answer to problems, many of which could otherwise be solved by mechanical and other forms of invention. Viewed in this light it is an alternative to technology, or a debased form of technology. The slave, observed Aristotle, is a living tool.

The institution of slavery presupposes a fairly advanced and militarily successful society. In such a society unpleasant, dangerous, repetitive and physically demanding tasks are undertaken by slaves, and so long as they are freely available there will, so the argument runs, be little incentive to invent or to employ labour-saving machines. Slavery, it is therefore alleged, retards invention and innovation. Furthermore, the association of manual tasks with slaves on the one hand and with practical invention on the other must devalue invention in the eyes of free citizens. A final possibility is that slave traders (presumably an influential element in the economy of a slave-based society) may be expected to do all they can to restrict or prevent the introduction of (slave) labour-saving devices.

Against this it may be argued that slave labour is not free. Slaves have to be trained, maintained and, when damaged, repaired if possible; a uniform supply of slaves cannot be assured; slaves have limited working lives and the measures necessary to restrict their liberty and to compel them to work are an added cost. To complicate matters, there were, no doubt, many and varied forms of the institution of slavery so that generalization may appear to be a risky procedure.

One argument, however, that slavery retards invention is immune against these and other objections. Slavery, if it means anything, means denial of individual freedom and that implies restriction of personal mobility. A slave with a good idea or invention will be in no position to hawk an idea around, seeking a sponsor, or a suitable place where it could be developed. Now the history of technics and technology shows clearly enough that freedom to travel has usually been an essential requirement of the successful inventor. In this respect the career of James Watt may be taken as the paradigm (see Chapter 7). The inventor who finds the incentive, the materials, the ideas and the necessary support all on the parental doorstep, metaphorically speaking, was – and is – the exception and not the rule. It may be that certain slaves at certain times and in certain places enjoyed considerable freedom, including freedom to travel. To the extent that they did so they were the less slaves and less typical of the institution. The argument can be expressed in general, or modern, terms: a society in which the mobility of labour is, to some extent, restricted will be less efficient, less innovative than one in which the mobility of labour is entirely unrestricted. We conclude that the institution of slavery must have retarded invention and innovation. This, it must be emphasized, is not to claim that it prevented them altogether.

Technology Moves West and North

During the 'dark ages' and in the early medieval period, which may be said to have begun in the ninth-century AD, slavery, in Europe, was progressively moderated into the form known as feudalism in which the serf was almost literally tied to the land, owing dues and service to the feudal lord. The rise of feudalism, in particular the appearance of the equestrian order, owed much, according to the late Professor Lynn White, to the introduction of the stirrup, a simple enough device that nevertheless made it possible for a knight to fight on horseback. In particular, he could use a lance by bracing himself against the stirrups to take the shock of the lance striking its target. Under feudalism the serf or vassal, and later the peasant, was allowed to retain a portion of the products of his labour for himself while the lord and his retainers were expected to give protection to the serf in violent times. In Hungary labourers working unpaid on behalf of the feudal lord were called *robotnik*, whence the modern word, by way of Karel Capek's play, *robot*. In the west, the lord would exercise additional monopoly rights by compelling the peasantry to grind their corn at his mill. Millers, as Geoffrey Chaucer[5] made quite clear, were often unpopular.

From the time of Charlemagne, western Europe became more stable and relatively prosperous. The Christian religion was established universally and the tide of Islam was halted in Spain and Italy before being steadily driven back. Towns became jealous defenders of their rights as they grew larger, and the runaway serf or peasant who, tired of the oppressions of the countryside, sought refuge in a town gained his freedom. As the old German saying had it, 'Stadtluft macht frei'.

It was during this period that Europe began to import inventions, directly from the Arab nations at those points in Spain and Italy where Europe and the Islamic world met. Many of these inventions, but by no means all, originated in China and India and were merely passed on by the Arabs[6]. That in itself was a considerable service. The Arabs had inherited and extended Greek astronomy and mathematics and had explored the realms of Greek philosophy. Confirming the extent of their astronomical work are the facts that many of the most conspicuous stars have Arabic names and that Ptolemy's treatise, the *Syntaxis*, became known in the west as the *Almagest*, a mild corruption of its Arabic name. The world is indebted to the Hindu mathematicians who made the enormously important invention of the zero sign in arithmetic, and to the Arabs who developed and transmitted the present system of numerals, including the zero, that replaced the clumsy Roman numerals with their time-wasting procedures for calculation. Many familiar mathematical and geometrical terms – algebra, algorithm, zero, nadir, etc. – are Arabic in origin. The Arabs developed the chemistry and metallurgy, pioneered in the Greek–Egyptian city of Alexandria, and made notable contributions to ophthalmic optics. Arab shipbuilders had invented the mizzen mast and Arab civil engineers built complex systems of water supply and irrigation. Finally, they made significant contributions to cosmetics and to the culinary arts (for example, they introduced coffee). In the end, however, the Arab civilization faltered and stagnated. Two very material reasons, apart from slavery, can be given. The Crusades (1096 to 1270) exhausted the Islamic nations while awakening their Christian enemies and rivals to the achievements of Arabs. And a series of invasions from the east in the thirteenth and fourteenth centuries went on to complete the ruin the Crusades had begun.

Imitation means innovation, which, in turn, often stimulates invention. This is the lesson of medieval Europe in its first phase; the lesson was later exemplified by Britain in the seventeenth century and by Japan at the end of the nineteenth century. Imitation is not, as may be supposed, an indication of inferiority. For an invention to be adopted – imitated – in a different community, that community must have reached about the same level of technical competence as that of the originating community; and, moreover, the adopting community must obviously be willing to learn. An example of what happens when this attitude is lacking is provided by China. The remarkable number of inventions that originated in China is proof enough of the genius of the Chinese people. And yet for centuries China refused to accept ideas or inventions from outside, with the result that Chinese technics eventually languished. On a vastly smaller scale the annals of industry in western Europe are full of firms that refused to innovate, to learn new ways or to copy their rivals and, accordingly, failed completely.

A willingness to imitate, or adopt, inventions made by foreigners is the first step towards the creation of an inventive and technically progressive society. We suppose that the process worked like this: town and village craftsmen – blacksmiths, coppersmiths, wheelwrights, millwrights, masons, carpenters, etc. – faced with demands to make imported inventions, began with slavish imitation, then learned how to use local materials or processes and how to modify the inventions for local needs; quite possibly for needs never envisaged by the original inventors. This means that the technicians made original inventions themselves; in any case they extended the range of their own skills and increased their own capacity to make inventions. But we do not know what basic drives initiated and maintained the – essentially social – impulse and the mechanism of technical advance.

From about AD 800 onward the two main areas of European technics were northern Italy and southern Germany; together these areas form what we may call the region of Alpine technics. Relative proximity to the Islamic world was probably one factor and the plentiful rivers – suitable sources of water power – another. The abundant forests that grew on the foothills and lower slopes of the Alps were a further asset. Apart from its use as a building material, wood provided the raw material for charcoal, essential in the metal refining industries. Medieval German technicians were particularly active in mining and metallurgy and this, it was much later pointed out, probably accounted for German excellence in the closely related science of chemistry, an excellence that has continued up to the present day. Italian technics were associated with art, architecture and mechanics. And, appropriately, modern Italy has a reputation for the manufacture of such things as high-performance automobiles. France, northern Germany, Scandinavia, the Low Countries and Britain remained relatively backward. In particular, Britain lagged until the end of the seventeenth century[7].

The pattern of medieval European technics differed in at least one respect from the pattern of ancient technics. No doubt this reflected very different social and political structures. The Europeans eschewed the extravagant and grandiose structures built to satisfy the needs, whims or ambitions of all-powerful rulers. European technics were initially on a much smaller scale. And so, for example, the fine Roman roads were allowed to fall into decay; for this neglect the moderns, everywhere concerned with their automobiles, and the Victorians, proud of their railroads, agree in condemning the medievals. But the medievals had no need of expensive roads. They had no great armies to maintain, no empire or continent to subdue, police and administer. And if the medievals built no fine villas with central heating and constructed no imposing aqueducts they did at least bequeath us their great cathedrals. And these were by no means the only innovations they passed on to succeeding generations.

Agriculture is the basic industry and here medieval Europe saw two important innovations before AD 1000: the rotation of crops and the wheeled Saxon plough that enabled heavy, fertile soils to be cultivated. A horse collar that allowed the animal to pull efficiently with its shoulders and nailed horseshoes were introduced. The windmill, unknown in the ancient world, was invented and by the time of the Norman conquest (AD 1066) about 6000 water-wheels were recorded as being in use in England. In the following century the applications of water power diversified to a variety of processes: to sawing wood, forging iron and driving fulling stocks as well as to grinding corn. The blast furnace appeared, the bellows being driven by a water-wheel, metallurgy was developed and methods of casting iron and bronze were improved. The crank was introduced and improvements were made in the transmission of power. In 1269 Petrus Peregrinus made the lodestone the subject of a theoretical study and the mariner's compass, possibly originating from China, came into use. Spectacles were invented towards the end of the same century and, coming from China, the manufacture of paper began[8]. (The Chinese secret of the manufacture of China ware (china) or porcelain, was not solved in Europe until the eighteenth century.) A variety of new chemical substances was adopted, due very largely to the Arabs. They included camphor, calomel, dyes, tincture, pigments, mordants and medicaments. The study of mathematics revived and Arabic numerals were adopted. These were only a few of the major innovations of the period, innovations that ensured that by the end of the middle ages the technical standards of western Europe were higher than those of any previous civilization.

The ancient arts of spinning and weaving thread are of unknown origin. For centuries the spindle-and-whorl were the only mechanical aids to spinning. The spinster, usually a young woman, held a mass of cotton or wool in one hand while a short length of thread, twisted between the fingers of the other hand, was fed to the spindle. The spindle-and-whorl was then set spinning so that as it was allowed to fall it drew out twisted thread from the hand-held mass. This remained the method of spinning thread until, in the thirteenth century, somewhere in Europe someone invented an entirely new technique for spinning: the so-called Great Wheel. A large-diameter, but lightweight wheel, free to rotate on a horizontal axle, drove a small horizontal spindle by means of an endless cord. As with the spindle-and-whorl a short length of thread was drawn out from a mass of cotton or wool and wound round the spindle. The big wheel was then set turning with the free hand. By drawing the mass of cotton or wool away from the fast-turning spindle a long thread could be spun, rapidly and easily. When an arm's length of thread had been spun it was turned through 90° so that the thread was wound on to the rotating spindle. By this means the productivity of the spinster was greatly increased.

Another and important advance in the technics of spinning thread was made two hundred years later, at the end of the middle ages. This was the introduction of the Saxon Wheel. Once again the inventor is unknown. The Saxon Wheel was considerably smaller than the Great Wheel and was driven by foot-operated treadles. Instead of driving a spindle the wheel now drove a horseshoe-shaped metal 'flyer'. This had a small metal loop at the top and further metal loops along one limb. The thread was taken through the topmost loop and down through the loops along the limb. As the flyer rotated it imparted twist to the thread, which was fed, after it had passed through the loops, on to a spool that was located between the limbs of the flyer. There was, therefore, no need for alternating operations of spinning and winding on: the spinning and winding on were continuous and simultaneous. The whole operation was much simpler and less tiring for the spinster and, at the same time, the operation was much more efficient than with the Great Wheel. It was to be three hundred years before the next great advance was to be made. The Saxon Wheel is, of course, the spinning wheel represented in children's stories, fairy stories and at least one opera and one ballet.

The Chinese, certainly, and the Byzantines, possibly, discovered the extraordinary flammable properties of 'black powder', a mixture of about 75 per cent saltpetre (potassium nitrate) 15 per cent ground charcoal and 10 per cent sulphur. It was first used in fireworks and as a military incendiary weapon with functions akin to the modern flame thrower. Later on someone, somewhere, made a remarkable discovery: if the black powder was forced into a confined space and a heavy stone – a projectile – put on top of it, the black powder would explode on ignition, throwing the projectile out with tremendous force. In this way gunpowder [*sic*] and the gun, or cannon, were invented. The first recorded mention of the cannon was made in 1318. These early guns were made by arranging rods of wrought iron to form a tube and binding them together with hoops of iron. They were siege weapons intended to reduce forts and castles with a power that the ancient battering ram could not rival. Standing off at a safe distance from the defenders, the gunners could methodically pound down the walls of the city, castle or fort. Even the great walls of Byzantium (Constantinople, now Ankara) could not resist the monster cannons, of two-foot bore, made on the spot by Hungarian engineers. In due course smaller cannons appeared, leading to the hand-held cannon, or hand-gun, that could easily be operated by one man. In turn, these led to long-barrelled arquebuses, matchlocks and then flintlock muskets. The long barrel ensured that the bullet was accelerated to the maximum velocity.

It was the musket that in due course and in spite of its disadvantages – weight, slowness of fire, expense of gunpowder, unreliability (particularly in bad weather) – defeated the bowman. Following the reinvention of military discipline a body of musketeers, protected by pikemen as they loaded or reloaded their muskets, could clear the field with one concerted volley (the pike was later united with the musket in the form of the bayonet). But a body of trained longbowmen could do the same, and far more rapidly. The decisive advantage of the musket was that it did not require a strong and highly skilled man to use it effectively. Physically strong and fully trained archers were always in short supply (and apt to be troublesome, in the way of monopolists). Virtually anyone could be a musketeer. Even armoured horsemen and mounted knights stood no chance against a clodhopper of a musketeer with a cheap musket.

The musket and the cannon evolved slowly through the centuries and gunpowder was made progressively more efficient. Indeed, the steady improvement of musket and gunpowder epitomize the process known as evolutionary improvement. Revolutionary improvement was not possible until the nineteenth century when the development of machine tools, improved metallurgy and new chemical explosive made possible the invention of the breech-loading rifle. This and the field gun remained the decisive elements of the battlefield until, within the last hundred years, the machine gun and the tank took over.

The Weight-Driven Clock

The dates of the inventions of a set of weights and of a foot rule are lost in remote antiquity. But the clock – Maxwell's third key feature of the whole system of civilized life – is beyond doubt a medieval invention. Proceeding from Maxwell's observation, it is certainly true to say that the technical, economic and social standards of a civilization can be measured by the accuracy and number of the time-keeping devices in use. For many millennia people had no need of clocks. It was only with the rise of civil societies possessing fairly elaborate social institutions that the measurement of time became important. Elaborate religious and civil ceremonies, military activities and commercial negotiations required more precise timing than the simple sequence of day and night could provide. The shadows cast by the sun were the simplest, most reliable and universal method of time-keeping: hence the sun-dial. This, of course, was useless at night but this was unimportant for so feeble were the available lamps and candles that most social activities ended with nightfall. Even in day-time the sun-dial had its limitations in northern latitudes and supplementary devices like calibrated candles, hour glasses and water clocks had to be used. The water clock was never satisfactory for it proved impossible to ensure a constant head of water and a sufficiently uniform flow. The first two were crude instruments and could only be used for limited periods of time. A complication was the custom of dividing the daylight time into hours whose duration would, in northern latitudes, vary with the time of year. This meant that sun-dials had to be calibrated with different sets of hours to suit the time of year.

The desideratum for a mechanical clock is a device that will cause an indicator, shaped like an arrow, to point to a series of numbers that correspond to the sun's position as it circles round the earth. The numbers will be those of the hours that it takes the sun to perform its journey: twelve for day-time, twelve for the night (the number twelve being derived, probably, from the twelve divisions of the zodiac). As the heavens are spherical the numbers should be engraved or painted on a circular disc.

So much for the problem in principle; in practice the problem was to settle on a suitable driving mechanism and, the nub of the matter, a suitable regulating mechanism[9]. A falling weight was the only natural driving force available, but unfortunately a weight accelerates as it falls so that a weight-driven clock would run faster and faster. The fall can be retarded by means of a brake so that when the weight reaches its terminal velocity it falls like a ball-bearing in treacle, with a uniform velocity. Unfortunately and inevitably the brake will become progressively smoother so that the terminal velocity will steadily increase. No solution was to be had that way and we are left completely in the dark about the steps by which some unknown genius – or geniuses – invented the escapement mechanism that solved the problem by

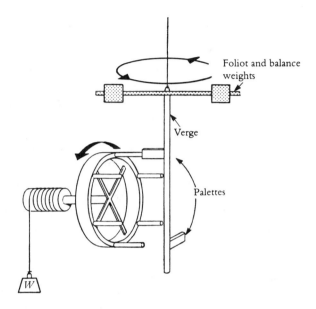

FIG 2.1 *Mechanism of the weight-driven clock*

making perhaps the greatest single human invention since that of the wheel.

The escapement works by releasing equal quanta of mechanical energy at equal, short intervals. Its mechanism is illustrated in figure 2.1. The wheel carries short, horizontal, uniformly spaced pegs and is driven by a weight at the end of a cord that is wound round the axle. In front of the wheel is a vertical rod – the 'verge' – carrying two small plates – 'palettes' – at slightly more than 90° to each other. The palettes are placed so that they can mesh with the pegs on the wheel. The verge is suspended freely by a rope or cord and at the top carries a short horizontal rod – the 'foliot' – with balance weights at either end; the position of the balance weights can be adjusted.

The operation of the escapement is simple. When a peg strikes the upper palette (as in the figure) the inertia – using the modern term – of the balance weights checks the rotation of the wheel. The driving weight then overcomes the inertia of the balance weights until the peg has pushed the palette out of the way and the driving weight can, momentarily, fall freely. The turning of the verge has brought the lower palette between the pegs so that the approaching lower peg, moving in the opposite direction, almost immediately strikes the lower palette. The swinging motion of the balance weights and the fall of the driving weight are checked until the latter, once again, pushes the palette out of the way. The driving weight momentarily falls freely and the upper palette meshes with the pegs again. The process repeats itself indefinitely with the foliot and balance weights swinging to and fro and the driving weight repeatedly checked in its fall by being compelled regularly to reverse the motion of the balance weights. By carefully arranging the proportion between the balance weights and the driving weight and the distance between the balance weights and the verge, the machine can be made to move with a predetermined regularly interrupted motion, the fall of the driving weight being periodically checked and restarted so that the average, or overall, rate of fall is uniform and effectively independent of friction.

There are two striking features of this invention. First, there is the brilliant mechanical, or kinematic, insight that the motions, in opposite directions, of the upper and lower segments of the escapement wheel can, by means of palettes at 90° to each other, be converted into the cyclic, to-and-fro motion of the balance weights. This was invent-ive genius of the highest order.

Second, and no less impressive, there is the mastery of the dynamical principle according to which the fall of the driving weight is *uniformly* checked by being made to impart the same quantity of accelerative motion to the balance weights. Understanding of the theory behind this was a long way in the future. Not until the century of Galileo and Newton could it be scientifically explained. This invention may, therefore, be described as 'precocious' for, like other medieval inventions, the principles governing its operation were far beyond contemporary scientific understanding. At the other end of the time scale, if we try to trace the antecedents of the mechanical clock we run up against a blank wall, for very little information is available. The first firm indication of the machine was in the year 1286, but who made it and for what specific purpose is not known. It may well have been an astronomical clock and descended from a line of models or instruments intended to imitate the rotations of the planets in their spheres. Its design may have resulted from the speculations of some millwrights who knew about gearing and the problems of uniform motion, and who, moreover, had astonishing insight into mechanical principles.

All we can suppose is that there must have been many attempts to devise a machine to indicate the position of the sun in its daily journey round the earth, and therefore to tell the time. Certainly many of the first clocks were astronomical ones, some of them of such elaborate design that the positions of the sun, the moon, the other five planets and even the motions of the tides could be displayed.

Clocks driven by falling weights were erected in the late-thirteenth and early-fourteenth centuries, being placed on cathedrals, churches and castles. Among the early pioneers of the art of clockmaking were Richard of Wallingford and the de Dondi family. After their time steady evolutionary improvement marked the history of the clock. In about 1500 the balance wheel and clockwork, or spring drive, were introduced and the watch was invented. But the application of spring drive raised a new problem. Coiled springs, unlike gravity, tend to run down and exert less and less driving force as they unwind. To compensate for this an ingenious method was devised of storing up energy – as it would now be called – at the beginning and then progressively releasing it as the spring unwound. This was called the 'stackfreed'. Another device to achieve the same end was the fusee. Later, in the seventeenth century, came Galileo's invention of the pendulum clock using the isochronous property of the pendulum instead of the inertia of balance weights. Later still came the dead beat and anchor escapements, notable improvements due to Robert Hooke. The apotheosis of the mechanical clock came in the eighteenth century with John Harrison's marine chronometer.

As the diurnal motion of the sun is, practically speaking, uniform and the same in summer and winter, it would have been a complicated business to calibrate clock faces with different hour marks, appropriate to the different seasons: widely spaced hours for winter nights and for summer days in northern latitudes. It is not surprising, therefore, that with the spread of the weight-driven clock the fixed, uniform hour – the astronomer's measure – replaced the old, variable one. More, by about the fourteenth century the clockmaker's art had progressed sufficiently for the hour to be divided into sixty minutes and, later, the minute to be divided into sixty seconds. Since those days the remorseless tick of the clock has marked the unending passage of time.

Astronomers, philosophers and theologians may well have been aware of the steadily flowing stream of time – Plato's moving image of eternity. The mechanical clock did more than exemplify this aspect of time, for without a precise, easily available and generally accepted time standard, the administrative, commercial and industrial arrangements of advanced civilizations would be impossible. Although we now have such sophisticated time-measuring devices as crystal oscillators and atomic clocks, we still, in the last resort, arrange all our affairs in the framework and by the measures of time that resulted from the medieval invention of the weight-driven clock.

The new industry of clockmaking was accompanied by the rise of a new and very superior class of craftsmen. These men became skilled in the design and manufacture of gears and in the different ways in which motion could be transformed and applied. There should, therefore, be a link, so far untraced, between the clockmakers of Europe and the forgotten Greek craftsmen who designed and made the Anti-Kythera mechanism. Inevitably, or so it seems, the clockmakers transcended their craft and before long they are to be found designing and superintending the construction of water-wheels; later still, in the early days of the English Industrial Revolution, clock- and watchmakers figured as key engineers for the construction and operation of textile machinery[10]. There was, in fact, a direct link between the medieval invention of the mechanical clock and the enormous industrial change that began in eighteenth-century England, a transformation that was to show the way to public affluence and now to hold out hope for the eventual ending of world poverty and famine.

Clockmaking was, for long, the pinnacle of mechanical arts and the training ground as well as the inspiration for practitioners in other branches of mechanics. So highly esteemed were the craft and its products that seventeenth-century pioneers of the Scientific Revolution and of the 'mechanical philosophy' conceived of the planetary system as a gigantic piece of celestial clockwork and God himself as the heavenly clockmaker.

Generally, the mechanical contrivances of the middle ages and the Renaissance tended to be over-complex, over-elaborate, as if their creators delighted in their new powers and had little regard for economy of operation and mechanical efficiency. This attitude changed radically in the seventeenth century but, in the meantime, it was not surprising that mechanics and magic came to be confused in the minds of the many and the distinct talents of the mechanic and the inventor ascribed to magical practices. Machines, after all, were commonly regarded as devices for cheating Nature and the machine maker consequentially regarded as a man of altogether superior knowledge and powers. Among the unsophisticated the legend of the mechanic–magician lasted a long time; it was perpetuated, for example, in the figure of Spallanzani and the tales of E.T.A. Hoffmann.

The last invention of the period to be discussed is a social and not a mechanical invention. However, its importance is so general that it merits a short discussion. It is the invention of mass adult education. We cannot date the foundation of the first universities. They evolved gradually from small groups of scholars who gathered round those cathedrals or cathedral schools that were refuges for men of learning. They were essentially Christian foundations and, indeed, the existence of universities implies a common culture based on a common religion. In its mature form the medieval university included four faculties. The junior faculty, that of Arts, provided a general education in the seven liberal arts: grammar, rhetoric and logic (the trivium) and music, geometry, arithmetic and astronomy (the quadrivium). After the Arts faculty the student could expect to go on to one of the three great professional faculties: those of Divinity, Law and Medicine. The language of instruction was, of course, Latin and examination was by defence of a thesis in public debate, the form of argument used being Aristotelian logic.

It might be assumed that medieval universities were, like modern universities, engaged in the advancement of knowledge; that research was one of their functions and the duty of every one of their members. This, however, would be a mistake. The Christian religion taught that humanity was in a fallen state; the common belief was that the golden age of secular learning had been in classical antiquity, since when knowledge and understanding had deteriorated. Accordingly, the classical authors in philosophy, literature, law and medicine were the ultimate authorities. From the thirteenth century onward the universities were at the peak of their importance and popularity. Although the philosophy first taught in universities had been Neoplatonism, through the efforts of St Thomas Aquinas Aristotle was enthroned as the master philosopher and the ultimate authority on logic and physics (defined as the science of change and therefore including biology). The books of Euclid accounted for geometry while Ptolemy was the authority for astronomy. Medicine was set out in the texts of Galen, Hippocrates and Celsus while jurisprudence was reinvigorated Roman law. Divinity meant, of course, the writings of the Church fathers. The modern university interprets its mission quite differently: research leading to new knowledge, as well as teaching, is assumed to be a defining characteristic. In medieval Europe the idea of progress had yet to be born; the ancients possessed all possible secular knowledge and common belief was in the fallen state of humanity.

Nevertheless the medieval university could hardly avoid having an influence on the advancement of knowledge and technics. The intensive study of Aristotle by the schoolmen refined that philosophical system and included the discussion of certain problems of motion (within the general question of change). These problems assumed greater importance when combined with the other endeavours – such as the efforts of artillerymen to improve their art – in the sixteenth and seventeenth centuries. In combination, these were to lead to the foundation of the science of mechanics.

Whatever effects the universities had on innovation and invention they established and consolidated a relatively homogeneous culture in western Europe with Latin as the common language from Wittenberg to Bologna and Salerno, from Oxford to Cracow. And they taught men how to think logically and to argue rationally. This was an information revolution. But the culture was even richer than this would suggest.

During the period when significant inventions and innovations were being made – indeed the universities themselves were innovations – knowledge that was basically static, knowledge of the sort the universities stood for, invited challenge. Hence arose what have been called the 'heretical sciences', those which opposed established learning. Alchemy and astrology (condemned by the Church) and natural magic were the leading forms of heretical science; Friar Roger Bacon was its best-known exponent. The transmutation of base metals into gold, the casting of a perfect horoscope and the construction of a perpetual motion machine were three notable objectives. Aristotle had insisted that if motion is to continue a force must be applied. Very well, the heretical scientist will ignore Aristotle's ruling. The heavens are in perpetual motion, but Aristotle insists that between heaven and earth there is an unbridgeable gap. The heretical scientist will try to close that gap. The needle of the mariner's compass points to the Pole Star around which the heavens rotate. Perhaps magnetism holds the key to a perpetual motion machine? The enterprise failed; all similar enterprises failed. Yet they were essential for the eventual establishment of the science of mechanics, founded as it is on the axiom of the impossibility of obtaining unlimited work from a 'perpetual motion' device or system.

The achievement of the 'heretical scientists' was that they broke the age-old 'taboo on the natural': the ancient belief that we cannot and should not even try to do what Nature does. That way disaster lies, as the legend of Prometheus shows. And the taboo still has force: we are familiar with the TV advertiser who proclaims that a particular product has 'pure natural goodness'. The operative word is 'natural'; it implies an excellence that human technics cannot attain. In overcoming this prejudice (and fear) the heretical scientists helped to open the way to modern science and technology. The revolt against medieval Aristotelianism that was a necessary feature of the Scientific Revolution of the seventeenth century was summarized by Whitehead in the form of a paradox when he described Galileo's science as fundamentally irrational (at least in Aristotelian terms).

In principle a monotheistic religion whose one God is both the creator of the world and whose concern is for the salvation of the righteous should be favourable towards science and technics. The believer who was also a discoverer or inventor had no reason to propitiate or to fear hostile and arbitrary spirits or gods who might be locally predominant. Such did not exist. At the same time Christianity rejected the enervating fatalism that had characterized many ancient religions. Associated with a revived Greek philosophy such a combination of religion and metaphysic should, in principle, positively encourage science and technics. This has been argued forcefully by Lynn White and others[11]. We can go further: in Christianity we have God in human form; God who met and associated with ordinary men and women, who knew about their everyday troubles. To such a God one could turn for understanding, protection and support more readily than to an impersonal deity, however benevolent. A religious faith of this kind must surely provide a strong shield for technologists working at the very limits of human endeavour, be they navigator, miner, inventor, metallurgist or surgeon. And in another, equally important, realm Christianity may be taken as favourable towards technics. The founder of the religion did not scorn manual labour as is confirmed by several of his sayings. Accordingly, certain monastic orders, notably the Carthusians and the Benedictines, were associated with physical labour and technics, including agriculture, iron working and medicine.

Western Europe underwent a fundamental change in the medieval period. This is confirmed by its characteristic art, its technology (including human dissection – a notable breach of the taboo) and its nascent science. Two facts are incontestable: Europe, during its highly creative centuries, was dominated by Christianity and no other civilization had surpassed the level of inventive and technical attainment that Europe had reached by the end of the middle ages. It is therefore plausible to assume that Christianity tended to encourage technical advance. The thesis, popular in the nineteenth century, that there was steady conflict between the two, a conflict that ended with the victory of science and technics, is rather less plausible now that we appreciate the enormous advances in technics that were made during the period known as the age of faith. Instances of the Church suppressing inventions, discoveries or scientific theories can, no doubt, be found. It would be surprising indeed if none had occurred. Against such instances must be set important cases, such as those discussed above and the many more not mentioned, that were not suppressed.

For the present we may imagine any student, at a modern university, looking anxiously at the clock as he or she picks up a sheaf of papers and rushes off to a lecture, perhaps on the philosophy of Aristotle or on geometry (probably non-Euclidean). This everyday event is as complete a testimony to the inventiveness of medieval Europe as one could propose.

New Worlds and
an Information Revolution

The English water mills recorded in the *Domesday Book* (1084) of William the Conqueror provided power for a population of about three million. Although we cannot estimate the average power of these mills, nor of the power available from horses and oxen, the figure does suggest that the gross horsepower per capita may have been relatively high. Water mills were to remain the main source of mechanical power in England throughout the Industrial Revolution and up to the third decade or so of the nineteenth century. Their great historical importance has been overshadowed by the dramatic successes of the steam engine in more recent years.

Water power was a strategic technology in the centuries that followed the millennium for, as we saw, it stimulated advances in other branches of technics. In the fourteenth century, however, the pace of innovation slackened. Historians are uncertain why this occurred although it is possible that the series of plagues that hit Europe in the middle of the century (including the 'Black Death' of 1348) may have been a cause. However, in the century that followed, the pace of innovation increased and has not slackened since then. Indeed, at the end of the middle ages and at the beginning of the Renaissance was made one of the most potent inventions of all time: that of printing by movable type. Although there have been several claimants to have been the first inventor and a number of presumed locations for the first invention, there are good reasons for locating the original invention in Europe in the middle of the fifteenth century and awarding the credit to Johan Gutenberg, a goldsmith of Mainz, in Germany[1].

In one very familiar form printing has existed from time immemorial. The royal seal, the signet ring, the punches used by gold- and silversmiths and, invented more recently, the bureaucrat's rubber stamp are all simple printing devices. However, before the printing of books, journals and newspapers became possible several ancillary inventions, or innovations, had to be made. A relatively cheap and effective raw material[2] – paper – had to be introduced (from China), printing ink had to be developed and the press had to be adapted (probably from the ancient art of wine making). Above all, the problem of making type cheaply and accurately had to be solved; and, too easily overlooked, a suitable alphabet had to be in common use. The Roman alphabet, universal in Europe, with its 26 letters happened to be peculiarly adaptable to mechanical printing.

It is reasonable to suppose that the demand for books was increasing during this era of technical progress and geographical discovery. And furthermore that this demand was being frustrated by the shortage, incompetence and restrictive practices of scribes and copiers. We know that by about 1400, in northern Italy, playing-cards and simple devotional pictures representing saints were being printed by means of stamps. But the cost of making a stamp carrying enough words, numbers and punctuation marks to constitute one page of the Bible would have been prohibitive. The problem was made more urgent by a requirement that every church should have a copy of the Bible. There was, therefore, an impasse and it must have seemed insurmountable to those concerned with book production. Yet the difficulty was resolved, and in solving it the waning middle ages left us with one of its greatest gifts: the printing press and the book as we know it.

Johan Gutenberg, or Gensfleisch (1394/99–1467),

certainly produced printed books before anyone else, and there are other facts in the case that make his claim to be accepted as the inventor of printing by movable type extremely strong. He was born in or near Mainz where his father was goldsmith to the archbishop, a post of some responsibility and, one supposes, profit. Johan followed in his father's footsteps, becoming, like many other Germans, a skilled metallurgical craftsman, a qualification that, as events showed, was to be an essential requirement for the inventor of what we should call the mass-production printing press.

Gutenberg's solution to the problem of printing fell into two quite distinct stages. In the first place he had to solve it in principle, theoretically as it were, and then, secondly, he had to make the invention-in-principle actually work; in other words he had to take it through what, today, would be called the process of development. The first, or theoretical, stage was simple. Instead of trying to carve out a separate and immensely complicated stamp for each page of print, or even for each separate word, Gutenberg proposed to make stamps for each individual letter, punctuation mark or other symbol. These small, individual units of type were made in large numbers and each one was put in an appropriate box: one box for all the capital 'A's, another for the small 'a's and so on. When he wanted to set up a page of type all he had to do was to select the necessary letters to make the desired sequence of words, set them in a frame and, when the page was complete, clamp them firmly together. He had now to ink the type face, using an 'ink ball', and press it against a sheet of paper to print off a page. After he had run off as many prints as he wanted he unclamped and dismantled the type, returning each small unit to its appropriate box to be used again when the next page was set up.

So much for the solution-in-principle; in practice Gutenberg would soon have run up against two serious and related problems that must have caused him a good

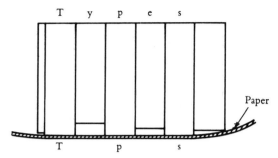

FIG 3.1 *Printing defects due to variations in height of type*

deal of time, patience, money and thought. To explain these problems let us suppose that the small units of type are clamped together as shown in figure 3.1. Some units, slightly longer than the average, project a little while others are somewhat retracted. Owing to these accidental variations in length some letters will be heavily printed while others will be faint or entirely missed, being masked by their immediate neighbours. For effective printing all the units of type must have exactly the same length; or, more precisely, the type face must be uniformly flat. Length, however, is only one of three dimensions and, as it happens, one of the other two is also critical. Let us consider what will happen if the thickness of the type,

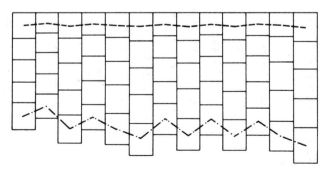

FIG 3.2 *Printing defects due to variation in thickness of type*

from the top of the page downward, varies beyond the permissible. It turns out that errors are cumulative, getting worse line by line, until after surprisingly few lines the individual letters are so mixed up that the print is incoherent, letters from one line being juxtaposed with those of another (figure 3.2).

The point is that these snags could hardly have become apparent before Gutenberg had started out on his invention. They must have arisen during the second, or development, stage. By using the greatest care and throwing away all those units of type that failed to meet the minimum standards of precision in two critical dimensions Gutenberg could have built up a stock of acceptable units of type. But the cost would have been prohibitive and the main advantage of the invention lost. This ruled out units of type made of wood or any other organic substance. Here was a cardinal dilemma.

We do not know how Gutenberg solved it; we do not, that is, know the train of thought and experiment that led to the final answer; but, in the event the solution-in-practice is no less brilliant than the original conception, the solution-in-principle. Virtually perfect accuracy could be obtained if all the type – of metal – was cast in the same mould. In this way the need for extreme precision was transferred from each individual unit of type to the common mould in which they were all to be cast.

As invented and developed by Gutenberg the type mould was adjustable, consisting of two separate, L-shaped members, free to slide over each other (figure 3.3). There were two good reasons for making the mould like this. In the first place, if the two sections can come apart cast type can be easily removed; and, in the second place, if the mould is adjustable in the third, non-critical dimension the width of type can be varied so that a small or lower-case 'i', for example, takes up less space along the line than does a capital or upper-case W. In this way a more pleasing and legible print is obtained.

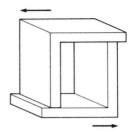

FIG 3.3   Type mould

The open, lower end of the mould is closed by the matrix, a piece of soft metal such as copper, carrying the impress of the particular letter to be cast. The letter had previously been punched in the soft metal by means of a hardened steel punch. When a sufficient number of units have been cast, a new matrix is substituted carrying a different letter and the process is repeated. The type metal must not, of course, stick to the sides of the mould or to the matrix. For this purpose an alloy of tin, zinc and lead was found to be suitable. It can now be appreciated how necessary it was for the inventor of printing by means of movable type to have had considerable experience of metals and in metal working. Gutenberg's training must have prepared him for this work.

A unit of type, removed from the mould, is shown in figure 3.4. The length is fixed by the groove and the

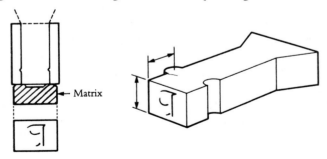

FIG 3.4   Matrix and type

corresponding ridge on the opposite side; these position all the units of type by effectively locking them together in the frame. The remaining problems were simple. The best composition for printer's ink was found, empirically, to be a mixture of lamp black and linseed oil; it was so successful that it remained in use up to the nineteenth century. The ink ball was made of leather and the principle of the screw press was taken, as we remarked, from the long-established art of wine pressing.

Gutenberg's printing press revolutionized the publication of books. The first known printed book, produced at Mainz in 1455, was the Gutenberg Bible. It has been estimated that more books were published in the first fifty years following Gutenberg (up, that is, to the beginning of the sixteenth century) than had been produced in the previous thousand years. A twentyfold increase in productivity is highly impressive and it would not be unreasonable to regard Gutenberg as the first production engineer. Furthermore, it is easy to overlook Gutenberg's achievement of complete interchangeability in manufacture; each little unit of type could be used over and over again and fitted between any other little units (see chapter 12). Finally, his invention brought about the first revolution in information technology. It was to be a long time – two hundred and fifty years – before another invention as dramatic and as important as Gutenberg's was to be made.

It is often claimed that printing was invented independently and possibly somewhat earlier in China and in Korea. In the case of China, the nature of the elaborate script made it impossible that Chinese printing could be of the same form as Gutenberg's. It seems likely that porcelain casts were made of each Chinese ideogram as it was required; this was the 'rubber stamp' solution to the problem that would have been impracticable for the Roman alphabet. The Koreans, on the other hand, are said to have invented printing by metal type during the course of the fifteenth century. It would be unwise to assume that the invention spread from Korea half-way round the world to Germany and was there copied by Gutenberg and his associates. On the contrary, the relatively advanced state of the metallurgical arts in Germany, the immense pent-up demand for books and the skills that medieval European inventors and mechanics had certainly acquired all combined to indicate that the time was ripe for the invention to be made independently in Europe. There is nothing implausible about simultaneous and independent invention. When two or more societies reach approximately the same standard of technical skill then the combination of felt wants, opportunities and stimuli provided by commerce, industry, administration and learning make it increasingly likely that the same things will be invented simultaneously and independently in each society. The more technically advanced communities become and the more communications improve, the greater the likelihood of simultaneous invention. It may now be said to be the rule rather than the exception. For example, radar was invented simultaneously and quite independently in half-a-dozen countries while the jet engine was invented simultaneously and independently in Britain and in Germany.

Most inventions have long prehistories. This appears to conflict with the commonly accepted view of the heroic, individual inventor, battling against scepticism and hostility to get his invention accepted. The explanation, which goes some way to reconciling these incompatible views, is that most truly revolutionary inventions are made by individuals working outside the craft or technique concerned; in which case much of the prehistory is not part of the history of the craft concerned. Gutenberg may well have been such an individual; there is no evidence that he ever earned a living, or served an apprenticeship, as a scribe or copyist. We shall come across other instances of outsiders making revolutionary inventions.

The mechanical clock and the printing press still determine the course of daily life to a quite remarkable extent and few of the individual inventions made subsequently can rival them in this respect. They were only two of the many innovations made in the medieval period, the motives for which ranged from the economic through the humanitarian – improving the amenities of life by inventing such aids as spectacles – to delight in ingenuity for its own sake or for ends that were mystical or magical[3].

Among the latter were automata and the highly elaborate public clocks of Europe's ancient cities and the long-sustained attempts to make a perpetual motion machine. Several factors were involved in the search for a perpetual motion device. The anti-Aristotelian heretical scientists agreed with Aristotle that everyday experience seemed to show that motion required a force but argued that a more exalted mechanics than that of Aristotle might be discovered and, on this higher plane, perpetual motion machines might be possible. Their endeavour failed but it made an indirect contribution to the development and understanding of clockwork.

From another point of view the idea of a perpetual motion machine was by no means absurd; on the contrary, the phenomenon was a matter of common experience. Apart from the (in practice, inaccessible) eternal rotation of the heavens, the tides on earth were in ceaseless motion, linked mysteriously to the motion of the moon; streams and rivers flowed on endlessly and the winds of the earth, although erratic, showed no signs of dying away to a sustained calm. Every tide-mill or water-wheel, every windmill, was therefore a perpetual motion machine of sorts. And if, therefore, an inventor could somehow master the basic principles of rivers, tides and wind he might hope to solve the problem completely and make a simple, self-contained perpetual motion machine. The lodestone – or magnet – seemed to hold out particular promise. These various endeavours, it must be stressed, were all made in the absence of any understanding of the hydrologic cycle, thermodynamics or magnetism. On the contrary, their repeated failure was to be a necessary condition for the achievement of scientific explanations.

Dreamers, speculators, 'heretical scientists' can perhaps be most fairly represented as one wing of a large movement, the other of which was peopled by severely practical inventors like Gutenberg and his many less well known contemporaries together with artists, architects and engineers. An informed historian is of the opinion that:

> From the end of the fourteenth century it was the superior group that Leonardo Olschki called the 'artist engineers' that took over from the academic philosophers as the pace makers of intellectual life. They were essentially a product of Italian urban society. Their achievement was to add to the logical control of argument of the philosophers a rational control of materials of many different kinds in painting, sculpture, architecture, engineering, canal building, fortifications, gunnery, music.[4]

This generalization is equally applicable to Gutenberg's Germany if we modify the list of technologies, putting rather more emphasis on mining, metal working and related chemical arts. Within the limits set by the available materials and power sources these arts were pushed to their limits and the geometrization that was such a feature of much late-medieval and Renaissance technics was to be one of the key factors in the Scientific Revolution of the seventeenth century. However, the inventions and innovations of the Alpine region – northern Italy and southern Germany – were not the only technical and quasi-technical factors that were to influence the great scientific movement. There was also the remarkable legacy of the discoveries – a truly unique episode in world history and one that shifts our attention from central Europe to the coasts of Spain and Portugal and the commercial ports of Italy.

At the beginning of the fifteenth century knowledge of the world outside Europe was still very limited. Early travellers by land, like Carpini, Marco Polo and Odoric, brought back information about Mongolia and China; seaward and in the other direction the Norsemen sailed beyond Iceland to Greenland and, in all probability, reached and even made settlements on the mainland of America (Labrador) (see Note at the end of the chapter). Much further south other voyagers rediscovered the Canary islands (known to the Romans) in about 1336 and, later on, Madeira, the Azores and the Cape Verde islands. The rest of the world was a matter of conjecture, fables – 'here be dragons' – and the hypotheses of Ptolemy's *Geography*. However, the dramatic series of discoveries that were to follow these tentative probings were to change the picture completely, filling in the enormous blanks on the map and establishing the true form of the world. No doubt there were economic motives for these discoveries: the opening up of trade routes that avoided the hazards of trading over land. They also owed a great deal to what can only be called the new philosophy of Nature; a down-to-earth and experimental attitude coupled with an unquenchable desire to find out at any cost. And the whole great adventure was made possible by advances in the arts of shipbuilding and navigation – the introduction of the three-masted ship, the development of the movable rudder to replace the steering oar and the invention of the mariner's compass.

Navigation, as opposed to the crafts of pilotage and seamanship, is a mathematical technology[5]. In the middle of the fifteenth century the Portuguese prince known as Henry the Navigator established a school of navigation – arguably the first technical college in the world – and encouraged his sailors to push further and further down the coast of Africa and out into the unknown Atlantic. In 1455 the constellation of the Southern Cross was discovered and in 1481 navigators crossed the equator to see, for the first time, stars wheeling round a pole in the southern heavens. In 1492 the Genoese sailor Columbus reached the Americas; two years later Bartholomew Diaz reached the Cape of Good Hope. In 1497 Vasco da Gama rounded the Cape and sailed on to reach India, thus incidentally disposing of Ptolemy's supposed southern continent by sailing across it. Finally, in 1517 Magellan's expedition set out on the voyage that was to end three years later when one of his ships got back to Portugal having circumnavigated the globe. In less than a century the Europeans had burst out of the confines of their little continent, as known from time immemorial, had discovered new continents and oceans – including the largest of all, the Pacific – and had gone right round the world. They had touched on the fringe of the Antarctic continent and at last knew beyond doubt the form and nature of their home, the earth. The laws of Nature were everywhere uniform; south of the equator, in the antipodes, stones did not fall down to the antipodean sky; they fell down to the ground. As far as physical phenomena were concerned everything was the same as at home. The great sphere of the heavens, with previously unknown stars and constellations in the southern half, completely surrounded the little globe of the earth.

It is hardly possible to overestimate the physical and – unprecedented – spiritual courage of the Portuguese, Spanish and Italian navigators. Apart from the obvious dangers of the sea there were imagined dangers and associated fears made worse by being obscure or unknown. The scholar, in the security and comfort of his study, could easily convince himself that the world was a perfect sphere and that the laws of Nature were everywhere the same. It could be a very different matter on high and unknown seas.

The social, political, economic and cultural consequences of the discoveries are too vast even to enumerate here. It is enough to note that the unambiguous establishment of the true form of the earth made it easy for men to conceive of it in realistic terms as a globe that, moreover, could be represented as such by a small model, just as a ship could be. And so, from about 1490, we find Martin Behaim, the cartographer of Nurnberg, making small model earths, or terrestrial globes, on which the progress of the discoveries could be marked. That terrestrial globes were relatively commonplace is proved by *The Ambassadors*, the painting by Hans Holbein (1497–1543). On the table between the two hard-faced, worldly, men are the tools of their trade: maps, legal documents, treaties – and a terrestrial globe. One key implication of this widely accepted view of the earth is straightforward. The earth's two near-neighbours, the sun and the moon, are clearly globes and they are certainly in motion. As the earth, too, is a globe an astronomer could, reasonably, take the next step and propose, if astronomical considerations required it, that the globular earth moves, orbiting round the sun. Proving it was a much more difficult matter.

The situation was explained by (Sir) Christopher Wren in the course of his inaugural lecture at Gresham's College in 1657. Referring to the voyages of Columbus and da Gama, he remarked:

> By these and succeeding Voyages, performed by the Circumnavigators of our Nation, the Earth was concluded to be truly globous, and equally habitable round. This gave occasion to Copernicus to guess why this Body of Earth of so apt a figure for Motion, might not move among the Coelestial Bodies; it seemed to him, in the Consequences probable and apt to solve the Appearances & finding it likewise among the antiquated Opinions

he resolved upon this occasion to restore Astronomy.[6]

In other words, there is no good reason why a ball, globe or sphere should not move. A proven spherical earth could therefore be conceived to be in motion. Astronomical considerations strengthened this hypothesis. Geocentric astronomy and the Ptolemaic system of epicycles and deferents could not be made to account acceptably for the movements of the planets. This was intellectually unsatisfactory; it made astrological prediction a hazardous affair; worst of all the precession of the ecliptic (due to the motion of the earth) made the determination of Easter uncertain. Reform was urgently needed. The heliocentric system[7] devised by the Polish astronomer Nicolaus Copernicus (1473–1543), imposed regularity on the heavens by assuming that the earth orbited round the sun. The retrograde motion and anomalies of the orbiting planets were shown to be due to the motion of the earth, catching up with and overtaking the outer planets (Mars, Jupiter and Saturn) or being overtaken by the inner planets (Mercury and Venus). The sequence of night and day was explained simply as due to the spinning of the earth on its axis.

The Copernican system, set out in *De revolutionibus orbium coelestium* (Nurnberg, 1543), 'restored' systematic planetary astronomy. But it made heavy drafts on accepted scientific understanding and on common sense. If the earth is spinning rapidly why is there not a tremendous gale blowing from the east? Why are we not thrown off? If it is orbiting round the sun why are we not left behind? Why is there no sensation of motion? Why is no stellar parallax observable? The first question could be answered by asserting that the air – the atmosphere – travels with the earth and must, therefore, be finite. The second and third questions, and related problems, were more difficult; to answer them a new system of mechanics was required.

The fourth question was relatively easy: just as on board a ship under way through a smooth sea, there need be no sensation of motion. The last question was not answered until the nineteenth century when, at last, stellar parallax was detected.

De revolutionibus was not the only major work to appear in 1543. In the same year Andreas Vesalius (1514–64) published, in Venice, his *De humani corporis fabrica*. Vesalius was born in Brussels and attended the universities of Louvain and Paris before going to Padua – the great medical university of the time – where he graduated MD in 1537, subsequently becoming Professor of Anatomy. In his book, a classic of human anatomy, he effectively refuted many of the assertions of Galen, whose dissections had been restricted to apes. The book had the additional merit of being magnificently illustrated. From a general point of view the most revealing illustration is the frontispiece, which shows Vesalius himself – a stocky, practical man – dissecting a body with a large number of students gathered round him. This confirmed that an almost universal human taboo had been overcome in Europe. A relatively minor social taboo was overcome by Vesalius himself. Before his time it had been customary for a learned professor to lecture from a pulpit above the table on which the body was dissected by a low, menial fellow called the 'demonstrator'. The professor would read passages from Galen and the other classical authors in the course of his lecture while the demonstrator revealed the organs mentioned in the texts.

Three years earlier Vannocio Biringuccio (1480–*c.* 1539) had published a book on a very different theme. This was his *De la pirotechnia* (Venice, 1540). Early in the sixteenth century a series of little handbooks on assaying had been published, anonymously, in Germany: the *Probier Bergbuchlein* and the *Nutzlicher Bergbuchlein*. Biringuccio's work was more ambitious. He explained how ores and lodes were discovered, the parting of gold and silver, assaying, and the properties of copper, zinc and other metals and alloys. He discussed the properties of mercury, sulphur, antimony, etc., and he gave what was probably the first account of a reverberatory furnace. He described how, in the steel-making process, a bloom of wrought iron (low in carbon content) should be added to molten pig iron (high in carbon) and the mixture constantly stirred, as in cooking, until the right mix of steel was obtained (this resembles the technique of 'puddling' iron; see chapter 7). Marble chips were added as a flux. The quality of the steel (intermediate in carbon) that was produced by this method would obviously depend very much on the skill of the workers. Needless to say Biringuccio knew nothing of the role of carbon in the formation of wrought iron, pig iron or steel.

De la pirotechnia contained an early account, with illustrations, of boring engines for cannon – perhaps the first published description of an industrial machine tool – and gave descriptions of cannon and bell founding as well as wire drawing and the manufacture of fireworks. There was only a brief mention of mining and little on geology and mineralogy. Biringuccio was contemptuous of the aspirations and claims of the alchemists. He eschewed speculation and was concerned solely with established facts. His book was published over nearly 140 years and went through many editions.

In contrast to, and complementing, De la pirotechnia was De re metallica (Basel, 1556), the author of which, Georg Bauer, had latinized his name to Agricola. Bauer (1494–1555), had received a wide education, including the classics, and had attended several of the north Italian universities. He qualified as a physician and was appointed town doctor to Joachimst(h)al, a town in north-west Bohemia with valuable silver mines[8]. De re metallica was the last and most comprehensive of Bauer's eight works on mining. It, too, went through many editions and remained a standard work for about a hundred and fifty years. The Latin of *De re metallica* is in contrast to the Italian of *De la pirotechnia*; while Agricola has many superb illustrations to Biringuccio's relatively few. The books complement each other in that Bauer deals exhaustively with mining and mining machinery as well as with metals and their treatment.

Bauer pointed out, in his preface, that mining had become an honourable and useful occupation. Although slaves had, in ancient times, been employed as miners this was also true of many other reputable occupations, including medicine. Having established the credentials of mining he began by giving an account of the mining geology and of the techniques and tools of surveying. He then went on to describe the different windlasses, winches, wheelbarrows and small trucks used for moving ores. A particular problem was that of moving heavy ore along the rough ground of the twisting galleries from the face to the shaft or entrance. One way of doing this was by means of a small truck, called a 'dog' (see plates). It was of rectangular form, made of wood and bound with wrought iron bands. It had a pair of large wheels and, in front, a pair of small wheels, close together. A stout iron pin projected below the rims of the small wheels and into a narrow gap between two smooth wooden planks running parallel along the ground; the two small wheels ran on these two planks. In this way the 'dog' was guided along the narrow galleries. *In the 'dog' we have the origins of the modern railroad.* Before the end of the century trucks running along wooden railways were in use in German mines.

Bauer next turned his attention to pumps for keeping mines clear of water, always a critical problem. He enumerated, illustrated and described seven different types of suction pumps including a multiple pump driven by a massive water-wheel and a combined suction and force pump. Other types of pumps that he described included bucket pumps and rag-and-chain pumps driven by water-wheels or animal muscle; windmills he did not consider.

A particular problem that he dealt with was how to boost the power of a water-wheel if, because of a summer drought, there was so little water in the stream that the wheel could not work the pumps. The solution he recommends is revealing. Assuming that the bed of the stream is steep and the water passes over the water-wheel, another, smaller, water-wheel should be placed just upstream from the main wheel. This small wheel can work a pump to raise water from the level of the main wheel and let it flow over that wheel thus adding to the water of the stream. Boosted in this way enough water will flow over the main wheel to enable it to work the mine pumps. Of course, initially human effort will be needed to help turn the main water-wheel so that pumping can begin. Once the small wheel can draw on water pumped up from the mine the system becomes self-sustaining.

What is significant about this is that Bauer did not envisage the problem in the same way that hydraulic engineers of the eighteenth and subsequent centuries would have done. They would have realized that more power could be got by increasing the head, or fall, of water exploited; and would then have proceeded to construct one of the hydraulic machines that they knew could do this (see below, chapters 6, 12). Bauer did not – could not – think like this; he was on the other side of the Scientific Revolution of the seventeenth century when the general principles underlying the solution to the problem were discovered. Instead, Bauer thought of finding a way to make more water flow over the main water-wheel. The result was the same in both cases; the approach was different.

The second half of *De re metallica* dealt with assaying and the treatment of ores. Throughout the book Bauer was, like Biringuccio, contemptuous of the alchemists. He accepted only established facts and his accounts are free of superstition and legend. His scepticism extends to the widely practised art of finding buried metals by means of a divining rod or twig. Divining was, in fact is, an ancient and reputable, or at least harmless, art. Bauer considered the matter carefully but decided against it. Divining, he said, was derived from ancient magic and was still practised by simple, ordinary miners. Bauer thinks it should be discouraged. Experience and practice should be enough to enable a miner to find exploitable veins.

It is disturbing to find, after this healthy scepticism, Bauer expressing the belief that if a magnet is smeared with garlic it loses its power to attract iron. To us this might appear to be simple credulity; more probably it was an instance of the familiar tendency of people in any age to accept uncritically beliefs – or superstitions – in fields outside their own areas of competence. The properties of magnets were incidental to the practices of mining and assaying.

The third in a trio of authoritative works on mining and metals to be published in the sixteenth century was Lazarus Ercker's *Treatise on ores and assaying* (Prague, 1574). Ercker, who was Chief Superintendent of Mines in Bohemia, wrote in German and his book was translated into English in the following century. There was nothing unusual about the language of Ercker's book. The engineers, administrators and managers of the mines of Bohemia and of neighbouring Slovakia (then part of Hungary) were German; the manual workers were Czechs and Slovaks. Metal working, like mining, was widely acknowledged to be a German technology.

Ercker, like Biringuccio and Bauer, provided an accurate factual record, substantially devoid of superstition and speculation. There were not yet systematic or scientific bases for metallurgy, mineralogy or geology but at least the knowledge had been assembled from which such generalizations could eventually be derived. The questions arising from these works are, first, to what extent were the rational attitudes of Biringuccio, Bauer and Ercker derived from, or a consequence of, the hard-won experiences of generations of craftsmen and technicians and, second, to what extent were they the consequence of the revival of learning in the middle ages? Here, we might recall the strictly rational approach to Nature taken by Aristotle in his three major biological works. Underlying our two questions is the very general one: how do we account for the remarkable courage of later medieval and Renaissance technologists in confronting Nature, in breaking the taboo on the natural, quite evident in the cases of the navigators, the anatomists and the miners? Most probably that courage derived in part from the teachings of a strictly monotheistic religion, in part from the knowledge of prior technical achievement and, to some extent, from the examples of the heretical scientists.

In the last quarter of the sixteenth century some interesting books on mechanical arts were published. These included Agostino Ramelli's *Le diverse et artificiose machine* (Paris, 1588), and Jacques Besson's *Theatrum instrumentorum et machinarum* (Lyon, 1569). These two works are significant in that they were published in western Europe and not in that cradle of European technology, the Alpine region. Besson included in his book an illustration, later often reproduced, showing a wood-turning lathe with a lead screw. The lathe was being used to turn a piece of woodwork, a chair leg perhaps, on which a screw pattern was being cut exactly following the pattern on the lead screw. Another book, written by V. Zonca at the beginning of the seventeenth century – *Nuovo teatro di machine et edifici* (Padua, 1607) – described and illustrated a silk-spinning mill in northern Italy (see plates). The spinning machines were driven by a water-wheel and twist imparted to the thread by a flyer. In the course of the eighteenth century this machinery was destined to become immensely important in a different context and in another country.

A hundred years after the invention of the printing press and the publication of the Gutenberg Bible a number of books were published that, in effect, indicated the state of technology and summarized the achievements of the middle ages. Over this period the pace of innovation did not slacken in Europe. At the same time the conquest and settlement of new worlds followed rapidly on the voyages of the navigators. In 1520 Hernan Cortes with his tiny army and his handful of cavalry entered Mexico City, having overthrown the kingdom of the Aztecs. The Conquistador with cannon and arquebus, military discipline and cavalry, had set about establishing vast new empires on the ruins of less successful cultures. The New World began at the same time as the first information revolution.

Technology and Understanding

We can now assert that, over the period known as the middle ages, European civilization underwent a fundamental and irreversible change. Although we are far from a full understanding of the causes and implications of this change, it is quite apparent that people looked on the world around them with different eyes from those of other and older civilizations.

This was first clearly apparent in the art and architecture of the Renaissance. However, it is quite impossible to mention more than a few of the many artists active at this time. Of them we mention only four. Leon Battista Alberti (1404–72), who studied at the University of Bologna, was a man of wide interests that included practical science and mathematics; these studies influenced his seminal writings on architecture. Of the artists, Paolo Uccello's (1397–1475) paintings show the influence of the revived geometry of the time, while Piero della Francesca (*c.* 1415–*c.* 1493), wrote a treatise on perspective. The paragon, from our point of view, was Leonardo da Vinci (1452–1519) who, besides being an artist of genius was also the great recorder (if not the actual inventor) of mechanical devices and an anatomist. These artists depicted – or created – a new world of perception and understanding. The people they painted were living people and animals, anatomically correct and not idealized abstractions; Nature was represented as it appears to our eyes and geometry provided the framework for art. Furthermore, the evidence of Renaissance art and architecture supports the view that humanity had become much less fearful of Nature: it was there to be overcome, to be mastered.

We cannot leave the subject of new worlds and the first information revolution without mentioning achievements in realms other than the fine arts and technics, for this was an incredibly rich period in human history. The religious movements known as the Reformation and the Counter Reformation are too vast to be discussed here and their relationships with technology have hardly yet been studied. On the other hand works of imaginative literature, particularly those of the Elizabethan dramatists, at the dawn of the scientific age, offer remarkable insights into contemporary thought and feeling. The concluding scene of Christopher Marlowe's play *Faustus* finds the doomed philosopher in an agony of terror and remorse as his allotted span of unlimited power runs out, after which his soul must be forfeit to hell. The clock strikes the half-hour; Faustus desperately hopes for some form of reprieve, even total annihilation, until, at last, the clock strikes the hour of midnight and the demons enter to carry him off to eternal damnation. What is perfectly clear here is the mechanization of time, which is not to be thought of as some meandering stream, or something subjective, but as mechanical, mathematical, relentless. Or, as Newton defined it many years later, 'Absolute, true and mathematical time of itself, and from its own nature, flows equably without relation to anything external.'

In his play *Tamburlaine*, Marlowe shows an acute

understanding of the concept of the secret weapon, so familiar in recent cold-war politics:

> I will, with engines never exercised,
> Conquer, sack and utterly consume,
> Your cities and your golden palaces,
> And with the flames that beat against the clouds,
> Incense the heavens and make the stars to melt,
> As if they were the tears of Mahomet.

This passage almost foreshadows thermonuclear war.

Shakespeare, it has been pointed out, was not interested in the new science of Copernicus, Kepler and Galileo that was being established at the time he was writing his plays and that was to influence, so strongly, later poets and playwrights. Their science was astronomical and cosmological. Shakespeare's concern was for the countryside: for trees, wild flowers and Falstaff's green fields. He is also deeply concerned, in both the plays and the sonnets, with time. Among many examples are:

> Come what, come may.
> Time and the hour run through the roughest day.
>
> *(Macbeth)*

> Come Desdemona, I have but an hour,
> Of love, or worldly matters and direction,
> To spend with thee: we must obey the time.
>
> *(Othello)*

> The Windsor bell hath struck twelve; the minute
> draws on.   *(The Merry Wives of Windsor)*

> The clock struck twelve when I did send the nurse
> In half an hour she promised to return.
>
> *(Romeo and Juliet)*

> Like the waves make towards the pebbled shore,
> So do our minutes hasten to their end.
>
> *(Sonnet 60)*

He can be guilty of a flagrant anachronism, in horology as in other matters:

Peace! Count the clock –
The clock hath stricken three,
'Tis time to part.

(*Julius Caesar*)

But clocks are not mentioned in *Anthony and Cleopatra, Coriolanus, Troilus and Cressida, King Lear, Cymbeline, Timon of Athens* and the repulsive *Titus Andronicus*, all of which are set in antiquity.

Finally, we remark that Puck's famous boast,

I'll put a girdle round the Earth,
In forty minutes.

(*A Midsummer Night's Dream*)

foreshadows the orbital flight time of space shuttles with remarkable accuracy.

The audiences, then, for Shakespeare as for Marlowe, can be assumed to have been entirely familiar with the notion of quantified time, divided mathematically into hours and minutes of equal durations and to which the affairs of men and women were rigorously subject.

Whatever the insights and understanding of the philosophers and theologians, it is clear that by the beginning of the scientific age this notion of time was widely accepted. It was certainly central to the science of mechanics, the great intellectual achievement of the seventeenth century. In its origin it owed much to the skills and intelligence of the clockmakers who were responding to the specific demands of the Church, of monarchs and of merchants. And the world picture owed a good deal to the navigators and globe makers who persuaded men to look on the world with new eyes. Clocks and globes were the tangible representations of a mechanized system of the world. The historian R. G. Collingwood wrote that the Renaissance view of Nature[9]:

. . . is based on the human experience of designing and constructing machines. The Greeks and the Romans were not machine-users, except to a very small extent: their catapults and water clocks were not a prominent enough feature of their life to affect the way in which they conceived the relation between between themselves and the world. But by the sixteenth century . . . the printing press and the windmill, the lever, the pump and the pulley, the clock and the wheel-barrow, and a host of machines in use among miners and engineers were established features of daily life. Everyone understood the nature of a machine, and the experience of making and using such things had become part of the general consciousness of European man. It was an easy step to the proposition: as a clockmaker or a millwright is to a clock or a mill, so is God to Nature.

Technology, therefore, had made a fundamental contribution to the interpretation of the world and to the rise of science in the seventeenth century.

A Note on Discovery

It is sometimes stated that the Norsemen, or Vikings, first discovered America. This claim rests on some misunderstanding of what is meant by discovery and is therefore unacceptable. The Vikings were sea rovers who at various times founded settlements on the islands and coasts of Great Britain as well as on more distant Atlantic islands such as the Faroes and Iceland. They settled also on the coast of Greenland and, almost certainly, on Labrador. There is no evidence that they had any conception of a continent and this is crucial to the problem. It seems likely that as far as they were concerned Labrador was just another island on which they could settle, at any rate for a time. The Vikings were skilful and courageous sailors but they were not navigators. That is to say, they were craftsmen and not technologists.

The Portuguese, Spanish and Italian navigators, on the other hand, had a hypothesis that the earth was a finite, spherical body. It was plausible, on this hypothesis, to believe that by sailing west one could reach Cathay and India. Columbus and his immediate successors failed to reach China or India that way but did discover, in a very short time, the islands of the Caribbean, the continents of South and North America and the Pacific Ocean. Columbus himself died still believing that he had discovered the western route to India. His was a fruitful error and fruitful errors are not uncommon in the annals of science. As Francis Bacon remarked, truth emerges more readily from error than from confusion.

Recently, on the occasion of the quincentenary of his voyage, Columbus has been strongly criticized for the alleged atrocities that marked the European colonization of North and South America. It is not clear, however, on what grounds a man can be held responsible for the actions of others who followed him and of which he can have known nothing. And it surely cannot be supposed that, had it not been for Columbus, the Americas would have remained undiscovered?

4

The Scientific Revolution

The great achievement of St Thomas Aquinas was to reconcile Aristotelian philosophy with the doctrines of the Catholic Church. Protestants such as Pierre de la Ramée, or Petrus Ramus, had, therefore, a reason to reject Aristotelian philosophy. Furthermore, increasing knowledge of ancient learning must inevitably have led to a realization of the superiority of contemporary factual knowledge and the enormous superiority of contemporary technics; sixteenth-century artillery would have ended the siege of Troy in a few weeks! At another level the renewed vitality of Roman law (it had never been extinguished) in medieval Europe implied, sooner or later, the idea of progress. Maine pointed out that the two dangers confronting law, and civilized society that is held together by law, in their infancy, are precocious critical intelligence, as in ancient Greece, and close identification with religion. The latter, he wrote, 'has chained down the mass of the human race to those views of life and conduct which they entertained at the time when their usages were first consolidated into a systematic form'. By marking off extensive areas of action and experience that were, by implication, not subject to religious interdiction, while at the same time allowing the individual the potential benefits of a monotheistic religion, the revival of Roman law may well have expedited the growth of science and technics over the past four hundred years.

Francis Bacon[1], Lord Verulam and later Viscount St Albans (1561 – 1626 *n.s.*), was an able lawyer and a skilful courtier who rose to become Lord Chancellor of England in 1618. Four years later he was found guilty of taking bribes, was discharged from office, exiled from court and heavily fined, although the fine, it seems, was not exacted. The details of his career suggest that he was a ruthlessly ambitious, none-too-scrupulous, man, who had on his conscience the betrayal of a friend and benefactor. But there was another, quite different, side to Francis Bacon. He was also the uniquely persuasive advocate and prophet of technology. He was the first man to lay down a comprehensive programme for science and technology and his influence in this respect extended over civilized Europe.

Apart from that brief flowering of the intellect during the times of the medieval scholars Occam, Grosseteste and Bradwardine, Britain had contributed little to the advance of civilization, so strongly centred on Italy and Germany. Britain was a land of warrior kings and turbulent nobles, not a land of artists, writers, craftsmen and inventors. At the end of the sixteenth century the situation changed completely; the age of Shakespeare needs no apology and Bacon was of that age. What stimulated the British Renaissance can only be conjectured. The rapid expansion of overseas exploration and the immense possibilities it offered must have had something to do with it. Certainly the discovery of the world strongly influenced Bacon. He had also been greatly impressed by German achievements as described in, for example, Agricola's *De re metallica*. These, in turn, influenced his programme for the restoration or reinvigoration of technics and learning that he styles the *Instauratio magna* and that he urges so eloquently in the books he published between 1605 and the posthumous *New Atlantis*.

Bacon's ambitions are for all humankind. He hopes that humanity's powers, so sadly reduced since the Biblical Fall, may, in large part, be recovered. His opinion of the, customarily revered, ancient Greeks is not very high. He dismisses their knowledge as verbal and barren; they had no history to speak of, only fables and legends; they knew only a small part of the world, being ignorant of Africa south of Ethiopia, of Asia east of the Ganges, and were entirely unaware of the New World. Accordingly Bacon holds that excessive respect for antiquity must necessarily obstruct the advance of science and he stresses the importance of the modern discoveries:

> . . . by the distant voyages and travels that have become frequent in our times many things in nature have been laid open and discovered which may let in new life upon philosophy. And surely it would be disgraceful if while regions of the material globe – that is, of the earth, of the sea and of the stars – have been in our times laid widely open and revealed, the intellectual globe should remain shut up within the narrow limits of old discoveries.

There are, he argues, four major hindrances to the advancement of science and technics. These he designates the Idols of the Tribe, of the Den or Cave, of the Theatre and of the Market Place. The Idols of the Tribe are humanity's psychological and physical limitations; the tendency, for example, to seek simplistic explanations and to put forward sweeping generalizations; human beings are, accordingly, handicapped in their efforts to understand and control Nature. The Idols of the Den comprise the limitations imposed on humans by their education and the society to which they belong. The Idols of the Theatre are those great intellectual systems – medieval Aristotelianism is a good example – that control and circumscribe our thoughts. Finally, the Idols of the Market Place are the ambiguities and difficulties resulting from the nature of the words and the languages that we have to use to communicate with one another.

These four Idols constitute, in effect, the limiting framework within which invention, creation and discovery are exercised. When it comes to a working scientific method Bacon rejects Aristotelian induction as demonstrably sterile. In its place he proposes a reformed inductive method that bears the strong imprint of his experience as a lawyer. The investigator studying a natural phenomenon takes all the evidence concerning the appearance of the phenomenon and then all the evidence bearing on its non-appearance when it might reasonably be expected. The investigator should now carefully sum up the evidence on both sides, after which the important causal factor will be revealed. This procedure, he holds, does not require exceptional talent or originality.

With the advancement of knowledge must come new opportunities for invention. New discoveries can lead to inventions that would otherwise be inconceivable, or that would be ridiculed as quite impossible[2]:

> If, for instance, before the invention of the cannon one had described its effects in the following manner: there is a new invention by which walls and the greatest bulwarks can be shaken and overthrown from a considerable distance, then men would have started to contrive different means of increasing the force of projectiles and machines by means of weights and wheels and other modes of battering and throwing. But it is unlikely that any flight of fancy would have hit on the fiery blast, expanding and developing itself so suddenly and violently, because none would have seen an instance at all resembling it in the least; except perhaps in thunder and earthquakes which they would have immediately rejected as the great operations of nature, not to be imitated by men.

In the same way, Bacon argues, no one whose knowledge of textile threads was limited to those spun from animal fur or vegetable fibre could imagine silk threads produced so copiously from the silkworm; nor could anyone ignorant of the remarkable properties of the lodestone conceive of the mariner's compass. It is therefore very probable that many things yet undiscovered and beyond our present imagination may be brought to light and so provide bases for new and very radical inventions. An important implication of this, that Bacon clearly recognizes, is that major inventions are often made from outside the technology concerned. The military and siege engineers of antiquity, for example, could never have invented the gun from the bases of their own distinctive discipline. On the other hand there are many inventions that, unlike cannon, silk and the mariner's compass, do not appear to depend on knowledge of the scientific properties of things. Many of the common items in daily use are of this type and the all-important invention of printing 'at least involves no contrivance that is not clear and almost obvious'.

Bacon, that is, distinguishes between science-based and what we may call empirical inventions. The former depend on the advance of knowledge and can only be made when, for example, the explosive properties of gunpowder, the magnetic properties of the lodestone or the life-cycle of the silkworm are clearly understood. Of course, this implies a simple idea of science; not so much the sophisticated, highly conceptualized science of today as a stock of general knowledge. Nevertheless the distinction between scientific and empirical invention seems valid enough. In the first category we should, today, place the gas turbine, computers, plastics, television, antibiotics and body scanners and, although one may question Bacon's over-simple classification of the printing press, in the second category things like barbed wire, the zip fastener, the sewing machine, the lawn mower and the revolving door. Of all the inventions that Bacon discusses three seem to be of particular importance:

Again we should notice the force, effect and consequences of inventions which are nowhere more conspicuous than in those three which were unknown to the ancients; namely, printing, firearms and the compass. For these three have changed the appearance and state of the whole world; first in literature, then in warfare and lastly in navigation; and innumerable changes have been thence derived, so that no empire, sect or star appears to have exercised a greater power and influence on human affairs than these three mechanical discoveries.

It will, perhaps, be as well to distinguish three species and degrees of ambition. First, that of men who are anxious to enlarge their own power in their country, which is a vulgar and degenerate kind; next that of men who strive to enlarge the power and empire of their country over mankind, which is more dignified but no less covetous. But if one were to renew and enlarge the power and empire of mankind in general over the universe such ambition (if it may be so termed) is both more sound and more noble than the other two. Now the empire of man over things is founded on the arts and sciences alone for nature is only to be commanded by obeying her.

The final words should be noted.

Bacon provided a workable and reasonable classification of inventions as well as a generous and truly international ideology for the advancement of science and technics. He observes that: 'The true and lawful goal of the sciences is none other than that human life be endowed with new discoveries and powers' (*Novum Organum*). Furthermore, in his posthumous *New Atlantis*, he suggested a specific organization for the advancement of technics. In the utopia he described there was to be an institution called Solomon's House, the purpose of which was to realize the goals that Bacon had described in his earlier works:

For the several employments and offices of our fellows: we have twelve that sail into foreign

countries, under the names of other nations, for our own we conceal, who bring us the books and abstracts and patterns of experiments of all other parts. These we call merchants of light.
We have three that collect experiments which are in all books. These we call depredators.

And so on to the number of eighteen more salaried fellows whose duties included experimental investigations, assessing results and using the knowledge gained to make further practical inventions. There were also novices and apprentices. Apart from the language it looks very modern in spirit, like a large enterprise with overseas sales and technical representatives, research and development department and policy-making executives. Interestingly, the duty of the largest group of Bacon's fellows was to bring back useful inventions from abroad. This was, no doubt, a reflection of the backward state of technics in England at that time.

It could be objected that this brief account presents an altogether too favourable interpretation of Bacon's ideas and that they have been selected with all the wisdom of hindsight. Bacon, it is claimed, missed the distinctive intellectual trend of the seventeenth century, which was the application of mathematics to the interpretation of Nature; he failed to understand the significance of Copernicus and he sneered at William Gilbert's important work on the magnet[3]; his reformed inductive method was not to be adopted by physical scientists; and, although he was sceptical of alchemy and astrology, he regarded witchcraft as the height of idolatry.

The general answer to these criticisms is that, beyond question, Bacon's ideas and ideals inspired those elements in Britain – mainly but not exclusively Dissenters from the state Church of England – that were to take the lead in technical innovation and to inaugurate the Industrial Revolution in the following century. To the charge that

Bacon had failed to recognize the importance of mathematics in contemporary science the reply could be made that medicine, chemistry, iron working, agriculture, river navigation, mining were extremely important non-mathematical arts and sciences that were of particular concern to the Baconians. For the rest it is probably sufficient to add that until well into the nineteenth century it was common – and justifiable – to regard Lord Bacon, as he was usually called, as one of the founders of modern science.

Bacon does not seem to have been particularly interested in mechanical technics; he accordingly omits the mechanical clock from the list of revolutionary inventions. His philosophy is basically *organic* rather than mechanical. Hence his precept that if we wish to command Nature we must first learn to obey her, advice that tends to go against the grain of our mechanically ordered civilization; mechanical solutions to problems are usually easier and quicker than organic ones. (Whether they are better is now a matter of urgent debate.) In Bacon's day mathematics and mechanics had little to learn from the progress of the great discoveries. Explorers might bring back new plants, animals, minerals and accounts of unknown cultures but they did not bring back new mathematics or mechanics. For this reason the wider significance of the great discoveries seems to have been overlooked in histories of the physical sciences. Bacon's inductive method has often been criticized by historians of science[4] but their criticisms, from our point of view, are misplaced. Bacon, we emphasize, asserted the autonomy of science, including technics, and he stressed that they were truly international activities for the benefit of all humankind. They are essentially co-operative and could be practised by people of modest ability, as experience in the modern world has so amply confirmed. He analysed the conditions for scientific advance and he proposed specific institutions to ensure this. His writings influenced Colbert and Leibniz and were partly responsible for the foundation of the Royal Society of London and the Académie Royale des Sciences, for the compilation of the *Encyclopédie* and for much of the idealism of the age of reason. Not the least of his services was to find an acceptable role for science and technology in an age dominated by conflicting religious creeds. Science and technology were henceforth to enjoy the autonomy that juridical law had known since the middle ages.

Two years after Bacon's death William Harvey, a graduate of Padua, published *Exercitatio anatomica de motu cordis et sanguinis in animalibus* in which he proved the circulation of the blood in animals. This was, beyond question, a major British contribution to science; paradoxically, Harvey was, like William Gilbert, an Aristotelian.

The Importance of Galileo

The career of Galileo Galilei[5] (1564–1642) stands in instructive contrast to that of Bacon. An able mathematician and a gifted musician, Galileo studied at Pisa University where he became Professor of Mathematics; he next moved to Padua before becoming mathematician and philosopher to the Duke of Tuscany. At different times he described himself as mathematician, engineer and philosopher and he could fairly claim to be all of these. Unlike Bacon, he was immediate heir to the mathematics, architecture, engineering, art and music of the Italian Renaissance. Again, unlike Bacon, he wrote in the vernacular, Italian, appealing directly over the heads of the academics to the enlightened men of the age. Like Bacon, however, he fell into official disfavour (this time with the Church). And his scientific methods and achievements complement the social analysis and propaganda of Bacon.

Galileo was a disciple of Archimedes. Like Archimedes, he was convinced that the laws of Nature were both simple and essentially mathematical. But, he argued, their simplicity is usually masked by contingent circumstances. Fortunately he possessed a remarkable gift for abstracting the heart of a problem from diverse and complex phenomena. He intuited that a body set moving would remain in motion at an unchanging velocity unless something – usually friction – slowed it down and stopped it. Friction is inescapable on earth: the air itself offers resistance to motion. This insight led him to the conclusion that all falling bodies should increase in velocity (or accelerate) by exactly the same amount, irrespective of their weights. If a falling branch, for example, falls more quickly than a leaf, it must be because of the greater air resistance to the leaf. An exact analogy is with an egg and an egg-shaped lump of lead falling through water. The latter falls much more rapidly than the egg, which has little weight in water and therefore little power to push the water aside. It followed from Galileo's argument that in a void or vacuum all bodies would fall with exactly the same acceleration. His arguments in support of this are among the classic statements of science. For a long time he believed that the increase in velocity of a falling body was proportional to the distance of fall or, generally, the distance through which it is accelerated. A stone falling twenty metres does more damage than one falling two metres; and, as every soldier and gunsmith knows, the longer the barrel of the musket, the greater the velocity of the bullet. Eventually Galileo realized that the increment of velocity was simply proportional to the time of fall, or continued acceleration, and he confirmed this experimentally.

We are, however, concerned with Galileo's contributions to mechanics and the science of machines rather than his contributions to what we should now call physics. Usually he is credited only with the inventions of a thermometer, the pendulum and the astronomical telescope, but his contributions went much further than this. He transformed our understanding of machines and the way in which their performances were to be assessed; furthermore, as we shall see, he made possible the establishment of the key concepts of work, power and energy.

We suppose that, up to and indeed for some time after Galileo, the value of a machine would be judged by normative standards: Was it solidly built and of good materials? Would it serve its purpose with some margin for emergencies? Was it aesthetically satisfying? The answers to such questions would determine the quality of the machine. On the speculative plane many looked on machines as ingenious devices for cheating Nature, for getting something for nothing. After all, Nature herself is prodigal in the ways in which she dispenses her powers: think of the rivers, the winds, the tides. And not far from the thoughts of some would be the old dream of a true, self-contained 'perpetual motion' machine.

After centuries of fruitless speculation and experiment many engineers and inventors had come to realize that Nature could not be cheated; there must be some basic principle limiting what one could get from Nature. In the same way long experience and disappointed hopes had discredited the claims of alchemy, as Bacon knew. Galileo was able to articulate this collective understanding and lay down the ground rules whereby the performance of a machine could be rigorously assessed mathematically. First, he showed that a machine is a device for applying the forces of Nature – wind, water or animal muscle – in the best and cheapest way. And he realized that a machine that can only just cope with its load is not, whatever common sense suggests, doing the most work in a given time. In the same way a man who can easily carry his burden will do more work than one who can only just carry his. Galileo has decided that the criterion is the amount of work done – however it is evaluated – in a given time.

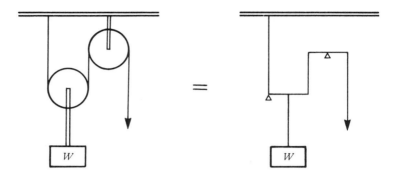

FIG 4.1 Galileo's lever principle applied to pulleys

The two components of any machine are the motive
agent and the mechanism by means of which the applied
force or agent can be made to do the required job. Now
Archimedes is famous for his observations on the lever as
well as for his studies on hydrostatics and specific gravity.
When he made his remark about a lever that could be used
to raise the world he was, in fact, talking about a machine;
a very simple machine, but a machine nonetheless.
Galileo developed Archimedes' approach and showed that
all machines – systems of pulleys, gears and the inclined
plane – are equivalent to systems of levers (figure 4.1).

The lever is, as Archimedes demonstrated, in
equilibrium when the product of the weight times the
distance from the fulcrum is the same on both sides.
Neither Archimedes nor his successors were able to go
beyond this to consider the case of more complex
machines in motion. The difficulty was a familiar
experience, the implications of which seemed to block all
further progress. No one would dispute that it is much
easier to prevent a vehicle, a light car for example, rolling
down-hill than it is to push it up-hill, even very slowly.
In general, for all machines, appreciable force or effort is
required just to set them in motion. And this force or effort

varies from machine to machine, from circumstance to circumstance. Galileo, with his remarkably clear head and his capacity for going right to the heart of the matter, saw that the inequality between the equilibrium force and the force required to set the machine in motion was not a principle of Nature; it is merely a consequence of the imperfections of all machines. A good machine will move at a uniform, unchanging velocity when the force and the load, including residual friction, are in equilibrium. A perfect machine, free from friction, distortion and other defects, will accelerate (slowly if loaded) under an extremely small force until it reaches an infinite velocity. The implications of this can be understood by considering the case of the lever and remembering that what is true of the lever is true of all machines.

Let us suppose that a light weight A is at a distance AC from the fulcrum of a lever, C being the fulcrum. A heavier weight B is at a distance BC from the fulcrum. According to Archimedes if A × AC = B × BC, the lever is in equilibrium. Now suppose that a negligibly small weight is added to A; then, according to Galileo, the 'machine' will start to move very slowly and the 'load' B will start to rise. A will descend a distance Aa while B rises Bb in the same time. Since, by simple geometry, Aa has the same ratio to Bb as AC has to BC, it follows that A × Aa = B × Bb. In other words, the product of the driving weight, or force, multiplied by the distance it moves equals the load multiplied by the distance *it* moves in the same time. And this must be true of all perfect machines.

It is impossible to overestimate the importance of Galileo's profound insight. It means that if we know the strength or force of any motive agent we can calculate the load it can shift or the *work* it can do applied to any perfect machine. By taking account of the load represented by friction and other defects we can calculate the performance and the efficiency (work done divided by driving force) of any real machine. Galileo did not cause the Industrial Revolution but he helped to make it possible. Before him machines could only be judged qualitatively; after him they could be assessed quantitatively. Franz Reuleaux explained another aspect of the same fundamental point[6]:

In earlier times men considered every machine as a separate whole, consisting of parts peculiar to it; they missed entirely, or saw but seldom, the separate groups of parts that we call mechanisms. A mill was a mill, a stamp a stamp and nothing else, and thus we find the older books describing each machine separately from beginning to end. So, for example, Ramelli (1588), in speaking of various pumps driven by water-wheels describes each afresh from the wheel, or even the water driving it, to the delivery pipe of the pump. The concept 'water-wheel' certainly seems tolerably familiar to him, such wheels were continually to be met with, only the idea 'pump' – and therefore the word for it – seems to be absolutely wanting. Thought upon any subject has made considerable progress when general identity is seen through the special variety; this is the first point of divergence between popular and scientific modes of thinking.

It is, at first sight, surprising that the idea of a pump could have presented such difficulties at the end of the sixteenth century. But it should be remembered that apart from a few points of similarity there was little common ground between machines that served very different purposes. How could one compare a fulling stocks with a mine pump, a corn mill with the bellows for a blast furnace, a saw mill with a pump for supplying water to a mansion? They did different things and were designed and made by quite different craftsmen. The only question, common to all, was whether each machine worked well and the only answers were that it was a good machine or it was not. After Galileo, however, all machines have the common function of applying 'force' as efficiently as possible. This means that the performance of machines can be quantified; ideally, the product of the driving weight, or 'force', multiplied by its velocity equals the product of the load times its velocity. Once this has been accepted a rational science of machines becomes possible and the design and function of each component can be studied in order to optimize efficiency.

The development of Galileo's idea in relation to technics – or, as we should now say, technology – was begun in the seventeenth century and continued in the eighteenth century. Different measures of the 'effect' (the load multiplied by the velocity) of machines such as water-wheels were used or suggested by engineer–scientists like Mariotte, Parent, Desaguliers and Beighton until, in 1784, James Watt standardized the unit of 'horsepower' as the capacity to raise 33 000 pounds one foot high in one minute (or, remembering the lever, 3300 pounds ten feet high in one minute, and so on). The establishment of a science of machines was more gradual and, according to Reuleaux, did not really get under way until the work of men associated with the Ecole Polytechnique, founded in 1794.

Galileo begins his classic work *Two new sciences* (1638) by remarking that the machines and devices in use at the Venetian shipyards offer many stimulating lessons for the student of mechanics. He notices that, owing to the scale effect, a ship that is supported only at the bow and stern will break in two whereas a small model of the same vessel can easily be held at the bow and stern without breaking. These reflections lead him to develop his immensely fruitful ideas on the strength of materials: the first of his two new sciences, the second being dynamics.

The strength of a beam, that is its capacity to withstand a longitudinal force, is plainly proportional to its area of cross-section. For if the strength is due to longitudinal fibres, as in the cases of wood and wrought iron, then

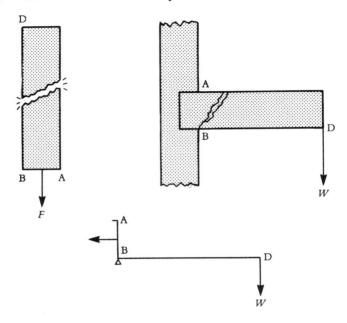

FIG 4.2 *Galileo's lever principle applied to beams*

the number of fibres must be proportional to the area of cross-section. The breaking force, *F*, of the beam shown in figure 4.2 will, therefore, be some constant, call it *k*, multiplied by AB × BC, the area of cross-section; the length of the beam is BD. The problem is: what load, *W*, can the beam carry when it is horizontal with one end mortised into a wall? For the sake of simplicity it is assumed that the load is concentrated at the unsupported end, D.

Characteristically, Galileo abstracted the essentials of the matter by reducing the problem to that of the lever. He assumes that the 'force', *F*, opposing the effect of the load, *W*, acts in the middle of the beam. By the law of the lever, therefore, at the breaking load ½AB × *k*(AB × BC) must equal *W* × BD. Or *W*, the maximum possible load, must be slightly less than ½AB² × *k*(BC/BD). At a load slightly greater than *W* the beam will crack at the fulcrum B. The depth of the beam, AB, should, therefore, be as great as possible. This is why the steel girders used in big buildings are of the familiar 'I' section; the metal is better used to increase AB at the expense of BC.

In the case of a cylindrical beam the area of cross-section is proportional to the radius squared, or r^2. In this case ½AB is also *r* so that the maximum strength of the beam is $k \times r^3$. To make the beam as rigid as possible we must increase ½AB, or *r*; but this means that for the same weight of material we must hollow it out to form a cylinder. This is why, in Nature, reeds and straws are hollow. However, for a given weight of material and a given length, there is a limit to the extent to which the strength of a beam can be increased by increasing its radius (or its web). At this limit the beam will buckle; that is, it will bend and distort before it breaks. Galileo did not consider this case.

As it happens Galileo did not allow for the unavoidable elasticity of beams. In fact the bottom layers of loaded beams are compressed, not stretched, and bending takes place about a fulcrum that is inside the beam. This was not a matter of great moment for Galileo's theory did not become important, practically, until very much later.

The Rise of Western Europe

The seventeenth century has been described as the century of genius; it was also a century of war. According to G. N. Clark there were only seven complete years in which there was no war, somewhere in Europe. Of these wars the most destructive was the Thirty Years War (1618–48) that began as a dynastic and religious quarrel and ended as a struggle for supremacy between Spain and France. Spain, the country that had played a leading role in the discovery of the world, was ruined while triumphant France was to become the dominant power in western civilisation. The German states, over whose territories the war was fought, were devastated. Although reliable facts and statistics of major disasters are usually difficult to establish and exaggeration is common, there is little reason to doubt the scale of the disaster. The consequences go far to confirm it, for Germany ceased to be one of the two main centres of technical achievement. Italy, too, lost its place in the van of European civilization. The opening of the sea route to India and the growth of Atlantic trade with the Americas resulted in the Mediterranean becoming a backwater. The centres of art, learning, science and invention shifted decisively to France, the Low Countries and, a little later, to England.

The first half of the seventeenth century was not a time of profoundly important inventions. War, as we shall note again, is not necessarily and always a stimulant to invention. The barometer, the telescope and the thermometer were the leading inventions of the time: appropriate enough for a century of immense scientific progress, but only one of the three had an application to war. Following Galileo's suggestion the pendulum, with its natural period determined only by its dimensions and by gravity, replaced the foliot and balance weights to regulate the weight-driven clock. There was a revival of interest in the powers of fire, or heat. Hero of Alexandria's simple steam toys had been invented at a time and place when and where there was no urgent, clearly defined, demand for continuous power. Conditions, at least in the long term, were more propitious when della Porta (1606) and Solomon de Caus (c. 1611) proposed methods of harnessing the 'impellent force of fire'. In both cases steam pressure raised water by pushing it up a pipe. About twenty years later Giovanni Branca invented a simple reaction turbine worked by steam. None of these inventions, or the several others to which they gave rise, was successful. They may, however, have stimulated a train of thought, an aspiration, that proved much more fruitful at the end of the century.

The 'Scientific Revolution' of the seventeenth century had no more articulate and persuasive an advocate than René Descartes (1596–1650). He brought into sharp focus the ideas and assumptions of Galileo and other leaders of the scientific movement. A brilliant mathematician, he was also the first truly original philosopher since antiquity. The world picture he presented is, with few modifications, that of modern physical science and, more generally, the rationalism that he advocated is now part of western culture. For Descartes the external world is characterized by matter, or more strictly extension (the defining property of matter) and motion. There are no colours, no sounds, no scents in Nature; these are put there by our senses and are therefore illusions. All that are unambiguously outside ourselves are inert particles, characterized by extension and in motion. In between the complex systems of particles that form a human body there flits the tenuous ghost that is the soul. The systems of particles essentially comprise a machine: levers operated by pulls and pushes. Here, the analogy of Renaissance mechanisms – clocks, automata and the like – is very close. We humans differ from animals in that we have a soul, or mind, that somehow controls the machine. How that soul or mind can do this, since, by definition, the ghost is not material, it has no extension, is a difficult question. This, the body–mind problem as it is called, has been at the heart of western philosophy ever since Descartes first posed it.

Harvey's account of the circulation of the blood was most acceptable to Descartes; it was consistent with his mechanical interpretation of the body. He notices, like Harvey, that the heart is hot and, he adds, it is because the heart is hot that the incoming venous blood expands[7]. The valve systems of the heart ensure that the expanded blood is forced out to the arteries. And in this way the circulation of the blood is maintained. It has been said that Descartes thought that the heart worked like a steam engine. This, of course, was not the case; the steam engine was not invented until more than sixty years after Descartes put forward these ideas. But it is perfectly possible that Descartes, a very well informed man, knew all about the inventions of della Porta, de Caus and others.

Descartes' philosophy was comprehensive. The motions of the planets round the sun must, he argued, be explained mechanically. He cannot accept Renaissance notions; sympathies, affinities, attractions cannot possibly explain the continued orbital motion of inert bodies. How, he would have asked, can dead, inert bodies like planets be *attracted* to anything. Attraction, sympathy, affinity are sentiments felt by humans. Descartes' universe is permeated by an ether composed of extremely small particles. Hot bodies, such as the sun, set up vortices, or whirlpools, in this ether and these carry the planets round with them, rather as bathwater carries soap bubbles round with it as it flows out of the bath. The earth has its own internal heat and so has a smaller vortex that carries the moon round and is the cause of gravity on the earth's surface. If there were no vortices the planets and the moons would sail off, on straight lines, into space and there would be no gravity. The whole picture was satisfyingly mechanical; there was nothing occult, or spiritual, about it.

The mathematical–mechanical science of Nature that Descartes strongly advocated must be, he believed, very much in the Baconian tradition, for the benefit of all humankind. This duty, as he sees it, leads him to accept that it will be possible to:

> . . . acquire a type of knowledge that will be of the greatest use to mankind and that in place of the Scholastic philosophy (Aristotelianism) we can establish a practical one where by understanding the forces and actions of fire, water, air, the heavenly bodies – all physical things about us – as clearly as we understand the skilled trades, we shall be able

to apply them in a similar way to appropriate uses and so give us command over nature.

It is one thing to assert that the advance of science leads to increased command over Nature; it is, however, much more difficult, if not impossible, to predict where, when and how that command will come about. In general terms more than two hundred years were to elapse before social change and the progress of science were to make systematic applied science, as envisaged by Bacon and Descartes, feasible.

The greatest scientific achievement of the century – indeed of all time – was, by universal consent, expounded in Sir Isaac Newton's *Philosophiae naturalis principia mathematica* (1687)[8]. Early in the century Johan Kepler had published his three laws of planetary motion, of which the most suggestive was that the planets travel round the sun on elliptical orbits, with the sun at one focus. At about the same time Galileo, with his telescope, had shown that the moon was of the same 'earthy' nature as the earth and had inferred that this was true of all the planets. He had also demonstrated that the trajectory of a projectile – a cannon ball, a stone or a javelin – was, assuming no air resistance, a parabola. The fact that planets and projectiles both follow trajectories that are conic sections suggested to the versatile Robert Hooke (1635–1703) that there was a certain common factor. And Hooke, together with Wren and Halley, speculated profitably on the nature of that factor.

To these insights, Newton added his concepts of universal gravitation, of mass and of accelerative force together with the principle of inertia proposed by Galileo and perfected by Descartes: a body in uniform motion will continue to move in a straight line unless acted upon by some external agency. And he added Descartes' notion of 'quantity of motion', or momentum (mass multiplied by velocity). The principle of inertia Newton enshrined in the first two of his famous laws of motion; the third law he proposed himself. After the famous three laws, or axioms, he appended, as a corollary, a formal statement of Stevin's parallelogram of forces.

There were serious mathematical and physical difficulties about Descartes' vortex theory; it could not, for example, account for Kepler's laws. Newton therefore dispensed with the ether and relied solely on the ideas that the force of gravity acted across space, always towards the centre of the gravitating body (the sun, planet or moon), and that it diminished with the square of the distance. With this interpretation of gravity as 'action at a distance' and with the Galileo–Descartes law of inertia, Newton, the supreme mathematical genius, solved the problem, as old as astronomy itself, of the motion of the planets; what the world would call the riddle of the universe. He had also laid the basis for a rational mechanics and he expressed the hope that his method of philosophizing – attractive and repulsive forces acting on point masses (atoms) across empty space – could explain the other phenomena of Nature. As for the ether he did not repudiate it even though he made no use of it in his *Principia*. It was to live on in physics until the beginning of the twentieth century when relativity finally disposed of it.

Newton's prestige became enormous in the eighteenth century and his influence profound. Yet the *Principia* is a conservative work. The problems of the planets solved, there was nothing further to do in that direction and astronomers of the following generations turned to the sidereal universe. And Newton had a challenger in G. W. Leibniz who argued, as did Christiaan Huygens, that the true measure of the 'force' of a moving body was the mass multiplied by the *square* of the velocity. *Vis viva* ($m \times v^2$), as Leibniz called it, was proportional to the extent to which a moving body could overcome resistance. A bullet travelling with doubled velocity will penetrate four times as deep into a tree trunk or a bank of clay; with tripled velocity it will go nine times as deep and so on. The *vis viva* of a moving body can also be measured by its weight multiplied by the height to which it can rise against the attraction – or resistance – of gravity, or, what amounts to the same thing, the height from which it must fall to attain that velocity. The measure of weight times distance is consistent with Galileo's insight into the work done by machines. It is hardly surprising that the engineers of the eighteenth and subsequent centuries found the Galilean–Leibnizian concepts more useful than Newton's momentum.

Robert Hooke[9] was a more versatile genius than Newton. Apart from his work on astronomy, his advocacy of the undulatory theory of light and his famous book, the *Micrographia,* he was able to make a number of highly original inventions. He invented the wheel barometer, the indexing machine, the spiral balance spring for watches and, it is believed, the anchor escapement. In addition to these and other inventions, he found that the extension of an elastic body is proportional to the tension producing it ('Hooke's law'). And he pointed out (1675) that, for stability, an arch should have the form of an inverted catenary. (A uniform chain or rope, suspended freely from both ends, hangs in the curve called a catenary.) Twenty-two years later David Gregory announced his conclusion that an arch is stable only if a catenary can be drawn completely within its outline.

The experiments and discoveries concerning the earth's atmosphere made in the second half of the seventeenth century might seem less impressive than the heroic achievements of Newton and his peers. They were to prove just as fruitful as the new mechanical science. Galileo had noted that a suction pump cannot raise water more than ten metres. He ascribed this to the breaking of the column of water under its own weight. His disciple, Evangelista Torricelli, gave the correct explanation. Torricelli realized that it was the pressure, or weight, of the earth's atmosphere that enabled the suction pump to work. This pressure, acting on the surface of the water, drove it up the pipe to the pump when the motion of the piston formed a vacuum. But the pressure of the atmosphere could not support the weight of a column of water more than ten metres high. To prove his point Torricelli used mercury, almost fourteen times heavier (denser) than water. He filled a glass tube with mercury and up-ended it with the open end immersed in a pool of mercury. The level of the mercury fell until it was 72 cm, or 1/13.6 of ten metres, above the pool of mercury. The empty space left above the mercury in the tube was void, or a vacuum. Torricelli had invented the familiar barometer; and the science of meteorology became possible. He had done more than that. Shortly afterwards Blaise Pascal took a barometer to the top of a mountain, noted the fall in the barometer and, knowing the height of the mountain, was able to estimate the height of the finite ocean of air surrounding the earth.

Some years later, when it was clearly recognized that air was a kind of fluid, Otto von Guericke[10] (1602–1686) invented the air pump and the unusual properties of the vacuum became clear, even if the explanations were in the distant future. A vacuum cannot support life or combustion and the sound of a ticking clock cannot be heard through a vacuum. On the other hand the recognition of the fluid-like properties of air soon led to the discovery of the Boyle–Mariotte law relating the pressure and the volume of air. And the recognition that there was a finite ocean of air round the earth suggested to Father Lana Terzi[11] (1672) the possibility of an air-boat, or balloon, or airship. Alas, the first successful lighter-than-air machine was a long time in the future!

Against the background of these truly unparalleled advances in science France and England were following quite different paths. The contrasts are informative. From 1661 onward J. B. Colbert (1619–83), a quintessential technocrat, consolidated a national tendency towards centralized administration and greatly enhanced the role of the state in industry and technology. On the apparently reasonable assumption that wealth and prosperity must follow from the manufacture of expensive commodities, Colbert encouraged the development of luxury industries, such as Gobelins tapestries and Sèvres porcelain. He supported the activities of the new Académie Royale des Sciences (founded in 1666). He realized the importance of good communications and began the construction of a modern road network. Road and bridge building on a national scale required experts with appropriate skills; in a word they required engineers. However, the only engineers in those days were military men, concerned with siege engines and the like. Colbert therefore instituted an Académie Royale d'Architecture, one duty of which was to advise on the construction of canals, roads and bridges. The architect and the engineer were often, in those days, one and the same man. France has many fine rivers and the improvement of navigation was another item on Colbert's programme. He was particularly interested in canals. Impressed by large-scale projects, perhaps his most imposing memorial is the great Canal du Midi, or du Languedoc. The idea of a canal linking the Atlantic and the Mediterranean was projected in 1633 but it was not until 1666 that its construction was begun under the engineer Pierre-Paul Riquet de Bonrepos. This magnificent project was completed in May 1681. The canal was about 240 km long, had 100 locks and rose to a height of more than 200 metres. With a long tunnel and three large aqueducts, it was the greatest feat of civil engineering since Roman times and it was to play a part, albeit an indirect part, in the English Industrial Revolution.

In France there was a command economy and an enforced religious uniformity resulting from the revocation of the Edict of Nantes in 1685 (in itself an aspect of the policy of centralization). However, a discussion of the policy of mercantilism is outside the scope of this book.

In any case Colbert's achievements were, in large measure, eroded by the wars and other follies of the vain if visually impressive Louis XIV. In England, the land of a hundred religions and one sauce, as Voltaire was later to put it, changes were taking place that were to lead to *laissez faire* and a minimum of centralized control. The rise of the puritan sects before and during the Commonwealth, 1649–60, was associated with great interest in Bacon and his ideal of science applied to social welfare. There were schemes for new universities and for the radical reform of Oxford and Cambridge. Aristotelianism was to be deleted from the syllabus and the teachings of Descartes and van Helmont put in its place. The members of the 'invisible college', mainly associated with Oxford University, such as Boyle, Halley, Hooke and Wren, were interested in practical, Baconian science; it was largely due to them that the Royal Society of London was founded in 1660 and granted its Charter in 1662. Isaac Newton, was not, however, interested in practical science.

The relationship between the tenets of puritanism and the ideology of private enterprise was explored in a well-known work by Max Weber. Professor R. K. Merton, in effect, extended the thesis to cover science and technology. After a careful study of the religious affiliations of early Fellows of the Royal Society, Merton found that a significantly high proportion of them were Dissenters from the Church of England. The extent to which this may be related to a distinctive Calvinist ethic and the extent to which it may have been due to other sociological factors – the exclusion of all 'Dissenters' from government and official office, for example – may be debated. Nevertheless, the prominence of 'Puritans' and other Dissenters in science and technology, particularly the latter, is beyond question.

In spite of their zeal Englishmen, whether Puritans or not, felt that their country lagged behind Europe, particularly Germany. The English had, for many years, admitted the supremacy of the Germans in mining and in metal working. The English iron industry was stagnant and a high proportion of the iron used in the country was imported from Europe. The tram track or railway had, we remember, been invented by German miners in the sixteenth century. By the second half of the seventeenth century Holland had, in many ways, taken the leadership from the German states; Dutch merchant ships dominated sea-borne trade; Dutch warships ruled the seas; in banking and commerce Holland was the most important country in Europe. And England recognized Dutch leadership in hydraulic engineering when Cornelius Vermuyden was commissioned to begin the drainage of the fenlands of East Anglia. At the beginning of the century the Dutchman Simon Stevin was Europe's most distinguished engineer; towards the end of the century Christiaan Huygens was not only a leading man of science but also an important inventor. So highly did the English rate the technical superiority of Continental Europeans that Daniel Defoe, a shrewd observer and an informed journalist, took it for granted that the English were no good at making original inventions but very good at copying and improving on other people's ideas. This is exactly the opposite of the view that the British now take of themselves.

PART II

The Industrial Revolution

Reason and Improvement

By the end of the seventeenth century the most active centres of technological innovation were to be found in western Europe, in France and the Low Countries and, slightly later, in England and Scandinavia. During the years between Galileo's first publications and the appearance of Newton's *Principia* there were several changes that radically affected the technological process.

Chronologically, the first of these was the movement to reform the patent system in England[1]. It is said that the first patent was granted in Florence in the fifteenth century and the practice was soon adopted in other north Italian cities and states. In sixteenth-century England patents were mainly monopolies in manufacture and trading awarded to Court favourites or in return for a bribe. The Statute of Monopolies of 1624 swept away many of the gross abuses of the system while it retained the practice of granting an inventor, or, generally, an innovator, letters patent safeguarding, initially for 21 years, his monopoly of his invention, or the invention or craft that he had imported from abroad. The Statute did not eliminate all abuses and it proved difficult to reach a satisfactory legal definition of what constituted an invention; accordingly the laws were progressively refined and improved over the following centuries. They were extended to Scotland following the Act of Union in 1707; in 1790 the infant republic of the United States of America instituted its own patent laws and in the following year Revolutionary France adopted a patent system inspired by the English Statute of Monopolies. It was not until the middle of the nineteenth century that a still fragmented Germany began to set up a system for the protection of inventions. It has been said that the patent system is nothing else but the idea of progress appearing in law; for it is plain that a static society would have no need of patents.

Prior to effective patent laws the only protection the inventor enjoyed was the uncertain one of secrecy. This necessarily discouraged invention and retarded innovation; it also had the incidental effect of surrounding the inventor with an aura of mystery and myth that has not been entirely dissipated, even today. The vast majority of inventions that have enabled humankind to rise from barbarism to civilization were made by men and women who remain entirely unknown. With the arrival of the patent system the inventor became a public figure; as modern jargon has it, the inventor entered the public domain. From the late seventeenth century onward the historian finds that the life and the individuality of the inventor are progressively more clearly defined than they were before. The inventor, or innovator, joins the gallery that includes the philosopher, the artist, the poet, the churchman, the warrior and the monarch. But it was not only the patent system that made inventors and their achievements known to the public and that expedited innovation. Journalism began in the seventeenth century[2]. In 1631 Theophraste Renaudot launched the *Gazette de France* in Paris; it was followed by the *London Gazette* (1665, 1666). In 1702 the first daily newspaper appeared in England. Specifically concerned with science and technology were the *Journal des Scavans*, founded in January 1665, and the *Philosophical Transactions of the Royal Society*, founded three months later. Hardly less momentous was the appearance of the first true encyclopedias: John Harris' *Lexicon technicum, or an Universal English Dictionary of Arts and Sciences* (1704) and Ephraim Chambers' *Cyclopaedia, or Universal Dictionary of Arts and Sciences* (1728) in two volumes. These works, characteristic products of the

Baconian school, eschewed curiosities, fables, superstition and unverified assertions.

The Baconian programme, the comprehensive philosophy of Descartes and the success of Galilean–Newtonian–Leibnizian mechanics determined the character of the centuries to come. The common cause must, we believe, be sought in the unique nature of medieval European society, in particular in the willingness to confront and challenge Nature to the limit. What brought about or encouraged this spiritual – and often physical – courage is a profoundly difficult question. It is enough to record that by 1700 the foundations of modern technology had been laid. Appropriately, the word 'technology' had been coined before the end of the century; and it seems likely that the word 'inventor' was beginning to be used in the way in which it is understood today.

The activities that are essential to life were greatly changed, particularly in England, during the eighteenth century. Farming was, perhaps, the first of these to show the effects of the new ideology. Socially these were an acceleration in the rate of enclosures and the rise of the 'gentleman-improver'. The old rural community had been the village with its common grazing land and its open fields, strips of which could be cultivated by individual peasants. The system was inefficient: it ruled out economies of scale; cultivable land was wasted on footpaths between the strips and human energy was dissipated in going from strip to strip. Enclosures to form large, single-purpose fields changed this but progress, where many individual rights were involved, was necessarily slow.

The exemplary 'gentleman-improver' was Jethro Tull (1674–1740), who had trained as a lawyer and had made the grand tour of Europe before settling down to farm in the south of England. In 1735 he published his influential work, *Horse-hoeing Husbandry*. He had noticed, during his tour through France, that rows of vines were kept highly productive by ploughing between the rows. Ploughing cleared the ground of weeds and stirred up the soil between the plants thus, he believed, making it easier for the roots to absorb the (hypothetical) atom-like particles of food in the ground. Tull developed, refined and generalized the vine cultivator's technique with such success that Lord Ernle[3] described him as 'the greatest individual improver that British agriculture had ever known'. He experimented (inductively) to find the depth at which to plant cereals and the density at which they should be sown to give the best yields. He used a microscope in his studies and devised and built his own apparatus. Finally, as a result of his experiments, he invented the seed-drill in which seed was poured down a notched barrel and fed into the earth at the right depth and at the right spacing; after this the earth was spread back mechanically over the seed. This was an advance over the old, random method of planting seed by broadcasting it by hand.

Tull, an educated man, can be classified as a disciple of Bacon, firmly committed to the doctrine of improvement. His method was Baconian, his philosophy scientific. But so great is the diversity of soils, of topography and climate (down to microclimate) that innovations could not diffuse rapidly in agriculture. And as life depends, very directly, on successful husbandry, experiments were not undertaken lightly. Agriculture is, understandably, one of the most conservative, perhaps the most conservative, of industries; it was only the great land-owners who readily adopted Tull's (and other men's) inventions, for they could afford the risks of innovation. Prominent among those who propagated Tull's doctrines and added innovations of his own was Lord Townshend (1674–1738), a gentleman-improver and an active politician. Townshend advocated improving the fertility of the land by marling (adding clay and lime) together with a rotation of cereals, roots, clover and grasses. His enthusiasm for the turnip, in particular, earned him the sobriquet 'turnip Townshend'. The great merit of his practices was that livestock could be properly fed throughout the year so that they were not half-starved by the end of the winter. Animals that had been well fed the year round ensured that the land was well manured so that the next crop yield was heavier, providing yet more food for more animals. This was, quite literally, an excellent example of feedback. By these and other innovations great land-owners radically improved agriculture in England during the eighteenth century. The small farmers copied them in due course.

In France, on the other hand, the nobility, the great land-owners, felt little or no incentive to improve their estates. According to Arthur Young[4] the causes were the dominance of Paris, the King and the Court, coupled with a grossly inequitable system of taxation that oppressed the peasantry while exempting the aristocracy and the Church. Young put much of the blame for this lack of an effective policy on Colbert, who had neglected the interests of agriculture in favour of those of industry.

The iron industry, outwardly so different, showed interesting similarities to farming. As the type and quality of iron ore varied from place to place, the same kind of obstacle to the spread of improvements applied in iron working as in agriculture. 'It might work there but not with the stuff we've got', would be a plausible objection to any new idea. Production units remained relatively small and location was determined by the distribution of natural resources. Small ironworks were to be found wherever there were local sources of iron ore, of wood for making charcoal and of water power for working bellows and tilt hammers. Transport difficulties further restricted the scale of operations. Most iron used was still wrought iron, hammered out of blooms to make nails, bolts, locks, hinges, spades, forks, scythes, brackets, tools, guns, ploughshares and, of course, horseshoes. Agriculture was a big market for ironware. Cast iron, brittle and of low tensile strength, had only limited use for things like large cooking pots and cannon. The great obstacle to the development of the iron industry was the need to rely on charcoal, which was slow and expensive to make and which could carry the weight of only a limited amount of iron ore and limestone, used as a flux, in a blast furnace. Experiments with coal as an alternative were, *pace* the claims of Dud Dudley in the mid seventeenth century, unsuccessful: the sulphur contained in coal ruined the iron.

In 1708, Abraham Darby (1677–1717), a member of the Society of Friends (Quakers) and an ironmaster from Bristol, leased an abandoned furnace at Coalbrookdale, on the river Severn. In the following year he succeeded in smelting iron using *coke* in place of charcoal. Darby had studied the techniques of Dutch iron workers, then recognized as the best in the world, and he had had experience in non-ferrous metal smelting. Coke could carry greater charges of ore and limestone than charcoal and in burning it generated higher blast furnace temperatures so that the melted iron was more fluid. With more fluid iron a variety of thin-walled items could be cast and this, Professor Hyde[5] suggests, was the reason for Darby's move to Coalbrookdale, where, incidentally, the local coal yielded a coke that was suitable for iron smelting. This raises the intriguing possibility that Darby had succeeded in smelting iron using coke before 1709, possibly in Bristol. In 1707 he had patented a process for casting iron pots in which sand instead of loam was used, and it would be reasonable to suppose that the next move would be to find a way of raising the temperature of the molten iron. It is significant, Hyde remarks, that Darby jealously guarded his 1707 patent but made no effort to patent his coke smelting process. Only some half-dozen coke furnaces were built in the following fifty years, the reason being, according to Hyde, that charcoal smelting remained cheaper for most purposes. Quakers were prominent in the English iron industry at that time and it is, in principle, likely that Darby would have told some of them of his success in coke smelting.

In medieval Europe industry had been mainly located in towns. The various guilds apportioned work, supervised apprenticeships and generally exercised restrictive practices, some of which were, no doubt, in the interest of the consumer. During the seventeenth century and, increasingly, in the eighteenth century industry migrated to the country where water power resources were to be found, where there was wood and, in the case of England, coal for fuel and where there were no guild restrictions on production or labour. Holland, on the other hand, was conspicuously lacking in water power resources; indeed, it might be said that the country had negative energy resources. And this, for all the genius of the Dutch people, must have limited economic growth and technical innovation.

Migration to the country and dependence on water power raised other problems. There was the question of communications. The techniques of road building had been forgotten, the need for an efficient road network had vanished and could hardly be imagined after the universal Roman empire collapsed. Revival, depending on complex economic, military and political factors, was extremely slow. In France and in England the maintenance of roads was for long the responsibility of local communities, who were supposed to provide conscript labour to do the work. The labour was, understandably, half-hearted, the direction uncertain and the technical knowledge conspicuously absent. The system was a failure. There was no national policy in England, although from 1663 onward 'turnpike' trusts were authorized for piecemeal improvement. The geography of England was such that a small ship was the best means of transporting goods to many places. Cattle, sheep and pigs could be driven to town, even to London, along the drovers' roads that were little more than open tracks.

There were many changes, in practice and in spirit, during the eighteenth century. The framework of a national communications authority began to evolve in France[6]. Road and canal building was, with varying degrees of dedication, continued by Colbert's successors. In 1716 the Corps des Ponts et Chaussées was founded. Under an inspector-general the staff included an 'architect senior engineer', three inspectors and 21 assistant engineers who could call on the services of masons, carpenters (most bridges were made of wood) and other craftsmen. Here we should note that the distinction between architects and engineers that now seems obvious was, in those days, still far from clear. This is confirmed by the title of Bernard Foret de Belidor's authoritative work, *Architecture hydraulique* (four volumes, 1737–1753), which dealt with roads, bridges, buildings, even the Newcomen engine (see below), besides hydraulic engineering. Although, therefore, architecture and engineering could hardly be distinguished in those days, they were to become increasingly differentiated by the advance of technology.

Buildings, whether palaces, forts or churches, were constructed by craftsmen whose skills had been handed down through the generations. As the properties of the traditional materials – stone, brick and wood – were common knowledge a skilled craftsman could tell, easily enough, what he could do and what he could not do with any particular item. There was, apparently, little need for any more sophisticated techniques[7]. But the spirit of technological innovation was not to be excluded. Philippe de la Hire made use of the parallelogram of forces in an effort to deduce the weights of the voussoirs – stones tapered to allow for curvature – of a stable arch; it being assumed that the voussoirs were perfectly smooth (1695). This idealized conception led to the absurdity that the lowest voussoirs would have to be of infinite weight. Seventeen years later he, more profitably, used moments to find the conditions of stability for a semicircular arch.

La Hire did not question Galileo's assumption that the neutral layer was on the under surface of a beam. Edmé Mariotte had, however, recognized in 1686 that even the hardest materials are, to some extent, elastic under stress and he placed the neutral layer in the right position. Unfortunately he made a slip in his calculation of the relative strength of the beam. The correct interpretation was given in 1713 by Antoine Parent (1666–1716). Parent saw that the neutral layer was not necessarily in the middle of the beam, for the elasticity in tension and the elasticity in compression might be different. And he recognized that for equilibrium in a loaded beam the total force of compression must equal the force of tension. Unfortunately Parent was effectively ignored and for many years it was a matter of individual choice whether or not to accept Galileo's location of the neutral layer.

The Foundation of Power Technology

Parent, as it happened, was interested in water power as well as in structures. That available water power should be used as efficiently as possible was quite obvious when mines had to be pumped out and when towns or palaces had to be supplied with drinking water. There had been, however, no yardstick of efficiency until the Galilean science of mechanics was widely accepted. As this yardstick was, itself, a radical innovation whose consequences have been of incalculable importance, we must discuss its origins in some detail.

The scientific study of water power was initiated in 1704 when Parent published his calculation of the maximum power that can be derived from any given stream of water[8]. Torricelli had shown, by applying Galileo's dynamics, that the velocity of a stream of water, rushing out at the bottom of a tank, would be just sufficient to carry the stream, directed smoothly upward like a fountain, to the level of the surface of the water in the tank. This was the dynamical equivalent of the well known and long-established principle that water finds its own level. Further, Torricelli showed that the square of the velocity of the stream of water was proportional to the depth of the water in the tank (i.e. v^2 is proportional to h; in modern usage, $v^2 = 2gh$; cf. Leibniz's *vis viva*).

Parent considered the case of a perfect, friction-free, 'undershot' water-wheel in which the stream strikes the flat blades, or paddles, and therefore acts by impact. If the wheel is so heavily loaded that it cannot move, no work is done; if it is unloaded so that it spins round with the same velocity as the stream, again no work is done. *Somewhere between these extremes must be a velocity at which the maximum work is done.* What is this velocity and what proportion of the total 'effort' of the stream can be converted into useful work or, as he calls it, 'effect'?

When the wheel is turning at a uniform velocity the force of water on the blades must, assuming that there is no gearing, be equal to the weight being lifted; if it is greater the wheel will accelerate, if it is less the wheel will slow down, in both cases until the velocity is uniform. The force, or pressure, of water on the blades is proportional to the quantity of water and the relative velocity (the excess velocity of the stream over that of the wheel) with which it hits the blades. Since the quantity of water is proportional to the relative velocity it follows that the force, or pressure, is proportional to the relative velocity squared. The force, or pressure, that the stream exerts on the blades corresponds, therefore, to a head or height of water less than that of the stream. The power, or the work done in unit time, of the machine is equal to the force, or pressure, of the stream on the blades multiplied by the velocity of the wheel, which is proportional to the speed at which the load is raised.

The full effort (energy) of the stream is equal to all the water rising to the head, or height, from which it must fall to gain the velocity it has. By using the newly invented differential calculus Parent deduces that the velocity for maximum power, or work done in unit time, is one third that of the stream and the maximum power is only four twenty-sevenths of the full effort (energy) of the stream (see Note at the end of the chapter). This, Parent admits, is a surprisingly small fraction. But, as he points out, the stream still has considerable velocity – and therefore effort (energy) – *after* it has passed the wheel. And, he adds, the impact of the water is reduced as soon as the wheel starts to move. His argument also implies that some water must slip by without striking a blade, and the faster the wheel moves, the more water slips by in this way. [But this is not necessarily the case. Since he is dealing with perfect water-wheels he is free to postulate a wheel with so many blades that virtually no water can slip by without striking a blade. Furthermore, the common practice in his day of dealing in simple proportions masked the fact that he should have equated v^2 to $2h$ and not to h. Making these two alterations we find that for maximum power the blades should move with one half the velocity of the stream and that the maximum power is one half that of the full effort (energy) of the stream.] Parent did not consider the alternative case of the 'overshot' wheel in which the water, passing over the top of the wheel, fills buckets and therefore acts by weight rather than by impact. Presumably, he felt that such a wheel could not be better than an undershot wheel.

Parent's paper was most influential. It inaugurated research into the efficient exploitation of energy resources, research whose importance today need not be stressed. And it introduced – if only by implication – a fundamental concept for both science and technology. His yardstick was the extent to which the initial situation could be recovered. Only four twenty-sevenths of the driving water could, by means of a perfect water-wheel driving a perfect pump, be restored to the height from which it had fallen. This criterion of 'reversibility' was, very much later, to become a basic concept in the nineteenth-century science and technology of thermodynamics. It is ironic that Parent's correct views on the strength of a beam should be ignored while his flawed theory of water power was accepted. Fifty years were to pass before his four twenty-sevenths rule and the associated theory were taken up and fruitfully criticized.

Another early eighteenth-century innovation that had to wait many years before it was fully exploited was Thomas and John Lombe's introduction into England of Italian silk spinning machinery. In 1702 Thomas Cotchett had built a 'mill' using Dutch machinery for spinning silk thread. It was on the river Derwent at Derby and was powered by a water-wheel. It was apparently unsuccessful. A few years later John Lombe went in disguise to try to find out how the Italians were able to spin silk by machinery. The disguise was unnecessary as the information was freely available. In 1717 his half-brother Thomas Lombe began the construction of a mill near Cotchett's original building. It used Italian machinery. Unfortunately, the machines were complicated for the time and, financially, the enterprise was a doubtful success. At least it confirmed that textile threads could be spun by machines driven by water power. The next step would be to spin cotton and woollen threads mechanically. This proved to be far more difficult. The silk fibre is very long and is adhesive; cotton and woollen fibres are very short and are not adhesive. More than fifty years were to pass before cotton and woollen threads could be spun mechanically. Nevertheless, Cotchett and the Lombes had pioneered an entirely new type of mill, a new application of water power, in England.

The First Working Steam Engine

The acute problem of how to keep mines free of the water that seriously limited the depth at which they could profitably be worked stimulated inventors to try to harness the 'impellent force' of fire. The importance of the earlier suggestions of della Porta and others should not be exaggerated in this context. At best they showed the experimental spirit of a progressive age. And it would be unwise to regard their proposals as direct precursors of the steam engine. To be successful a steam engine had to satisfy three basic requirements. First, that it was a safe and reliable mechanical means of harnessing the 'impellent force' of fire; second, that it could satisfy a clearly recognized need; and, last, that it could be shown to be able to meet that need more cheaply than any established source of power: animal muscle, wind or water. How this was achieved is a curious and instructive story. Studies of the vacuum and of atmospheric pressure following Torricelli, Viviani and Pascal might seem irrelevant to the problem of power, but this was not so. Otto von Guericke's book *Experimenta nova Magdeburgica de vacuo spatio* (1672)[9] describes his invention of the air pump and the experiments he carried out with it. He produced vacua on a large scale and the consequences of his work were far reaching. The illustrations of the book were excellent but they were also dramatic and inspirational. Perhaps the best known is a double spread showing two teams, each of eight horses, trying to pull apart two hemispheres of about one-foot diameter that had been put together and from which all the air had been exhausted. The picture shows that the horses – they must have been feeble animals – were unable to separate the hemispheres, such was the weight – the pressure – of the atmosphere holding them together (about 750 kilograms). Another picture shows how this new agency could be put to practical use: a vertical cylinder with a piston is exhausted of air – a boy or girl of eleven could work the pump – so that atmospheric pressure will drive the piston down. By means of a rope, fixed to the piston and passing over a pulley above the cylinder, a very heavy load can be lifted.

Here was the clearest possible hint: find a simple and efficient way of exhausting air from the cylinder and you have a new source of power.

Almost simultaneously Christiaan Huygens and the Abbe Hautefeuille independently suggested exploding a charge of gunpowder to expel the air, but the method would have been too expensive and the practical difficulties were too great. (Curiously, a very similar method was used in the first really successful gas engine, the precursor of all subsequent internal combustion engines.) A much more fruitful idea was put forward by Denis Papin (1647–1714), in 1690. This was to boil water at the base of the cylinder. The piston would rise on the steam until it reached the top at which point the fire could be removed, the cylinder would cool down and the steam condense. Atmospheric pressure would then drive the piston down. This was, in principle, very close to the answer.

A different idea was developed by Thomas Savery (1650?–1715), of Devonshire, in England. Savery was a merchant, an inventor who also held an official position under the British Admiralty that required him to travel to various seaports, including Plymouth and Dartmouth, where there were depots and agents. In 1698 Savery was granted a patent for an engine to raise water by the agency of fire. In 1699 he demonstrated a working model of his invention at a meeting of the Royal Society and in that year an Act of Parliament extended his patent monopoly to 31 years. In 1702 he gave an account of the engine in a booklet entitled *The Miner's Friend*[10]. John Harris described the engine enthusiastically in his *Lexicon technicum* and other writers, in France and England, followed suit.

A pipe from a boiler carries steam, via a stop-cock, to the top of a large iron vessel; another pipe, from the base of the vessel, carries the steam to a long, vertical pipe. Above and below the point where the steam pipe joins the vertical pipe there are upward-opening valves. The lower end of the vertical pipe, not more than ten metres below the vessel, is immersed in the water that is to be raised; the upper end is bent over a trough. Steam from the boiler flows through the upper valve and blows off at the top of the vertical pipe. When this happens the stop-cock must be closed and cold water poured over the iron vessel to condense the steam inside, leaving a vacuum. Water, pushed by atmospheric pressure, will rise up the vertical pipe to fill the vessel. The stop-cock is now opened and steam, at high pressure, drives the water out of the vessel and up the pipe to the trough above. Savery insists that steam must not be blown off at the top; this, he says, would be wasteful.

Savery, who was elected a Fellow of the Royal Society, was a man of affairs. He was familiar with Papin's ideas, although the extent of his understanding of contemporary scientific knowledge is uncertain. The idea of condensing steam to form a vacuum was, therefore, not original but the distinctive form of his engine raises some interesting questions. The technique of pull–push, of sucking up the water and then blowing it up the pipe, was novel; the most obvious analogy was with a water-power-operated suction and force pump that Papin described in a paper to the Royal Society in 1685 (see plates), and which he suggested could be used for draining mines; admittedly, the analogy is clear only with hindsight. The development of Savery's ideas is therefore an open question and more research on the subject is desirable even though a complete answer may not be possible. Savery was boastful and his invention was widely publicized. He claimed, echoing von Guericke, that a boy of thirteen or fourteen could operate the engine.

There is evidence that about four engines of the type described in *The Miner's Friend* were built. It can be inferred that at least one engine was worked for a time, but how satisfactorily is not known. Savery made provision for replenishing the boiler with fresh water. The working boiler was joined by a pipe, fitted with a stop-cock, to an auxiliary boiler in which high-pressure steam was generated. The pipe extended almost to the bottom of the auxiliary boiler so that, if the stop-cock was opened, the greater steam pressure in the auxiliary boiler would drive hot water into the working boiler. Such a device bears the mark of practical experience. Sporadic efforts were made in France and England to improve the engine and to make it automatic, or self-acting, but without notable success until, towards the end of the century, a greatly modified and simplified form of the engine found a particular but limited market.

The question is, then, what were the main shortcomings of the engine? It is clear that the use of steam at anything very much above atmospheric pressure would, at that time, have been virtually impossible; the requisite metallurgical and manufacturing techniques to build high-pressure vessels had not been developed. The height of the 'push' phase was therefore severely limited. And Savery's insistence that steam must not be blown off at the top of the vertical pipe raises the question: how did he propose to get rid of the air that, unavoidably, would come over with the steam from the boiler (it had been well established by Savery's time that air is dissolved in water)? The progressive accumulation of air in the vessel would seriously reduce the efficiency of the engine unless blown off through the vertical pipe. It is, of course, possible, making the large assumption that the engine worked reasonably well, that engine men ignored Savery's advice.

The fuel consumed just to heat up the iron vessel to the temperature of high-pressure steam was fuel wasted; only the fuel required to generate the volume of steam to fill the vessel and then to push the water back up the pipe was used economically. As the iron vessel together with the water in contact with the steam had to be heated up once every stroke, or cycle, the engine was plainly extremely uneconomical. And, with these severe shortcomings, it would hardly merit discussion here had it not been the precursor of the first unambiguously successful steam engine. Furthermore, the Savery engine has its place, usually unrecognized, in the annals of the steam engine by virtue of its designated purpose. Savery clearly understood that the new source of power could meet the near-desperate need of the mining industry in Europe and, particularly, in Britain for some means of coping with the flood waters that so seriously restricted (relatively) deep mining. And even if the Savery engine was inefficient, the inventor himself deserves credit for loudly and effectively calling attention to the possibilities of steam power. This was not to be the last time that a well publicized false start played a part in launching a new technology, (see chapter 16).

Thomas Newcomen (1664–1729) came, like Savery, from south Devon. He was, by trade, an ironmonger. His few surviving letters and documents show that he was well educated and a man of some property. We know also that he was a devout Baptist and that he was a trustee of the Baptist meeting house at Bromsgrove, south-west of Birmingham (Bromsgrove was then a centre of religious dissent in Britain). Two families closely associated with the Baptist community in Bromsgrove were the Potters and the Hornblowers. The Potters were to collaborate with Newcomen in launching his new steam engine while the Hornblowers later played a prominent part in the development of Newcomen's engine.

Newcomen and his assistant, John Calley, erected the first successful steam engine in the world in 1712 at Coneygree coal mine near Dudley Castle and therefore not far from Bromsgrove[11]. As Mr Allen has pointed out it is inconceivable that Newcomen knew nothing of Savery and his ideas. Both were members of prominent local trading families and Savery regularly visited Dartmouth. This apart we know nothing of the invention and development of the Newcomen engine, which, in principle, was closer to Papin's model than to Savery's engine. Our knowledge of the Coneygree engine is based on a drawing of 1719, a drawing of 1717 of a different engine and later accounts. It appears to have been a well developed machine and the inference must be that a great deal of research and development work, together no doubt with numerous set-backs, had preceded its successful debut. It was certainly a large machine, standing some seventeen metres high. Working normally, it made about twelve pumping strokes per minute, and developed, it has been estimated, just over five horsepower (in modern measure). Unlike five or more horses it could work 24 hours a day, seven days a week without respite.

The structure and action of the Newcomen engine are easy to understand. A large brass cylinder with a piston is fixed vertically above a boiler. A strong chain links the piston to the arch head of a long beam. At the other end of the beam, which is pivoted in the middle, is another arch head connected by a second chain to the pump rod that goes down the mine shaft. The curvature of the arch heads ensures that the chains are always vertical. Steam from the boiler is admitted to the cylinder and the piston rises, not by steam pressure but because of the counterweight on the other end of the beam. When the piston reaches the top of the cylinder the steam is cut off and, simultaneously, cold water is sprayed *into* the cylinder so that much of the steam is condensed, a partial vacuum formed and the excess atmospheric pressure drives the piston down. When the piston reaches the bottom of the cylinder the condensing spray is turned off, simultaneously the steam is turned on and the next cycle begins. The operations of turning on and off the steam and the condensing water are carried out automatically. A long wooden rod, the so-called 'plug rod', hanging from the great beam, is fitted with plugs that engage two very ingenious systems of levers. These control the steam and water valves so that their actions were both simultaneous and – most important – instantaneous. Steam must not enter the cylinder while the condensing spray is on, or vice versa!

Although the principles of the engine are simple there were a number of design details that had to be settled before the engine could work satisfactorily. Condensing water and condensed steam were removed from the cylinder by means of the 'eduction pipe' at the end of which was a leather flap, forming a non-return valve, submerged in the 'hot well'. The air that came over with steam from the boiler had to be expelled from the cylinder; if it was not removed the engine would eventually stop, 'air-logged'. This was arranged by allowing the steam, as it rushed into the cylinder at the start of a cycle, to carry off the air from the previous cycle down the eduction pipe and out through the 'snifting valve', so-called because it made a noise like a man 'snifting' with the cold when the engine was working. The piston, it was found, could not be made to fit the cylinder tightly enough to prevent air from entering and ruining the vacuum; accordingly a pool of water, fed from a small tank, was maintained on top of the piston. This served to complete the seal. At the same time experience showed that it was uneconomic to try to condense all the steam by cooling the cylinder right down. Far too much steam would have been required to heat the cold cylinder up again every cycle and the engine would have worked too slowly (cf. the Savery engine). Better, it was found, to work more quickly with a warm cylinder, making use of only about half the available atmospheric pressure but consuming less steam and therefore less fuel.

When a one third scale model of the earliest engine was constructed some years ago (it is now in the Manchester Museum of Science and Industry), the technicians of the Department of Mechanical Engineering, UMIST, who built it, found that the pool of water on top of the piston was absolutely necessary, even though they 'cheated' by using modern machine tools to ensure a tight fit between piston and cylinder. They found also that the correct setting of the plugs on the plug frame was essential and that the diameter of the pipe carrying the condensing water was critical: too wide and the partial vacuum in the cylinder would suck in water so rapidly that the cylinder would flood, stopping the engine; too narrow and the condensing action would be unsatisfactory. Other problems encountered and solved concerned the size of the steam orifice, the settings of the snifting valve and the eduction valve and the correct weight of the pump rod. In fact several months were required before highly competent technicians, using modern tools and equipment, could make the engine work satisfactorily. Little imagination is needed to picture the difficulties and the disappointments Newcomen must have had before he got his engine to work satisfactorily in 1712. This is a case in which the historian must admit that practice counts for more than documentary evidence.

A novel feature of the Newcomen engine was the use of an internal spray of cold water to condense the steam. This was a great advance on the Savery engine in which water was poured over the cylinder, cooling it right down. It is not, *a priori*, obvious that a spray of cold water will result in efficient condensation. It is likely, therefore, that the explanation given by a reliable witness, Mårten Triewald, is correct. Triewald, who probably knew Newcomen and gave the first eyewitness account of the working engine, said that the advantages of internal condensation were discovered by accident. The practice had been to condense steam by Savery's method when a sudden, marked improvement in performance was found to be due to a crack in the cylinder that admitted a jet of water. (If the crack was big enough to admit an effective jet of water, why, it must be asked, did air not enter as well and so stop the engine, air-logged?) This accidental discovery does not, however, detract from Newcomen's achievement.

The discoveries and ideas that led to the Newcomen

engine were due to Italians, Frenchmen, the German von Guericke and the Dutchman Huygens. With the Frenchman Papin it might be said that the problem of ·harnessing the 'impellent force of fire' had been solved. The Englishmen Savery and Newcomen then stepped in and took the profit arising from this Continental work. The story might therefore be said to accord with the modern complaint – whine might be a better word – heard in certain countries, and particularly in Britain, that *we* make the inventions or discoveries but the foreigner makes the profit out of them. As remarked before, virtually no major inventions are without a prehistory. On the other hand the difficulties of turning an idea or an invention into a practical reality often demand as much (or more) originality as (than) the initial invention. The case of the Newcomen engine confirms this. When full account is taken of the difficulties Newcomen overcame and of the final success of his engine, it must be granted that he was an original genius of the highest order whose success was eventually to transform the world.

Other Considerations

One of the notable changes of the period was the replacement of traditional Latin by the vernacular for learned publications in both France and England. At the individual level, however, communication was along very different lines from those of today. There were no professional institutions, no recognized qualifications and, apart from patent law, few or no legal requirements. Thomas Newcomen, we may suppose, was known to members of the Baptist Church as a God-fearing, skilled and reliable man and that was sufficient. It may indeed be that church networks played a more important part in the diffusion of technical knowledge than has hitherto been recognized.

Marking out in time the era when the bases for modern

power technology were laid and the English agricultural revolution accelerated was the construction of two memorable palaces. In 1688 Louis XIV's magnificent Palace of Versailles was finished; in 1722 Blenheim Palace, built for Louis' opponent, John Churchill, Duke of Marlborough, was completed. Blenheim is far smaller than Versailles; its facade is only slightly more than a fifth as long as that of Versailles. Both buildings have certain features in common and some instructive differences. Both were built of stone, brick and timber, materials familiar from time immemorial. There was little or no scope for the application of Galileo's new science of the strength of materials. Traditional skills and the limitations of the materials played their parts in determining the forms of the buildings. Another point in common was that Versailles and Blenheim were intended to glorify individual men, not God. And to that extent they marked the onset of a more material, less spiritual age.

The surroundings of the two palaces have, on the other hand, nothing in common. The gardens of Versailles were laid out by André le Notre (1613–99?), the leading landscape gardener of the age. His style was highly artificial. 'The chief object', wrote an impartial authority[12], 'seems to have been to subject nature to the laws of geometry and to practise geometry, architecture and sculpture upon lawns, trees and ponds.' The surroundings of Blenheim were designed by Lancelot 'Capability' Brown (1715–83) following a style created by William Kent. Here the aim was to bring out the natural landscape, to co-operate constructively with Nature, as Bacon would have recommended. It would be rash to draw conclusions about that elusive factor, national character, from this comparison. Observation shows that, with its neat, disciplined rows of identical flowers and shrubs, and lawns with edges cut straight or in perfect circles, the typical English suburban garden of the present day resembles nothing so much as a miniature Versailles. But, admittedly, it would require something like genius to apply Brown's principles successfully to a small suburban garden.

Ladies and gentlemen of sensibility during the succeeding age of reason found pleasure in a landscape that showed the taming hand of civilized people. To be fully appreciated a landscape had to be a work of art. They regarded with distaste the uncivilized, barbarous wildernesses of stone, rock, ice and snow that were the Alps and the other high lands of Europe. On the other hand, the great majority of people had little or no time in which to form aesthetic judgements of this sort; for them life was hard and short and their expectations modest.

A NOTE ON PARENT'S THEORY

Let us suppose that the velocity of the stream is v, the velocity of the wheel V and the load being lifted W.

Then the quantity of water striking the blades per minute is proportional to $(v - V)$ and the relative velocity with which the water strikes the blades is also proportional to $(v - V)$.

When the wheel is turning at a uniform rate, $(v - V) \times (v - V)$ must be proportional to W. Assuming that there is no gearing, the velocity with which the weight W is raised is V so that the work done in one minute is $W \times V$ and this is:

$$\text{proportional to } V \times (v - V)^2$$

Differentiating this expression with respect to V and putting it equal to zero we find V is equal either to v or to $(1/3)v$. Substituting $(1/3)v$ for V in this expression gives $(4/27)v^3$ for the maximum effect.

But the quantity of water flowing per minute is proportional to v and the height to which it can rise is proportional to v^2 so the available effort is proportional to

v^3. The efficiency of the water-wheel is, therefore, proportional to 4/27 of the available effort. If, in any specific instance, the actual value of the available effort is calculated, the factor 2 must be taken into account and the efficiency of the perfect water-wheel becomes 8/27.

Progress in Practice

Newcomen engines were housed in ordinary, far-from-revolutionary, buildings of stone and wood. In that respect no more advanced technology was required than for the palaces of Versailles and Blenheim. It would be true to say that the engine was itself a structure, a working building, rather than an engine in the sense that would be understood today. From this point of view it was a variant of the mill and most of the skills required to construct a Newcomen engine were the same as those required to construct a water mill. A 'Proprietor' (of the patent), a coal-viewer (or mining engineer), a surveyor or an instrument maker might design an engine; local masons, bricklayers, carpenters, millwrights and blacksmiths would build it. It was fortunate for England that contemporary skills could meet the requirements of the design and that there was a sufficient number of skilled men to make the engine a feasible proposition[1].

The success of the Newcomen engine was acknowledged without delay. It triumphantly satisfied the three major requirements: it was reliable, relatively economical and met a clearly recognized need. If more water had to be raised to a greater height all that was required was a bigger engine with a bigger cylinder. With a Savery engine the same requirement meant a bigger engine and an increasingly dangerous higher steam pressure. Nevertheless Newcomen's engine could not be patented, for, as the law then stood, Savery's patent covered all engines that harnessed the 'impellent force of fire'. Technological knowledge was not sufficiently advanced to recognize the unique merits of the Newcomen engine and the line between invention and improvement was not well enough drawn to enable the differences between the Savery engine and the Newcomen engine to be clearly seen: both, after all, were 'fire engines'. One may as well say that Newton's *Principia* was an improved version of Ptolemy's *Syntaxis* (or *Almagest*)! In fact, by modern standards – or even those of fifty years later – the Newcomen engine included at least four major recognizable patents.

Between 1715 and 1733 the rights were held by a Committee of six Proprietors, of whom Newcomen was originally one. The exhaustive researches of Drs Kanefsky and Robey show that over these years 94 Newcomen engines were built in Britain, many being in the north-east coalfield – the largest coal mining area in Europe – the midlands, the north-west and the north Wales coalfields and in the non-ferrous metal mining area of Cornwall. One of a number of engines not associated with mining was that at York Buildings (1726), near Charing Cross in London, which was used for water supply. It was sometimes compared with the great water-wheel at Marly (1705) that pumped water up to the gardens at Versailles. A member of the Committee of Proprietors, John Meres, undertook the construction of an engine at Passy, in Paris (1726), to pump water from the Seine; it, too, was described as a triumph of hydraulic engineering. At about the same time Isaac Potter from Bromsgrove was associated with the installation of an engine in Vienna to supply water for ornamental fountains. More useful than these were the engines that a Colonel John O'Kelly and his assistant were to build in the coalfield near Liège, in Belgium; the first of these was at work by 1723.

In what was then northern Hungary and is now Slovakia was a group of valuable non-ferrous metal mines owned by the Austrian state and controlled from Vienna[2]. As in Bohemia the managers and engineers were German speakers; accordingly the three leading communities were known, at least in the west, as Schemnitz, Koenigsberg and Windschact rather than by their Slovakian names, Banská Štiavnica, Nová Baňa and Štiavnické Bane. Isaac Potter was persuaded to move on from Vienna to Koenigsberg, a few kilometres from Schemnitz, where he built the first (1721–22) of what were to be seven engines (see Note at end of the chapter). Potter's engine was reported (probably optimistically) to have been twice as powerful as the original Coneygree engine and to have raised water from a depth of 152 metres (in modern measure it would have developed about ten horsepower).

The Newcomen engines in the Schemnitz area were unquestionably successful. There was one serious drawback: they burned wood, between nine and thirteen cubic metres a day, and wood was used in great quantities in the smelting processes as well as for the construction of all mine machinery and structures. So great was the destruction of the forests that restrictions had to be placed on the working of the engines; they could not all be worked at the same time.

Wood was also the fuel used in what was perhaps the most celebrated and certainly the best documented of the early Continental engines. This was the engine erected at Dannemora iron mines, in Sweden, by Mårten Triewald (1691–1747), and the subject of an exemplary monograph by Professor Lindqvist[3]. Triewald came to England in 1716 where his interest in science and technology was aroused by Edmund Halley and, particularly, J. T. Desaguliers. Next year Triewald went north, to Newcastle, and there, in collaboration with Samuel Calley, the son of Newcomen's associate, he worked on the erection of four Newcomen engines. In addition he made a name for himself as a public lecturer in Edinburgh and Newcastle before returning to London where he met the elderly Newton himself. Triewald must have congratulated himself on his years in England, 1716–25.

In the meantime news of the new engine had reached Sweden by 1720. In the next few years John O'Kelly tried to persuade the Swedish mining authorities that he should build a Newcomen engine to drain a mine at Dannemora, about 100 km north of Stockholm. A Colonel de Valair, about whom little is known, made a similar proposal. Neither was successful. The task fell to Triewald: he had had extensive practical experience, he had acquired considerable scientific knowledge and he was a Swedish national of good repute. By 1728 the engine was completed. Although it demonstrated its capacity to drain the mine it was not a success; it was always breaking down until in 1734 it ceased work for the last time. The individual components of the engine have long since disappeared but the engine house still stands and is carefully preserved. It must be the oldest relic of a Newcomen engine in the world.

Lindqvist gives convincing – and revealing – reasons for the failure. Swedish technology, he points out, was based on wood and the standard of iron working skill in Sweden was markedly below that in England (this is surprising). Essential parts that in England would have been made of iron were made of wood. Furthermore, Triewald increased the diameter of the cylinder by 90 per cent in order to get more power; but he failed to increase the scale and strength of the other components to match the increase in power. To make matters worse there were only two men – of whom Triewald was one – in Sweden capable of running the engine. Neither of them spent much time at Dannemora: Triewald had interests in Stockholm; the other wanted to be in Uppsala. As Lindqvist remarks, if a technology is to be successfully transferred from one community, nation, or culture to another the resources and skills available in the recipient must match those of the donor. The remarkable speed and success with which the Newcomen engine was adopted in England – Kanefsky and Robey conclude that over a thousand were built in

England in the course of the eighteenth century – confirm the comparatively high standard and wide distribution of technical skills in the country. If Newcomen and John Calley supervised the building of the very first engines there were soon others like the Potters, the Hornblowers and Stonier Parrott competent enough to continue the job. And there were enough far-seeing entrepreneurs to keep the engine builders busy. There was no engineering industry at the time. And no formal qualifications were demanded of engine builders for the sufficient reason that there were no training courses, academic or practical. The schools and universities of Britain can claim no direct credit for this revolutionary invention.

The modern reader may, at first, be surprised that the Newcomen engine, the precursor of all subsequent heat engines[4] including internal combustion engines, aroused so little public interest when it was launched on the world. The comparison with the wild excitement caused by the first balloon flights at the end of the same century seems striking. But, of course, no one could then foresee the future development of the engine. And, although there were many engines, they were usually located well off the beaten track. They were to be found in remote, uncongenial areas like Cornwall, Cumberland, the north-east coalfield, the Black Country, Lancashire and Yorkshire. And from the outside all that could be seen was an unimpressive stone or brick building with a chimney and the end of a massive wooden beam, sticking out of one wall, moving slowly, ponderously, up and down. The big water-wheel under London Bridge was far more impressive: it had cranks, pumps and big beams moving up and down; so too had the even more imposing Machine de Marly. Although Bacon would have been delighted with it, for most people the Newcomen or fire engine was just another new pumping machine and, in particular, an advance in the art of mining.

The progress of the Newcomen engine in mainland

Europe was restricted by the relative scarcity of coalfields. In the Schemnitz district the engines had to be supplemented by water power. Accordingly, a senior engineer, J. K. Hoell, or Hell, who had built one of the Schemnitz Newcomen engines, invented and built a distinctive hydraulic engine, the so-called column-of-water engine, (in French, the *machine à colonne d'eau*; in German, the *Wassersaulenmaschine*). The column-of-water engine was plainly – and was acknowledged at the time to be – inspired by the Newcomen engine, of which it could be said to be the hydraulic version. Water pressure, instead of atmospheric pressure, acted on the piston. The water was brought by pipe from a conveniently high spring or reservoir and applied above the piston, the top of the cylinder being closed so that the piston rod had to be taken through a stuffing box. For the rest the engine was much like a Newcomen engine with working beam, arch heads, plug rod and valve mechanism. Its great advantage over a water-wheel was that the latter could not use a head of water greater than its diameter while the column-of-water engine was limited in this respect only by the pressure that its pipes, cylinder and valves could stand. Furthermore, there was very little wastage of water and, in theory, an efficiency approaching 1, or 100 per cent could be reached. That is to say, all the potential effort of the head, or fall, of water could be converted into useful effect. Later versions of the engine incorporated a refinement that extended the analogy to Watt's expansively worked steam engine (see chapter 7).

Hoell's first column-of-water engine was erected in 1749. In all, he built nine of these machines. Although one or two accounts of similar machines were given to the Académie Royale des Sciences before 1749, these were of experimental models and there is no evidence that they worked, or that Hoell knew anything about them. Hoell's machines certainly worked satisfactorily. Almost certainly they were independent inventions on his part, springing from his extensive knowledge of the Newcomen engine. The column-of-water engine was not Hoell's only notable invention. His air machine, or 'Heronic engine', of which he built four, was widely reported and greatly admired for its ingenuity and simplicity. Water was carried by pipe from a high spring or reservoir to a strong air vessel at the top of the mine shaft. After the water had compressed the air in the vessel sufficiently the supply was turned off and the compressed air admitted to a second pipe that led down to another vessel at the bottom of the shaft. The compressed air then drove the flood water that had filled the lower vessel up a third pipe to the top. This was probably the first instance of the successful use of compressed air in industry.

A curious phenomenon associated with the working of the Heronic engine greatly puzzled the scientific world of the day. It was noticed that after the air had completed its cycle by driving the flood water up to the top (and necessarily expanding in doing so) it was extremely cold when it was expelled from the machine as the next charge of water flooded in to the lower vessel; so cold was the air that even on the hottest August day it was accompanied by snow flakes. One theory put forward half-heartedly and quite wrongly was that the waters in the Schemnitz mines were so impregnated with mineral salts that they constituted a freezing mixture.

The air machine was simple and cheap to build. Its disadvantages were that, unlike the Newcomen engine or the column-of-water engine, it was never made self-acting; it required two men to operate it, working all the time. In addition, it was slow and the depth from which it could pump water depended on the head of driving water available and the limited strength of the pipework. The great advantage of the column-of-water engine was that, at Schemnitz, it cost half as much to run as the Newcomen engine. It was therefore suitable for all places where fuel was expensive and adequate heads of water available.

C. T. Delius described and illustrated it in his textbook, *Anleitung zu der Bergbaukunst* (1773). Apart from the French translation of this book, Gabriel Jars gave an enthusiastic account of the engine in his celebrated *Voyages metallurgiques* (1780). It became well known all over Europe and in the second half of the eighteenth century many such machines were built, particularly in Germany and France. They were not as numerous in England where coal was cheap almost everywhere and Newcomen and, later, Watt engines predominated.

As we noted, French inventors had, earlier in the eighteenth century, conceived of hydraulic engines working on the same principles. In the second volume of Belidor's *Architecture hydraulique* (1739) there is a short description of a machine with two pistons, one driven by water pressure, to 'raise water above its level', and mention of a complex machine invented by Denisard and de la Dueille. In 1741 de Gensanne described a small hydraulic machine he had made; it bore striking resemblance to a Newcomen engine. De Gensanne was associated with a lead mine near Rennes, in Brittany, where he had been working since 1738. Unfortunately his machine was subsequently destroyed in an accident, but not before the assessors of the Académie Royale des Sciences had reported favourably on it. The most interesting feature of de Gensanne's account was his calculations of the effect to be obtained from a given quantity of water falling fifteen feet and used, first, to drive an undershot water-wheel and then his machine. He showed that his machine was far more efficient than a water-wheel whose effect he calculated by using Parent's four twenty-sevenths rule.

The account, with illustration, of de Gensanne's machine was not published until 1757; in any case it and the machines described by Belidor were hardly more than experimental models. The credit for the independent invention of the column-of-water engine and for pioneering its use in practice, under the hard working conditions of mines, belongs to Hoell, as French engineers generously conceded. De Gensanne's idea marked, however, a stage in the public refutation of Parent's argument. The process was taken further when, some years later, the Chevalier de Parcieux was called on to arrange the water supply for Mme la Marquise de Pompadour's château. The problem was that the reservoir for the château was 163 feet above the small river that was to supply it and this meant that when the river was low there would not be sufficient power, according to the four twenty-sevenths rule, to supply enough water. De Parcieux reasoned that falling water, acting by weight, should, in principle, be able to raise more than four twenty-sevenths of its own weight. The water in the buckets of an overshot water-wheel should be able to raise a weight not much less than its own; such an arrangement is exactly analogous to a weight attached by a string passing over a pulley to another, slightly lighter weight. No one could doubt that the heavier weight would raise the lighter one. De Parcieux observed that if an overshot water-wheel drove an identical wheel in reverse so that it acted as a scoop-wheel, the system would, if both wheels were friction-free and perfect in all other respects, enable all the driving water to be returned to the source. This was an explicit recognition of complete reversibility, or the recoverability of the initial situation, as the limiting condition; it had only been implicit in Parent's work. For the present, however, we must assume that the water supply to the château was installed to the complete satisfaction of Mme la Marquise.

Finally, in 1767 the Chevalier de Borda published the correct general theory of the undershot water-wheel, showing that the maximum effect was one half that of the effort of the stream. He pointed out that it should be possible to minimize the waste of *vis viva*, due to turbulent impact of the water on the blades, and so increase the useful effect, by suitably curving the blades to point up-stream. In this way, he argued, the efficiency of an undershot wheel could approach 100 per cent; moreover, such a wheel could be run at high speed. De Borda had laid the foundation of the theory of the turbine. The difference between the circumstances and the ways in which the performance of water-wheels was investigated in France and in England in the middle decades of the eighteenth century is instructive. In France the approach was theoretical, mathematical; in England it was practical. The English approach was exemplified by the works of John Smeaton (1724–92), the founder of British civil engineering[5]. Born into a middle-class family, his original intention was to become an attorney but natural gifts soon asserted themselves and he trained as an instrument maker, a scientific craft of increasing importance at a time when the growth of overseas trade entailed improved navigational aids. But it was as what was later called a civil engineer – as distinct from the long-established military engineer – that Smeaton made his reputation. His constructions, many of which are still standing, ranged from bridges in the north of Scotland to the fine lighthouse he built, and which stood from 1755 to 1879, on the difficult and dangerous Eddystone rocks, twenty kilometres off the coast of Cornwall. (Smeaton's tower was dismantled in 1879 and re-erected on Plymouth Hoe where it now stands.) With no successful precedent and no adequate theory to guide him Smeaton had to rely on his engineering intuition. He chose the tree trunk as the model – broad at the base and tapering to the top – for a strong tower best suited to stand the force of Atlantic gales and seas. This was to be the model for all subsequent offshore lighthouses.

Smeaton's wide practical experience convinced him that Parent's four twenty-sevenths rule was wrong. He knew, he said, of actual water-wheels more efficient than four twenty-sevenths , or 14.8 per cent. He therefore resolved to measure the efficiencies of water-wheels experimentally. For this purpose he built a small wheel, about a metre in diameter, supplied with water from an adjoining tank. The water, after passing under the wheel, was recycled to the tank by means of a pump; the level, or head, of water in the tank being kept as constant as possible. The rate of flow of water was increased, step by step, and for each step the load on the machine was systematically varied and the corresponding work done was recorded. After allowance had been made for frictional losses Smeaton concluded that the maximum efficiency of an undershot, or impact, water-wheel was about one third. The inference that he subsequently drew was that, ideally, the maximum efficiency was a half, for which the velocity of the wheel would have to be one half that of the stream. He then studied the performance of the wheel when worked 'overshot'; that is, when water arrived on top of the wheel and filled buckets so that it acted by weight rather than by impact. A repeat of the same series of experiments indicated a maximum efficiency of two thirds. This time the inference was that a perfect overshot wheel, revolving extremely slowly, would have a maximum efficiency of unity. The inferiority of the undershot wheel Smeaton ascribed to the effort lost as the stream flowed away at half its initial velocity plus that lost by the stream 'distorting its figure' on impact, i.e. lost in generating turbulence. He did not wonder whether the effort lost in turbulence could be recovered.

The authority and skill he showed in these experiments were acknowledged when he was awarded the Copley Medal, the highest distinction of the Royal Society, to which he had been elected. The experiments, together with those on windmills, were published in the *Philosophical Transactions* for 1759. From this time onward the overshot water-wheel was acknowledged to be twice as efficient as the undershot wheel. And, Smeaton recommended, if the river was relatively flat, with no convenient fall, so that an overshot wheel would be expensive to install, the best compromise would be a 'breast' wheel. In this case a weir should be constructed to raise the river to the level of the axle of the wheel and the wheel should be faired in to the concave downstream face of the weir. The water would therefore act as much by weight as by impact.

Smeaton's researches, in effect, greatly increased the power resources of the country. The traditional water-wheel, as represented in sentimental and nostalgic art (examples are lovingly preserved), provided enough power to grind the corn, cut the wood and hammer out the iron needed by small local communities. Before Smeaton the industrial water-wheel had been confined largely to mining. He made it much more efficient and able to give sufficient power to drive a factory. And this was most opportune for an imminent major technological innovation was, for its full development, to demand all the power – and more – that was available.

Smeaton later turned his attention to the improvement of the Newcomen engine. He constructed a small engine – a model standing only about four metres high – of which he systematically varied each component in turn and noted the effect of each variation on the performance of the engine. From the data he recorded he was able to specify the optimum design for a Newcomen engine. It is reported that as a result he was able to double the efficiency of the engine, his measure of the efficiency of an engine being the volume of water it was able to raise one foot high for the burning of a bushel (or sackful) of coal in the furnace. This, adapting the term Parent and others had used for the work done, he called the 'effect'. Smeaton built very large engines indeed; one, at Chasewater in Cornwall, had a cylinder two metres in diameter. It developed, in modern measure, 76 horsepower. An improvement he introduced was to substitute for the simple chains, used to couple the beam to the piston and the pump rod, chains of a form similar to those developed for the fusees of watches. These chains consisted of flat iron plates united by round pins and are now commonly used on bicycles.

The advantage of such chains for use in Newcomen engines is obvious.

The Textile Industry

When John Kay invented his fly shuttle in 1733 he initiated a revolution in textile manufacture and, beyond that, gave an impetus to the nascent Industrial Revolution. Before the fly shuttle was invented the weaver had to pass the bobbin carrying the weft thread across the loom by hand; this was a slow business and it limited the width of the fabric that could be woven. With Kay's invention the bobbin was carried in a shuttle tapered at both ends and fitted with small wheels (figure 6.1). By jerking a cord the weaver could, without leaving his seat in front of the loom, project the shuttle at high speed from its shuttle box − its 'launch pad' − to the shuttle box on the opposite side of the loom. With another jerk at the cord the weaver could speed the shuttle back to the first shuttle box. This not only speeded up weaving but it increased the width of the fabric that could be woven. The resulting much improved productivity of weavers put pressure on the spinners to speed up the production of thread.

In 1738 Lewis Paul, the son of a refugee Frenchman, and

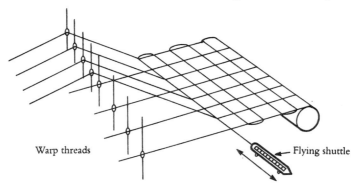

Warp threads

Flying shuttle

FIG 6.1 *Flying shuttle*

John Wyatt patented a spinning machine. It was modelled on the silk-spinning machine that Lombe had brought back from Italy. Wool in the form of a roving, or a tube about the thickness of a thin pencil, of untwisted parallel fibres, was fed between a pair of rollers and thence, via a flyer – adopted from the Saxon Wheel – that imparted twist, to a bobbin on a spindle. Paul and Wyatt set up a spinning mill, driven by two donkeys, in Birmingham in 1741. Later, three other mills were built but none of them was successful. 'Leisure preference', it is said, led to labour troubles, for the habit of disciplined work had not yet been inculcated and there were serious problems with machinery that was complex for the time. In 1748 Paul took out a patent for a carding machine to comb out the fibres mechanically. This, too, was not particularly successful.

Where Paul and Wyatt failed Richard Arkwright (1732–92) succeeded, triumphantly. He, more than anyone else, carried through the textile revolution that was to open the way to a new industrial world. Arkwright, born of working-class parents in Preston, Lancashire, was apprenticed to a barber and wig maker before setting up in the same craft on his own account. His work must have given him useful insights into the properties of natural fibres, and the experience combined with his natural aptitude for practical mechanics must have prompted him to study the problems of a growing textile industry that was still using methods developed in the later middle ages[6]. Although the sequence of Arkwright's major invention is unclear, having been obscured by rival claims and a lack of documentary evidence, the achievement itself is simple and easy to understand (figure 6.2).

In place of Paul's single pair of rollers Arkwright used four (later three) pairs. The first pair, to which the roving was fed, rotated relatively slowly, the next pair rotated faster and the last fastest of all. As a result the roving was drawn out, or attenuated, before reaching the flyer that

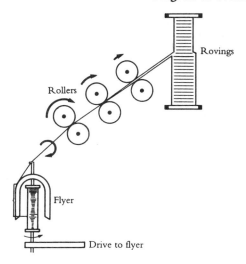

Rovings

Rollers

Flyer

Drive to flyer

FIG 6.2   *Arkwright's invention*

imparted the twist and wound the thread on to the bobbin. Arkwright showed his mastery of the process by making the separation of the pairs of rollers equal to the average length of the fibres. Had they been closer together some fibres would have been snapped; had they been further apart an unsatisfactory thread would have resulted. Furthermore, he realized that the pressure between the rollers was critical, so he weighted them to give just the right pressure on the roving passing through.

Such, then, was the famous 'water frame', patented by Arkwright in 1769 (see plates). Together with his partners, John Smalley, David Thornley and, later, Samuel Need and Jedediah Strutt, Arkwright decided, in 1771, to establish a horse-driven spinning mill in Nottingham. Before that mill had commenced work they had begun to build a water-powered mill at Cromford in Derbyshire. It is not clear why they chose Cromford. It was not on a good road, there was no canal and the river Derwent was not navigable at that point. On the other hand there was abundant water power and the families of the men who worked in the nearby lead mines offered a potential and docile labour force. Cromford mill began working in 1772; driven by an overshot water-wheel it presaged the modern factory system.

The water frame required well formed rovings to produce the best results. The next requirement was, therefore, to improve the preparation of the rovings. In 1775 Arkwright patented his carding engine, effectively a rotary comb in the form of a drum with numerous short metal teeth protruding from it. As the drum rotated the cotton wool fed into it was combed out into a continuous film with the fibres parallel to each other. In the patent specification Arkwright described several ways in which this film could be formed into a roving; one of them amounted to, or was held to amount to, a covert attempt to extend the run of his 1769 patent. Jealous competitors were later to challenge the patent and it was revoked in 1785. This was not entirely surprising. Arkwright was essentially a driving businessman, a tycoon who happened also to be an ingenious inventor. He understood, as no man had done before, the nature and potential of textile manufacture and he saw clearly how he could harness the new mechanical inventions to create what was, in most respects, a new industry. The skilled manual process of spinning was broken down into a number of separate stages, each one of which could be mechanized so that skill was no longer required and production enormously increased. The new industry was epitomized by the cotton mill. The first one at Cromford was spacious and stood five storeys high. It became the universal prototype. It had, as Arkwright's biographer, the late Dr Fitton, put it '. . . no counterpart in English architectural history and became the basic design in industrial architecture for the remainder of the eighteenth and throughout the nineteenth centuries'.

The Arkwright manufacturing empire spread all over England and Scotland and its creator became the richest man in Britain, quite possibly in Europe and the world. He showed that the way to industrial growth was by manufacturing cheap fabrics for the multitude; not by making expensive silks for the wealthy few. Sir (as he became) Richard Arkwright was not a man of science; he never bothered to become a Fellow of the Royal Society although he could easily have afforded to buy up that august body. His inventions and those associated with the early textile industry involved no principles or materials that would have puzzled Archimedes. Science played no part in their formulation. Nor did it play a part in James Hargreaves' rival and roughly contemporaneous invention of the spinning 'jenny', a simple device that was basically a spinning wheel on its side and driving not one spindle but a whole array of them.

The textile revolution of Kay, Paul, Arkwright, Hargreaves and the others indicated the potential of mass-production industry; it created enormous wealth consequent on high productivity and, in addition, it stimulated innovation and invention in a number of other industries; indeed, it brought new industries and technologies into being. Only the great mining industry had done (and was to do) as much. The rapid expansion of the industry, for others soon joined in once Arkwright and his partners had shown the way, exerted pressures for innovation in construction, in machine making, in transportation, in power generation and transmission and in bleaching, mordanting, dyeing and printing. Bottlenecks that were hardly noticeable in the days of the individual spinner and weaver became intolerably constrictive as the transformed industry expanded. The first generations of cotton and woollen mills, from Cromford onward, were powered by overshot or breast wheels on Smeatonian principles. The early textile masters had to squeeze the last ounce of 'effort' out of the last drop of water falling the last inch. The history of the new textile industry confirms, as we noticed in the case of the Newcomen engine, that there was a widespread distribution of practical skills and entrepreneurial spirit in eighteenth-century England. And, as Arkwright's biographer suggests, there were few institutional or formal obstacles to a working man making a huge fortune and rising almost to the top of the social tree; always provided that he had ability, determination, a modicum of good luck and robust good health. Cotton, as the saying goes, was king.

Cotton fabrics, when they leave the loom, are an uninspiring greyish colour. Before they can be sold they have to be cleaned, bleached, mordanted and dyed or printed. The old method of bleaching had been to expose the fabric in an open field for about six months to the sun, wind and rain. This was an inefficient and, apparently, a rather risky process for stealing from bleach-fields must have been common: it was made a capital offence in eighteenth-century England. A somewhat faster, but still expensive, process was the use of sour milk (lactic acid) as a bleaching agent. Here, at the very end of the processes of textile manufacture, ancient procedures, archaic technics, restricted the effectiveness of the latest technology. The resolution of this problem involved the creation of a new industry: the chemical industry[7]. As a first stage sulphuric acid was used in place of sour milk as a bleaching agent.

Sulphuric acid, or oil of vitriol, was used in relatively small quantities for special purposes such as cleaning and parting metals, the preparation of pharmaceuticals and in a variety of small trades. The acid was prepared by burning a mixture of sulphur and nitre (potassium nitrate) over water in a large bell-shaped glass vessel (*per campanum*), the process having been imported into England from France and Holland. Manufacture on a much larger scale began when Joshua Ward, a pedlar of quack medicines, and John White set up their works at Twickenham, near London, in 1736. Economies of scale led to a substantial reduction in price and the acid became available for large-scale industrial use as well as for the small-scale preparation of pharmaceutical products. A second substantial advance in the manufacture of sulphuric acid was made in 1746 when John Roebuck and Samuel Garbett, making use of an observation by J. R. Glauber that sulphuric acid does not attack lead, built a large cubical wooden frame and lined it on the inside with lead sheets. There was a pool of water at the bottom of the lead chamber over which the mixture of sulphur and nitre was burned. It was still a slow process: repeated charges of sulphur and nitre had to be burned before a relatively concentrated acid could be obtained. Production on a much larger scale was now possible. The continued reduction in price resulting from these advances was such that Francis Home, Professor of Materia Medica at Edinburgh University, could recommend the use of sulphuric acid as a bleaching agent in place of sour milk. Home's book *The Art of Bleaching* (1756) is described as the first scientific study of bleaching. It is more than that: it is one of the first indications of the arrival of Scotsmen at the forefront of technology.

The Iron Industry

While Arkwright was busy founding his textile empire the iron industry was undergoing quantitative and qualitative changes. The price of charcoal having, according to Hyde, increased rapidly, it became more economic to use coke to smelt iron. More coke furnaces were built from the 1750s onward while charcoal furnaces went out of use. The most spectacular individual event was the opening in 1760 of the great Carron Ironworks in Scotland. Significantly two of the three major shareholders were John Roebuck and Samuel Garbett. The furnaces were, of course, coke-fired. And, from Coalbrookdale came equipment and manpower to operate the plant as well as to teach the Scots the new technology of iron.

In due course Carron Ironworks became by far the largest single plant of its kind in Britain. It was famous for casting cannon, in particular the naval gun called the 'carronade'. The versatile Smeaton advised that, for blowing the furnaces, an air pump in the form of a large iron cylinder with a piston should replace the old-fashioned leather bellows. The pump was driven by a large water-wheel, the water for which was recycled by a Newcomen engine. The power supply was therefore substantially independent of the state of the stream or river. This meant that ironworks could be worked through the driest summers and could be located virtually anywhere where coal and iron ore were available. Moreover, the powerful blast from Smeaton's pump meant higher furnace temperatures and molten iron that was more fluid.

An indication of the growing cheapness of cast iron was the introduction by Richard Reynolds of Ketley in 1763 of iron L-section rails for the tramways that linked coal mines with rivers, canals and ports. Such tramways became particularly common in the north-east of England. The L-section rails were laid on massive flat stones with the vertical edge plates on the inside, an arrangement that kept the wheels of the coal waggons running on the flat plates. Cast iron rails not only lasted longer than wooden rails but could carry much heavier loads. However, before the full potential of railway transport could be realized, several related developments had to take place. The first of these was the establishment of a national canal system.

The beginning of the English canal network dates from 1761–64 when the Duke of Bridgewater's canal was built[8]; it linked his coal mines at Worsley, twelve kilometres west of Manchester, with Castlefield on the south-western edge of the town. Even before the canal was completed the construction of a national network had begun, so that by the last decade of the century the growing industrial areas of the north-west, the north-east and the midlands were linked to one another and to London by canal while London was linked to the Bristol Channel. Canal transport was, of course, slow but it was extremely cheap. It was reliable and it could cope with bulky loads that could not be carried by road. As engineering achievements the English canals were certainly less impressive than those of France (The young Duke of Bridgewater, when on the grand tour required of every young nobleman, had greatly admired the French canals. They inspired him to build his canal.) But the technological knowledge and skills required for canal building should not be underestimated. The route had to be accurately surveyed and adequate water supplies arranged, particularly at the highest point – the summit – reached by the canal. The water level of the canal had to be kept constant so the bed and sides had to be waterproofed by coating with suitable 'puddle' (a mixture of light loam with sand or gravel). A labour force had to be recruited, equipped, trained and paid. Finally, legal and political skills were needed since land and water rights had to be bought out or acquired by compulsory purchase. It was not fortuitous that the first geological map was made by William Smith, a canal engineer.

Dr John Roebuck's name is not as familiar as are those of other leading British entrepreneurs of the eighteenth century. He was, nevertheless, one of the most interesting. His career links Coalbrookdale, the lead chamber process and the Carron Ironworks; he was also to play a significant role in the early career of James Watt. Roebuck had been a medical student at Leiden; he was therefore a disciple of the great Dutch chemist and teacher Hermann Boerhaave.

The political, social, philosophical and economic roots of the Industrial Revolution no doubt go back many years before the mid eighteenth century. Of that revolution, the lead chamber process, the Cromford cotton mill, the Carron Ironworks and Coalbrookdale with its famous iron bridge can be taken as among the earliest material achievements. These were soon to be closely related and, in important respects, mutually stimulating. This can also be said of the work of Robert Bakewell (1725–95), a farmer in the English midlands. What Tull, Townshend and others did for the improvement of land and crops, Bakewell did for livestock. He systematically improved the breeds of sheep and other domestic animals, bringing out those qualities of meat and fat (fat was greatly relished by the English in those days) that he wanted to develop and reducing those, such as bone, of little or no economic value. Bakewell was careful to breed in and in. Crossing a good animal with an inferior one he regarded as adulterating the stock. Good must be mated with good and so on until the qualities he was aiming for had been established. So famous was Bakewell that land-owners and farmers came from all over Europe, including Russia, and from the Americas to learn from him. With one important exception, agriculture, unlike mining and textiles, did not stimulate invention or discovery in other industries. It formed a big market for the innovations of other industries, particularly the iron industry; it provided the food for the growing numbers of industrial workers and, as it became more efficient, it made farm workers redundant and therefore available for the new and expanding manufacturing industries. It was, we may say, a passive supplier and not a source of innovation. Bakewell's contribution, however, was an exception. His successes in animal breeding – it might almost be described as animal engineering – were much later to have a profound influence on the thoughts of Charles Darwin and Gregor Mendel.

Improvement, manifest in such diverse fields, was given formal and institutional recognition when the Society for the Encouragement of Arts, Commerce and Manufactures was established in London in 1754. This distinguished body can point to an impressive record of contributions to technology, industry, technical education and commerce. Its provincial and rather more specialized counterpart was the association of midlands industrialists that was called the Lunar Society[9]. 'More than any other single group', writes Professor Schofield, 'the Lunar Society of Birmingham represented the forces of change in late eighteenth century England.' The name of the Society came from the members' custom of meeting at the time of the full moon when travelling on unlit lanes, roads and streets would be relatively easy and safe. It began about 1765, the driving force being Matthew Boulton; the other members included James Watt, Josiah Wedgwood the potter, Joseph Priestley, Erasmus Darwin (grandfather of Charles Darwin) and Richard Lovell Edgeworth, inventor, reformer and father of the novelist Maria Edgeworth. Their interests were scientific, technological and industrial and their activities covered an area extending from Warrington in the north west, Derby in the north east and Coventry in the east. Distinguished visitors to Birmingham were entertained by the Society and their opinions and advice – discreetly – sought on matters of commercial, industrial or scientific interest to the members. The Society had no constitution; no minutes of its meetings were kept and no journal published. Although it was an entirely secular body, its influence was considerable. It owed nothing to the state, the established state Church or the universities of England. As Matthew Boulton grew older and less active and as political reaction set in as a consequence of the French Revolution, the Lunar Society faded away. Had it managed to survive those turbulent years it might have established itself on a permanent basis, in which case it might have evolved into something like the Société Industrielle de Mulhouse. The Royal Society of Arts, on the other hand, with its wider range of membership and dissociated from any overt political movements, has survived and is still very active today.

A NOTE ON JACOB LEUPOLD

In his comprehensive volumes, *Theatrum machinarum generale* and *Theatrum machinarum hydraulicarum* (Leipzig, 1724, 1725), Jacob Leupold ascribed the invention, as well as the erection, of the engine at Koenigsberg, in Hungary, to Isaac Potter. In his drawing of the engine the steam and condensing valves are shown as coupled together and operated by a simple chain connected to the 'plug rod' (which is, of course, shown without tappets, or plugs). Such an arrangement could not work. Leupold explained that he had not yet seen proper drawings of the engine. Interestingly, his illustration depicts the system of counterweighting at the pump end of the beam that was to become common practice in Britain and the rest of Europe.

Leupold also illustrated a simple two-cylinder engine worked by steam pressure. This was a good example of the solution-in-principle long before there were the technological means to make it effective.

The Birth of the Factory

The acceleration in invention and innovation during the eighteenth century is confirmed by several independent, objective criteria: the numbers, power and diversity in use of steam engines, the increased production of sulphuric acid and the increasing complexity of textile mills being three such indicators. Qualitatively, one of the best indicators was the publication of the great *Encyclopédie* . . . *des sciences, des arts et des metiers* of Denis Diderot and Jean d'Alembert, the mathematician. The *Encyclopédie* was, it is said, inspired by Ephraim Chambers' two-volume *Cyclopaedia* but was a far larger work of 28 volumes, published between 1751 and 1778.

The editors leave no doubt about their inspiration and whom they regard as the architects of the modern world. D'Alembert wrote of the men who had brought about the great instauration[1]: 'At the head of these illustrious personages must be put the immortal Chancellor of England, *Francis Bacon*, whose works are so justly esteemed, and the more esteemed as they are better known . . .' Bacon, it is interesting to note, is followed in precedence by Descartes, Newton, Huygens, Kepler, Locke, Barrow and Galileo. The *Encyclopédie* can therefore be said to have celebrated and confirmed the triumph of the Baconian philosophy in the west. It did more than extol science, technology and industry; it advocated, tacitly, the rational society, a society in which reason rather than tradition, privilege, faith, superstition and the other familiar features of the old (and, of course, the future) Europe prevailed.

The ideas it helped to propagate were those that led to the Revolution of 1789.

The same period saw the first major scientific advance to be made in North America. It was in the scientific study of electricity. Benjamin Franklin (1706–90), a man of wide interests and great attainments, proposed the 'one-fluid' theory of electricity[2]. A sound Newtonian, he explained many electrical phenomena in terms of a repulsive force between the atoms of a specific electrical fluid that was strongly attracted to and by matter. Franklin proved, with his famous kite experiment, that thunder-clouds are electrically charged. His theory predicted, and experiment confirmed, that if a sharp metal rod is connected to the earth ('earthed') and pointed at a charged body that body will become discharged. On this basis he invented his lightning 'rod'; for Franklin reasoned, on exemplary scientific grounds, that such a rod would discharge thunder-clouds overhead and thereby prevent lightning. The lightning discharge is, however, more complex than Franklin or any of his successors for many years could have known and the modern expression 'lightning conductor' is ambiguous; it can mean a device that conducts the brief but enormous current harmlessly to the earth. This, nevertheless, does not affect the validity of the invention. Before Franklin's invention many large buildings – churches and cathedrals were particularly vulnerable – were regularly damaged by lightning; after lightning rods were installed, damage was greatly reduced.

Franklin was to be one of the leaders of the revolt by Britain's North American colonies that began in 1775 and ended with the final defeat of the British in 1781. The American War of Independence and the Declaration of 1776 might have seemed a small affair, judged in comparison with the major preoccupations of the peoples of France and, indeed, of England. And yet this revolution, if revolution it should be called, was destined to have profound consequences for technology in later years.

However, for decades to come, American technology was to be that of pioneering rural communities, rather than that of settled industrial societies.

Many members of the Dissenting community in Britain sympathized with the rebellious colonists. Men like Josiah Wedgwood, the potter and member of the Lunar Society, had no reason to support a government and a king that discriminated against them simply on grounds of religious faith. It is likely, too, that the Quakers tended to support the American cause, although as pacifists they could hardly approve of violence. But business was business and even at Coalbrookdale it was unlikely that principles would be pushed to extremes.

Abraham Darby III, unlike the discreet, if not secretive, founder of that dynasty of ironfounders[3], was publicity conscious, as is proved by the numerous pictures that he took care were painted or engraved of their ironworks in its dramatic, even beautiful, setting. It may be that his awareness of the value of sound publicity owed something to the activities of the energetic businessman, Matthew Boulton, in nearby Birmingham. In 1779 Abraham Darby III completed the famous iron bridge over the river Severn, a structure whose publicity value, for that part of the world, is still considerable. Apart from its material – cast iron – there is nothing revolutionary about the bridge. It consists of five semicircular iron ribs carried on masonry abutments and supporting a roadway (now limited to a footpath) with a curved profile. The design is essentially that of a masonry bridge with all the members in compression. Iron would have permitted a flatter profile with ribs of greater radius, which would have meant a greater lateral thrust to counter the inward thrusts on the abutments, which, as it happens, have been a continuing source of trouble at Ironbridge. But criticisms of a radical innovation are usually out of place. Although iron had long played a relatively minor part in building – there are, for example, the wrought iron bars supposed to act as tie-bars in medieval cathedrals – the construction of the iron bridge marked – and publicized – the beginning of a new building technology in which iron was to play a far more important part than it had done before.

Of the traditional building materials, bricks and masonry are strong in compression, and timber is elastic and strong in tension (how else could trees stand up to gales?). Moreover, timber can easily be shaped from a primitive form that is already roughly suitable for beams, rafters or columns. And it comes in varieties that can serve different purposes, including the simply decorative. On the other hand, timber has familiar disadvantages. It is liable to rot, it is bulky for its strength and it is highly flammable. These deficiencies would have seriously retarded the development of mills and other industrial buildings where the weight and vibration of machinery had to be taken into account. Fortunately an alternative and superior building material was becoming available cheaply and in quantity – iron.

Another market for cast iron was in the construction of Newcomen engines. Early Newcomen engines had brass cylinders and copper boilers. The difficulty of casting and boring a right cylinder with a diameter far greater than that of any cannon but with much thinner walls was evidently too much for the iron industry and the crude tool-making skills of the first half of the eighteenth century. With the spread of coke smelting and with improved boring machines iron cylinders became cheaper than brass ones. In this, as might be expected, Coalbrookdale led the way while the ironmaster John Wilkinson, of Bersham, had developed a relatively efficient and accurate boring machine.

James Watt and His Colleagues

The invention of an efficient engine worked by steam and not by atmospheric pressure, and its development to drive machinery without using an intermediate water-wheel were due entirely to one man. Not only did the success of this engine increase the demand for iron, it enormously widened the range of application of heat power and, in doing so, transformed the conditions of life in industrializing countries. The inventor, James Watt (1736–1819), was, like John Smeaton, born into a lower middle-class family and, again like Smeaton, was trained as an instrument maker. As a child he showed unusual mathematical ability, a keen interest in Newtonian natural philosophy and great mechanical skill. After his training and helped by some family influence, the University of Glasgow employed him as instrument maker. In 1757 the University allowed him to set up shop within its precincts, advertising himself as 'Instrument Maker to the University'. His early and close association with a university at a propitious time sharply distinguished his career from those of Smeaton and other men involved in the power business. The small 'Glasgow College', with barely a dozen professors and virtually no supporting staff, was extraordinarily rich in talent: Adam Smith and Joseph Black were among the professors[4].

Black (1728–99) was one of the founders of the scientific study of heat. The idea of quantifying heat seems to have developed quite naturally from the established method of calibrating clinical and meteorological thermometers: a pound of water at the boiling point, 212°F, mixed with a pound of water at the freezing point, 32°F, gives two pounds of water at (by definition) 122°F, and so on in arithmetic proportions. Black noticed that the simple law did not apply if (say) a pound of iron at 212°F were to be added to a pound of water at 32°F. The same amount of heat that raises a pound of water by 1°F causes different temperature rises for different substances: Black had discovered the concept of (specific) heat capacity and he went on to discover the concept of latent heat, or the amount of heat required merely to bring about a change of state; for example to change boiling water into steam, or ice into water at the freezing point. All that he needed to make this discovery was the key idea of a quantity of heat and reflection on the long time taken for snow or ice to melt, or water to boil away while all the time heat was entering the liquid or the solid. Black put forward no theory as to the nature of heat although, at that time, there were two theories of heat; one, favoured by some seventeenth-century thinkers, was that heat was due to the motion of (assumed) atoms that comprise all material bodies; the other was that it was in some way related to 'fire'. Black, a friend of David Hume, was a cautious man and his attitude was non-committal, positivistic.

The story of Watt's key invention and its subsequent development is well documented[5] but certain details are often misunderstood. Called on to repair a model Newcomen engine belonging to the University, Watt noticed that the cylinder became very hot when full of steam but was considerably cooled in the condensing phase. Moreover, although the engine was supposedly a scale model of an actual working engine, it could not perform more than a few strokes before running out of steam. Watt inferred that too much steam was being used merely to heat up the cylinder only to be wasted in the condensing phase. He was familiar with the scale effect; Newton himself had pointed out (in the *Opticks*) that a small body loses heat more rapidly than a big one and this, Watt reasoned, could well explain the poor performance of the model engine. It might even be that the alternate heating and cooling of the cylinder was a major diseconomy in the working of large engines. On the other hand it could be that it was a minor diseconomy compared with other, remediable, sources of waste. Watt settled this point when he found that the volume of steam supplied to the engine was several times greater than that required to fill a volume equal to that of the cylinder; the extra steam was being used only to heat up the cylinder. It *was* a *major* diseconomy.

What was required was a substance that would absorb much less heat as the steam warmed it up. Black had found that wood had a much lower heat capacity than any metal. (Black did not distinguish between specific heat capacity and conductivity. In fact, he denied the concept of specific conductivity, which was not established until the end of the century.) Watt therefore made a cylinder of wood, treated it with linseed oil and baked it. Experiments showed that much less steam was required to fill it; an economy had been made. However, on adjusting the condensing jet to ensure that no more cold water was injected than was necessary to condense the volume of steam, he found that the power of the engine was seriously reduced. The temperature of the condensate as it left the cylinder indicated the cause; it was so hot that it would boil in a vacuum, and go on boiling until the vacuum was vitiated. Watt knew that Dr Cullen, Black's predecessor at Glasgow, had shown that tepid water will boil *in vacuo*.

Here was an insoluble problem. For economy, the cylinder and contents must be kept hot all the time; for power, the cylinder and contents must be cooled right down; the objective was to achieve maximum economy with maximum power. The solution occurred in a flash of understanding and insight while Watt was walking near the 'Golf House' on Glasgow Green on Easter Day 1765. If *two* cylinders are used, one, in which the piston moves, can be kept hot all the time while the other, in which the steam is condensed, can be kept cold all the time. A vacuum in the condensing cylinder, or condenser, would cause all the steam to flow under its own pressure from the working cylinder into the condenser, there to be condensed by the cold metal walls, the condensing cylinder being immersed in a large tank of cold water. There would therefore be a vacuum in both cylinders while the working cylinder remained hot, ready for the next influx of steam.

The broad details of the complete engine rapidly

sketched themselves out in Watt's mind. For maximum economy, the pool of water on top of the piston had to go, as cold air was no longer an acceptable motive agent. Hot steam, at atmospheric pressure, was the only acceptable motive agent, or 'working substance'. To ensure further economy the working cylinder was encased in a steam-filled cylinder insulated with wood to minimize heat loss. The working cylinder was closed on top, the piston rod passing through it by way of a stuffing box. Steam, entering above the piston, drove it to the bottom at which point the inlet valve cut off the supply from the boiler and, simultaneously, another valve, in a by-pass pipe joining the top and bottom of the cylinder, opened to allow steam to flow from above to below the piston, the exhaust valve to the condenser being, at the same time, closed. The pressures on both sides of the piston were now equal so that it could rise to the top of the working cylinder. The inlet valve then opened, the by-pass valve closed and the next cycle could begin just as the exhaust valve opened to allow the steam under the working piston to rush into the condenser, there to be condensed leaving a near-perfect vacuum. The condenser consisted of a series of metal pipes immersed in cold water, the steam being condensed by contact with the cold metal and not by a jet of water, as in the Newcomen engine. The air pump that necessarily replaced the snifting valve (steam could hardly be used to flush air out of the cold condenser!) served also to pump out the small amount of condensed steam each cycle.

The relations between Watt and Black have been the subject of some misunderstandings. There is no doubt that Watt learned of the fundamental concept of heat capacity from Black but he was never Black's student, nor is there evidence that Black made any suggestions that led to Watt's invention of the separate condenser. Watt, in his experiments to find the right amount of water to condense a volume of steam, stumbled across the phenomenon of latent heat. He found that if one part of boiling water (at

212°F) was added to thirty parts of cold water, the change in temperature of the mixture was barely perceptible to his fingers, while a small amount of steam (at 212°F), bubbled through cold water, soon made the water boil. (Watt had difficulty *measuring* the latent heat of steam by this method. See the Note and the end of this chapter.) Puzzled by this he consulted Black, who told him about his discovery of latent heat.

This has been the source of a persistent legend that originated with Watt's friend, John Robison. According to this, having been told by Black about the high latent heat of steam, Watt realized that economy of steam was important for the economical working of a steam engine. Although Watt himself denied it, this misleading story has been repeated in many accounts until recently. It is nonsense, for if the latent heat of steam had been negligible very much more steam would have been required to heat up the cylinder, or work the engine; if, on the other hand, the latent heat had been far bigger than it actually is, much less steam would have been needed and the Newcomen engine would have been thermo-dynamically more efficient, so that the addition of a condenser would have been unnecessary (whether the engine would have been *economical* would, of course, have been another matter). Latent heat, in short, has nothing to do with the problem, which is a matter of heat loss through the cylinder.

Watt patented his engine in 1769. The same specification included a description of a simple, direct rotative engine consisting of a revolving drum inside a fixed drum. The revolving drum had attached to it a vane or paddle extending to the inside surface of the fixed drum, which had a retractable baffle hinged to it that could be withdrawn to allow the vane or paddle to pass. Steam admitted on one side of the baffle drove the inner drum round while an exhaust port on the other side of the baffle allowed steam used in the previous cycle to flow into the condenser. A fly wheel carried the inner drum past the dead space when the baffle was retracted. This extremely simple engine was not a success, however, and despite the efforts of subsequent generations of engineers it proved stubbornly impracticable. Persistent attempts to make it work were abandoned with the advent of the successful steam turbine (1884).

The development of Watt's engine, first in partnership with John Roebuck, who went bankrupt, and then with the far-sighted, resourceful and successful Matthew Boulton (1775), has been described in a number of studies (see Bibliography). Here we may note that Watt expended much effort trying to perfect his surface condenser. And it is enough to point out that the new engine, while at least twice as economical as the best Newcomen engine, was extremely difficult and therefore expensive to construct. It was, in fact, at the very limit of the mechanical resources of an age when the only *industrial* machine tools were primitive lathes and inaccurate boring engines used mainly to bore out cannon. Cost factors and the market soon led to simplification; the outer cylinder – the steam jacket – was dispensed with and surface condensation was replaced by a simpler and cheaper jet of water. Even so, there was only one substantial market for the engine (see plates). This was in Cornwall, the 'county of fire engines', where there were many rich and potentially rich copper and tin mines but where flooding was critical and coal, imported by sea, was expensive. In Cornwall Boulton and Watt made their names, a great deal of money and learned all about installing and improving steam engines. They made the valve mechanisms and for the rest acted as consultants and licensees. They advised their customers to buy components from specialist makers – boilers from Coalbrookdale, cylinders from Wilkinson, pumps from William Jessop – and reserved for themselves the design of individual engines and the supervision of their erection.

Not until 1796 were the partners able to manufacture complete engines themselves.

Cornwall had its limitations and commercial hazards. The partners were well aware of the demand for a rotative steam engine to provide power for urban flour mills, breweries and the new textile mills. It was to this market that Watt turned his attention in the 1780s. To meet its needs he invented the double-acting engine in which steam was applied above and then below the piston so that the whole cycle was powered. As the piston now pushed, as well as pulled, the beam, a chain connecting the piston rod to the arch head could not be used; the connection had to be rigid. But a rigid piston rod jointed to a rigid connecting rod would vibrate and damage the stuffing box. The planing machine had not been invented so the almost obvious solution of constraining the motion of the piston rod between rigid and accurately machined guide rods was not feasible: it would have been impossibly expensive to have made accurate guide rods by filing by hand. Watt therefore devised what he called the 'parallel motion', a combination of hinged rigid rods that prevented the piston rod from vibrating. (Several different combinations of hinged rods can be made so that the end of one rod moves in a straight line. The pantograph of an electric locomotive is a simple and familiar example of such a combination.) Finally, he invented the sun-and-planet gear to get round what he believed to be a restrictive patent on the use of the crank in 'fire engines'.

Watt's career and achievements have a general significance extending beyond the history of the steam engine. He was not, when he made his key invention, a 'fire engine' man; he was very much an outsider. Newcomen's engine was well established, simple and reliable; it was cheap to build and operate, particularly at collieries where waste pit-head coal was literally dirt cheap and unsaleable. It could even be adapted to drive machines, for which purpose the best arrangement was to use it to pump water back over a water-wheel, as was done with Smeaton's blowing machine for ironworks. Towards the end of the century the same arrangement was used when the Savery engine was reintroduced in modified form by Joshua Wrigley. Only the suction, or pull, phase was used so there was no need for high-pressure steam, condensation was by internal jet and the steam and condensation valves were operated automatically by the motion of the water-wheel. It was, writes Dr Hills[6], 'an engine of extreme simplicity, because it had few working parts, of low cost because there was no cylinder which had to be bored accurately to fit a piston . . . [and it was] . . . as economical as its contemporary Newcomen engines'. It met the needs of mill-owners who wanted a small engine to supplement their water-wheels. Accordingly it enjoyed a certain popularity although it is not known how many were actually built. Hills emphasizes what a remarkable achievement it was on the part of the early textile engineers to harness their clumsy, lumbering engines to the spinning of fragile cotton threads.

In view of the problems of manufacture and the competition from rival sources of power, the launch of the Watt engine was unquestionably a major entrepreneurial achievement[7]. The first engines carry the hallmark of the scientific instrument maker; a perfectionist, unfamiliar with the rough world of eighteenth-century mining. Against this must be set the improbability that an established or traditional engine man could have made this remarkable invention. In other words, in the case of Watt's inventions we have another and very well documented example of the outsider bringing revolutionary ideas to an established technology. There have been many such examples since his day. A unique feature in Watt's case was his close association with Glasgow University and, in particular, his use of the new scientific ideas pioneered by Black. It must be stressed that no other early engineer, technician or craftsman could have known anything about the nascent scientific study of heat. Boerhaave had claimed the study of heat for chemistry and medicine; accordingly, from Black's time up to the first decades of the nineteenth century, the study of heat in Britain was the virtual monopoly of medical men and chemists and therefore far removed from the world of the power engineer. Moreover, Black preferred to announce his discoveries in the course of his well-attended lectures rather than to publish them in a journal. This hiatus had, as we shall see, its effect on the development of steam power. It is even possible that Watt had some understanding that *heat* was the motive agent for the action of his engine and its predecessor, although many years had to pass before this could be established. The invention of the Newcomen engine had been based on established scientific knowledge; the invention of the Watt engine was directly related to the latest, to *progressive*, science. This mode of technological progress has now been formalized through industrial and government laboratories.

It can be argued that Watt would have been content with a much simpler engine had he not had the instincts of a man of science, or a laboratory perfectionist. There were, during the period of Watt's monopoly, many attempts to pirate or to evade his patent. One way of getting round the patent was the so-called 'pickle-pot' condenser[8]. Part of the pipe connecting the cylinder to the boiler was enlarged in the rough shape of a pickle jar. Cold condensing water was injected into the relatively small 'pickle-pot' while the working cylinder could be kept relatively hot all the time. These engines, while not as efficient as the typical Watt engine, were more economical and not much more costly than an ordinary Newcomen engine. An additional refinement might have been to use hot boiler water to complete the seal on top of the piston. Watt, it seems, never contemplated compromises with his ideal engine beyond those enforced by the market.

It is a familiar saying that science owes more to the steam

engine than the steam engine does to science. One of Watt's inventions that tends to confirm it was that of expansive operation. He had realized that, when the piston reached the end of its working stroke, recoverable work would be lost if the steam at boiler pressure was allowed to rush – like a gale – into the condenser. He therefore proposed to cut off the steam supply after the piston had travelled a short distance down the cylinder; the trapped steam would continue to press against the piston although the pressure would steadily fall to a minimum when the piston reached the bottom of the cylinder. In this way as much work as possible would be squeezed out of the steam. This insight was to be a vital element in the foundation of the science of thermodynamics.

Watt rendered another service to the science later to be called thermodynamics when he adapted Smeaton's empirical measure of the efficiency of Newcomen pumping engines to the measure of his engines. He replaced Smeaton's measure of the *volume* of water lifted by the *weight* that his engines could raise for the combustion of a bushel of coal. This he called the 'duty' of the engine and it, too, was a factor in the establishment of thermo-dynamics. (The term 'duty' has long since passed out of use; but the principle is familiar to most people who, on buying a car, take into account the number of 'miles it can do to the gallon' or 'kilometres to the litre'.) Lastly, although not of scientific importance, Watt standardized the unit of horsepower. In 1783, when he began to sell rotative engines to millers, cotton manufacturers and others unfamiliar with mining practice, he described his engines in units of 'horsepower', about which everybody had some idea. He laid down that one horsepower represented 33000 pounds raised one foot in one minute. This is still the practical unit of horsepower.

Watt was fallible; he had his Achilles' heel. He remained throughout his long life strongly opposed to the use of high-pressure steam even though it would allow smaller, more powerful engines to be made and would enormously widen the market for the steam engine. He dreaded that a disastrous boiler explosion would discredit all steam engines. This danger, he might have realized, was not insuperable and it was the high-pressure steam engine that, from the time Watt's extended patent finally expired in 1799, was to dominate and utterly transform life in the century to follow. If Watt failed to foresee the future it is also surprising that Adam Smith, in his *Wealth of nations* (1776), made no reference at all to steam engines, much less to the radically improved version that his young colleague had invented.

There was one other attribute of James Watt that marked his individuality: he was a Scotsman. He was not only one of the greatest of engineers, he was the first great Scottish engineer. Before about 1700 Scotland had been on the cultural as well as on the geographical edge of Europe. A remote country with an indifferent climate, inhabited by violent, war-like men, Scotland's contribution to European advancement had been modest. The grim John Knox was the best-known Scotsman. In the realm of science there were John Napier, the inventor of logarithms, the mathematicians David Gregory and (after 1700) Colin Maclaurin, and George Martine, a physician who wrote an interesting book on thermometry. But after Watt and Black Scotsmen rapidly took their places among the leading engineers and scientists in the world. As the age of Watt and Black was also that of Adam Smith and David Hume, it is clear that this flowering of a national culture was – and remained – broadly based. What brought it about is an intriguing question. Perhaps it represented a collective, national response to the implicit challenge of English domination. Whatever the causes there is no doubt that the excellent Scottish universities had a lot to do with it. Indeed, many highly creative Englishmen were indebted, directly or indirectly, to the Scottish universities. It was not mere sentiment that caused John Dalton to dedicate the first volume of his *New System of Chemical Philosophy*, in which he expounded the scientific atomic theory, to the Universities of Glasgow and Edinburgh.

Boulton and Watt showed their skill at salesmanship when, in 1786, they opened the Albion Flour Mill on the south bank of the Thames, in the heart of London. The mill had two rotative steam engines and all operations were powered by these engines. The opening ceremony was a grand social occasion and the enterprise received the widest publicity. Technically, the mill was a great success. It heralded the end of the traditional local mill, driven by a stream or wind power, and it began the era of the large flour mill situated on the coast or navigable river. Financially it was a disaster for it was burned down in 1791 long before the capital debt had been paid off. Sabotage was suspected but no evidence was ever found to support the charge. Albion Flour Mill certainly publicized the new rotative steam engine and its fate emphasized the fire hazard.

Structures

The structural use of iron in Britain was, as Dr Arnold Pacey has pointed out[9], pioneered by the ironmasters; in France architects and mathematicians led the way. The British favoured cheap cast iron; the French preferred wrought iron, which lent itself to more versatile use. The first structural use of cast iron in Britain was to make the, relatively slender, columns in St Ann's Church, Liverpool (1770–72). In France J. F. Calippe suggested an iron bridge (1777); it was to have been of wrought iron with a 200-metre span (most bridges in France had been wooden). A model was put on display and this may have influenced the architect J. G. Soufflot (1713–81) to use wrought iron girders for the framework of a truncated pyramidal skylight over a stairwell in the Louvre Palace (1779–81). It suggested to other architects that iron might be used to

Floor

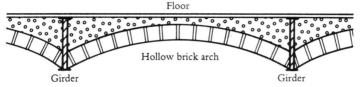

Hollow brick arch

Girder   Girder

FIG 7.1   *Girder with brick arch*

support the floors or ceilings of buildings. In 1785 J. V. Louis began work on the Théâtre Royale using long stressed girders with hollow brick arches between them for the floor and ceiling (figure 7.1). The intention was to make the theatre fireproof.

Wood had particular shortcomings as a building material for theatres. Its limitations of strength meant that deep, protruding balconies, or 'circles', could not be built, which is why eighteenth-century theatres and opera houses had tiers of shallow circles, only a few rows of seats deep. A further disadvantage was that theatres were, in those days of oil lamps and candles, particularly vulnerable to fire. Whatever the course of development in France might have been, the outbreak of the Revolution in 1789 put a stop to it and the next significant innovations were in England. These were in the sector of most rapid economic and technical development: the textile industry. Here the problems were much the same as in the theatre world. According to the late Dr Fitton[10]:

> The problem of building fire-resistant cotton mills was a pressing one for factory masters. The early timber framed structures were most vulnerable. Rooms were imperfectly ventilated; methods of dust extraction were primitive; oil from the wooden machines dripped on to the wooden floors; the machinery contained much wood; the lighting was by oil lamps and candles; and fire-fighting was inefficient. Few firms escaped loss at one time or another.

To the problem of fire we should add that the weight of ever more machines made a strong structure essential for the continued prosperity of the industry.

William Strutt, FRS, was the son and heir of Arkwright's partner, Jedediah Strutt. His interest in the possibilities of iron was aroused in a curiously indirect way. He had heard that Tom Paine, soon to be the famous author of *The Rights of Man*, was engaged in building an iron bridge[11], for which he had a patent, in Sunderland. Strutt, according to Fitton, shared Paine's political views and probably met him in Derby or Belper. Stimulated, perhaps, by the destruction of the Albion Mill, 'the most advanced industrial building of the day', Strutt began the construction of a fire-proof mill in the following year. Slender cast iron columns of cruciform section supported wooden beams that had wedge-shaped iron 'skewbacks' bolted to each side. These skewbacks served as bases to take the thrust of brick arches that spanned the spaces between the wooden beams. The bottoms of the beams were plastered, to reduce their vulnerability to fire, and the space between the crests of the brick arches, the spandrel, was filled with sand; quarry tiles were laid on the sand to make the floor.

The next innovation was made by Charles Bage. In 1797 he completed the Castle Foregate Flax Mill in Shrewsbury for the Leeds partnership of Marshall and Benyon. In this mill the beams, like the columns, were of cast iron. In 1799 construction of George Lee's Salford Twist Mill began; it was completed in 1801. This, too, had cast iron beams as well as cast iron columns, but it represented a further advance in that it was designed as an integrated whole: the steam engine was built as part of the structure. Another novel feature was that it was the first mill to be lit by coal gas light, a further instance of the association of the textile industries with the latest technology.

Although cast iron, easy to produce and easy to mould into any desired shape, continued to be the commonest form, the demand for wrought iron increased with the advance of industrialization[12]. Patents were granted at intervals throughout the eighteenth century for processes involving the use of reverberatory furnaces, most notably to John and Charles Wood (1760), the Cranage brothers of Coalbrookdale (1766), Peter Onions of Merthyr (1783) and, finally, to Henry Cort (1784) for his successful 'puddling process'. This last could fairly be described as the outcome of a sustained evolution to which a large number of men had contributed. Molten pig iron in a reverberatory furnace is kept stirred by long iron rods – each rod being used only for a short time – called 'raddles'. As fresh iron is brought to the surface little blue flames of burning carbon monoxide show the puddlers that the carbon content is being reduced. As this steadily falls from about 4 per cent down to about 0.1 per cent the melting point of the iron rises from about 1000°C to about 1400°C. It becomes less and less fluid, until, finally, it forms into a large plastic lump, too big to be removed through the furnace door. The puddlers have to be careful to prevent this happening. They have to form the iron into blooms of a convenient size that do not stick to each other or to the sides of the furnace. When this is done the blooms are lifted from the furnace and forged or rolled into bars in a rolling mill, the rolling mill having been invented by the versatile Swedish engineer Christoph Polhem[13] (1661–1751).

It has been said that puddling iron was the hardest work that men have ever done. No one who has seen puddlers at work doubts it. Not only was great strength needed to form and handle the heavy blooms, but the puddler had to endure the intense heat and glare of the furnace. It was, therefore, customary for puddlers to work as a team, taking turns to keep the iron stirred and assessing the changing composition of the melted iron. And as it was also a highly skilled occupation it is hardly surprising that it took some time for the puddling process to become widely established. It was perfected by about 1840 when the problems of finding a lining for the furnace that did not interact with the molten metal was finally solved. As far as I know the puddling process was last operated at the Atlas Forge, in Bolton, where it was used to supply a small, highly specialized market for wrought iron. The oil crisis of 1973 made the process uneconomic. Mild steel is a substitute for wrought iron, although advocates of wrought iron claim that mild steel lacks its finer qualities.

The substitution of iron for wood in the construction of textile mills was not as simple as replacing one type of wood with another and stronger one. In fact it entailed a fundamental change and it brought with it a number of difficult problems. Understanding of the properties of cast iron as a structural material was in its infancy. Adequate strength could be obtained by making the columns and beams as thick as possible, but iron was not dirt cheap and no business prospers by spending more money than it needs to: a truth that was as well understood and appreciated in the Industrial Revolution as at any other time, before or since. The desideratum was to obtain the necessary strength for the minimum amount of iron. Fortunately a simple set of rules was available.

In 1758 William Emerson had published the first edition of his *Principles of Mechanics*, a text book that, among other things, gave an elementary and uncorrected account of Galileo's theories of the strength of materials. The leading ideas were summarized and published in even simpler form by John Banks, a peripatetic lecturer, in his book *On Mills and Millwork* (1790, 1792). Now although Edmé Mariotte and Antoine Parent had, long ago, pointed out that Galileo's theory ignored the elasticity of a loaded horizontal beam, Emerson and Banks did not realize this. They assumed, with Galileo, that the bottom surface of a loaded beam was the neutral layer and was neither compressed nor stretched. English pure and applied mathematics, still dominated by Newton, was archaic judged by Continental standards. And even though some of the more advanced English mathematicians might have heard about Mariotte and Parent's works, Emerson and Banks were self-taught popularizers and not internationally minded scholars. Fortunately their shortcomings did not matter. As Pacey pointed out, Banks followed the English custom of reasoning in proportions so that as the error occurred on both sides it cancelled out and he arrived at the correct result.

The two rules, laid down by Banks to guide the beginner working with the new material, were that for maximum strength the 'web' of the beam should be as wide as possible: this indicated that it should be 'I' shaped in section. And, secondly, if we consider an unsupported beam mortised into a wall, the shape for a given strength using the minimum amount of material must be a parabola. Such a form was not, of course, suitable for a horizontal beam between two vertical columns and supporting brick arches. A compromise was to settle on a semi-elliptic form with a straight lower edge carrying a

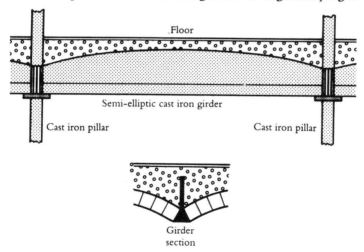

FIG 7.2   *Cast iron beam with cast iron pillars*

flange from which the brick arch could spring (figure 7.2). This was, with progressive refinements, the form of cast iron beam used in the great textile mills that spearheaded the Industrial Revolution. With this entirely new method of constructing entirely new buildings, an old problem, previously hardly more than theoretical, had arisen in a very practical form. This was the problem of scale, so brilliantly analysed by Galileo. Triewald, as we saw, failed to appreciate the importance of scale in the construction of his Dannemora engine. Reasoning from a model, or a smaller machine, to a full-size or larger machine requires – as Galileo pointed out – that attention be paid to the scale effect. Christoph Polhem was the first to reason correctly from models to full-scale machines but he was before his time. Smeaton and Watt used models in their researches but they were not concerned with the size and strength of their machines. Banks found that a wrought iron bar, one square inch cross-section and one foot long, broke under a load of 2190 lbs applied at the unsupported end. From this he was able to calculate the correct dimensions for full-size beams. This was satisfactory enough for early textile mills, but in the next century far more complex problems were set when engineers were required to build railway bridges to stand the vibration and weight of heavy, fast trains and when shipbuilders set out to construct huge iron ships.

A short discussion of some details of Strutt's Belper North Mill (see plates), in effect, summarizes some of the points made in this chapter. Although built at the beginning of the nineteenth century (1803–04) this mill illustrates neatly the achievements of the eighteenth century and indicates the promise of the future. The often reproduced drawing (by John Farey, from Rees' *Cyclopaedia*) shows a compact structure with minimal distances between the power source, a large breast-wheel and the textile machines. Power transmission is by overhead line-shafting and belt drive to the individual machines. The cast iron columns are exactly aligned, one above the other, to obviate shear stresses that would otherwise distort and weaken the structure: there was nothing casual about the construction of a cotton mill. On the different floors there are rows of identical machines; in the middle, for example, there are two rows of carding engines. These indicate an entirely new technology, the process of batch production that foreshadowed the automobile assembly lines of the twentieth century. More, and most significantly, the whole structure, the rows of machines, the method of distribution and allocation of power show careful planning and, above all, an awareness of *an integrated production system*. This was something entirely new.

To the right of the mill is the spiral staircase with, beside it, the shaft that accommodated the teagle, a device to carry passengers up or down to any floor they chose. This ingenious machine was, according to Rees' *Cyclopaedia*, invented by Henry Strutt in about 1790. (Fitton and Wadsworth, however, do not record a Henry Strutt in their Strutt family tree; they suggest that the initial 'W'(illiam) was misread as 'H'.) Whoever has the credit for the teagle it could not have been invented before multi-storey buildings were in use with power available on the topmost storey. Textile mills were the first buildings in the world to meet this requirement. Teagles are now, quite illogically, called elevators, lifts, *ascenseurs*, etc. (how can one go *down* in a lift or an elevator?).

Lastly, we should note the school room in the attic. The Strutts were devout Presbyterians and had a concern for the welfare of their employees. At the same time the demand for labour was such that provision had to be made for the children of workers when both parents were employed in the mill. The requirements of morality and of economic advantage therefore coincided most happily.

A NOTE ON LATENT HEAT

After his initial experiment Watt continued to be interested in the latent heat of steam. He wanted to discover if there would be any significant economy by working his engine with steam at different pressures and therefore at different temperatures. His experiments showed, however, that as the temperature of the steam increased the latent heat diminished, and vice versa, so that the 'total heat' ('sensible heat' plus latent heat) remained roughly constant. Watt's expression 'total heat' was used, much later, as a synonym for enthalpy in earlier books on thermodynamics.

Black's method of measuring latent heat was simple and exemplified the way in which proportions, rather than equalities, were used in eighteenth-century physical science. He measured the rate of rise of temperature when a pan full of water was brought to the boil and then measured the time it took for all the water to boil away. Assuming that the rate at which heat entered the water was constant, he used these two figures to calculate the temperature the boiling water would have reached had none of it been evaporated into steam at 212°F. This notional temperature was the latent heat. A similar simple experiment enabled him to find the latent heat of fusion of ice into water. It should be noted that it is not necessary to know the weight of water involved in these experiments as times are proportional to weights.

Watt's method of measuring latent heat was more complicated than Black's. It involved latent heat plus the effect of mixtures (i.e. the mixing of the cold water and the hot water of condensed steam). Watt made two mistakes in his calculations: he added the weight of the condensed steam to that of the cold water and he assumed that the steam, once condensed, was heated up from the temperature of the cold water. This gave results that were inaccurate, but not grievously so (see Robinson and McKie, in Note 4).

It must be pointed out that these elementary mistakes were a consequence of the method of calculation he had to use. The method of proportions gave no insight into equalities.

It was not until September 1783 that Watt learned of Lavoisier and Laplace's *Memoir sur la chaleur* in which the physical process was represented by *equations* and not by proportions. Putting m_1 for the weight of cold water, m_2 for the weight of condensed steam, t_1 and t_2 for the temperatures of the cold water and of the mixture, Lavoisier and Laplace's equation for Watt's experiment is

$$m_1 (t_2 - t_1) = m_2 (212 - t_2) + m_2 L$$

where L is the latent heat. This equation gives a clear insight into the physical process.

THE EXPRESSION 'SPECIFIC HEAT'

Black's 'heat capacity' was called *chaleur specifique*, or specific heat, by the Portuguese scientist J. H. de Magellan in his book *Essai sur la chaleur* (1781). It is now commonly called specific heat capacity.

8

Technology Becomes Autonomous:
The Logic of Improvement

The German states had dominated chemistry, the chemical 'arts', metallurgy and mining from the medieval period up to the middle of the seventeenth century. From that time onward the secular movement of the centres of technological activity westward to France, the Low Countries, Scandinavia and England coupled with – and surely hastened by – the disasters of the Thirty Years War led to Germany taking a subordinate role in what had been distinctively German sciences and technologies. The last big idea to come from Germany before the great revival in the early nineteenth century was to cause a prolonged debate, familiar to all historians of eighteenth-century chemistry. The phlogiston theory[1], put forward by J. J. Becher (1635–82) and G. E. Stahl (1660–1734), was designed to account for combustion and calcination. A burning body gave off phlogiston, leaving an ash or calx. Similarly, a metal was a calx plus phlogiston and therefore could not be a simple or elemental substance. It has been pointed out that the phlogiston theory accorded with the experience of iron smelters that heating charcoal (obviously rich in phlogiston) with iron ore (iron without phlogiston) yields the metal by restoring its phlogiston.

The theory proved a dead end and, as Professor Crosland remarks, it came to be forced 'to explain so much that it really explained nothing'. It is, however, doubtful if it seriously misled many people and it had the merit of directing attention to the key problem in chemistry, that of combustion. This was more than Newton's system could do, despite his clear hint in the preface to *Principia*. Progress in chemistry during most of the eighteenth century was achieved piecemeal through the accumulation of knowledge of various reactions and, following the discovery of the atmosphere, through great interest in 'airs', as gases were often called. Joseph Priestley (1733–1804), the polymath, was puzzled by the fact that, while a candle burning in a closed vessel soon destroys the capacity of the enclosed air to support either combustion or life, the earth's atmosphere appears to have been unaffected by the fires that have been burning since the beginning of time. This led him to the discovery (1774) that plants extract 'fixed air' (carbon dioxide) from the atmosphere and return 'dephlogisticated air' (oxygen) to it. This was the beginning of understanding of photosynthesis. And he discovered that calcined (oxidized) mercury, when heated, yielded dephlogisticated air and the original amount of mercury.

A.-L. Lavoisier (1743–94) was a chemist of genius as well as an affluent and versatile man. His practical interests included such matters as improving the street lighting in Paris, the manufacture of gunpowder, agricultural reform and, ill-advisedly, acting as one of the private-enterprise tax-gatherers. Lavoisier had both great experimental skill and, like Joseph Black, an awareness of the importance of exact quantification. Unacceptable ambiguities and inconsistencies in the phlogiston theory had accumulated by the time Lavoisier began his successful attack. He repeated Priestley's experiment using a burning glass to calcine mercury in a closed vessel and carefully measured the reduction in the volume of air when the calcination of the mercury stopped. He then removed the calcined mercury and heated it, recovering the mercury and a volume of 'dephlogisticated air' that he measured carefully: it equalled the original reduction in volume. And he found that when added to the reduced volume the quality of the original air was fully restored: it would support combustion and life.

There was no need for phlogiston. It added nothing and explained nothing. 'Dephlogisticated air', he decided, was an individual gas, an element that he called oxygen, and it was the chemical combination of oxygen with other substances that constituted combustion and calcination. Lavoisier continued his increasingly successful attack on phlogiston and propounded a new theory of chemistry that he set out in his *Traité élémentaire de chimie* (1789). Beside the oxygen theory of combustion he set out his list of chemical elements. In place of the three, four or five elements of antiquity he identified no fewer than thirty, including the newly discovered gases hydrogen, oxygen, nitrogen and chlorine. Lavoisier laid down that all thirty were to be considered provisional elements; it *may be* that, in the future, some of them would be shown to be compounds. In addition to the new list of elements, Lavoisier reformed the language of chemistry. He swept away the old, often obscure and misleading, names to substitute new names formed from, and therefore indicating the chemical composition of, each compound. This was chemistry as it is known today.

Some years, however, elapsed before the reformed chemistry of Lavoisier proved itself invaluable in industry. Great though the contributions of French *savants* and engineers were, it could be argued that the British policy of implied, if not openly avowed, *laissez faire* had triumphantly succeeded while the Continental policies of *dirigisme*, or Colbertism, had been more or less completely discredited. On the other hand, Britain had enormous advantages. Her geographical position was favourable for overseas trade while her possessions in India and the West Indies generated wealth that became increasingly available for industrial investment. She had extensive and valuable non-ferrous metal mines, a growing coal industry and a progressive agriculture. There were, in addition, important political and administrative advantages, but to discuss these would take us too far from our theme. Arkwright and his immediate successors had established an industrial society of which Henry Ford's Dearborn plant and the modern Japanese car factories are direct descendants. John Smeaton had created the profession of civil engineer in the English-speaking world. James Watt had, perhaps for the first time, allied current scientific research to technological invention and development; this is now associated with highly 'scientific' industries although, of course, under very different conditions from the days when Watt could incorporate in himself a research and development department, a sales engineer, a works manager, a training officer and a joint president of a major corporation. With some hindsight, and a degree of hindsight is surely allowable, it is possible to see weaknesses in the British position, weaknesses that would become more apparent in the nineteenth century. For one thing direct state action in matters technological is not necessarily and in all circumstances ineffective or damaging. The state's role in education, including technical education, is a case in point. Not since the turbulent days when Tudor merchants had founded the old grammar schools had the English bothered about education. Why should they have done? They were economically and numerically the dominant people on a militarily secure island. They did not need to teach their young how to be English. Foreigners were a long way off, kept at their distance by rough seas. In contrast, a state with extensive land frontiers must, to some extent, concern itself with the education of its peoples as it must concern itself about military defence. Scotland, a small country having a common boundary with England, laid great emphasis on education in the eighteenth century and afterwards. Scottish universities were, by common consent, the best, the most effective, in the world in the eighteenth century.

In France also interest in education was at a low ebb.

The once great French universities were moribund or, like Oxford and Cambridge, quiescent. France, after all, was the dominant European power with secure frontiers. The tendency in France towards centralization meant that the state had an interest in good communications, sometimes from the military and political points of view, sometimes as necessary for economic development. The Corps des Ponts et Chaussées was a state communications authority. Such a body, operating on a national scale, requires experts with appropriate experience and skill; in short, it requires engineers with officially recognized qualifications. There was still some confusion about the designation of these new officials – were they engineers, as in the army, or were they architects? – and, to start with, there were no clear rules concerning their training. The Corps had to recruit suitably qualified trainees who, initially, would have to work and learn under the direction of established officers. This was no more than the old and well-established system of apprenticeship: the young man learns by working under the direction of the skilled craftsman. It tends, however, to be a time-wasting procedure. Senior officers are not necessarily good teachers and, quite naturally, do not like having to interrupt important work in order to instruct juniors. It was found more satisfactory to rationalize training. The Ecole des Ponts et Chaussées was, accordingly, initiated in about 1747 under the direction of Jean-Rudolphe Perronet (1708–94), perhaps a leading practical civil engineer of the eighteenth century. It is impossible to be precise about the dates at which such institutions were founded; it would be wrong to imply that in a certain year an organization with staff, students and syllabuses was set up where, in the previous year, there had been nothing. Perronet himself was vague about the matter, referring sometimes to his office, sometimes to the school. Not until 1775 was the title Ecole officially bestowed. By then the number of students had grown from ten to sixty and classes were formally organized. Senior officers continued to be responsible for a modicum of teaching, much of which was taken on by older and more experienced students. In a few words, the Ecole evolved from a procedure of informal, *ad hoc* instruction into a formal school[2].

Perronet remained in charge of the Ecole for 47 years. In that time the system of technical education that was developed was probably unsurpassed anywhere in the world and was, in its own field, unique. And the Corps that it served represented, even before the Revolution, an entirely new administrative technological system that was to leave its marks – material, educational and political – on France and on much of the rest of the civilized world.

A number of similar Ecoles, serving various official military and civilian organizations, were established in France in the eighteenth century. The most successful was the Ecole du Génie (1747), serving the corps of military engineers. Others were less successful and proved ephemeral. Even decentralized, *laissez faire* England ventured into the field of technical education when the Royal Military Academy, Woolwich, was founded in 1741 for the training of army engineers and artillerymen. But it was in the German states and in Austria–Hungary that the other major contribution to technical education was to be made.

The mining areas of the German-speaking states did not suffer too severely in the Thirty Years War and one or two did not suffer at all. In any case the silver and gold ores were too valuable to be left in the ground and, following contemporary economic principles, the exploitation of the mines had a high priority. As early as 1720 J. F. Henschel established an analytical laboratory in Freiberg, aided by a state grant. It soon attracted students, the most famous being M. Lomonosov, who was sent by the Russian government. The mines of Saxony were largely state-controlled and the same pressure on skills and knowledge applied there as in the case of the Corps des Ponts et

Chaussées. A system of apprenticeship was supplemented and then effectively replaced by a Bergakademie, established in 1767 with two professors aided by three mine officials. The course lasted three years and was highly organized: students were required to wear a uniform. The most distinguished graduate of Freiberg Bergakademie was A. G. Werner (1749–1817) who, in 1775, returned to become a professor there. He is best known for his 'Neptunian' theory according to which all the surface rocks of the earth were sedimentary, having been laid down in a primal, all-embracing sea. It was probably Werner's success as teacher and scholar that attracted many students to Freiberg; among those who later became famous were Leopold von Buch, Alexander von Humboldt and James Watt, junior.

Another mining college, in its time at least as well known as the one at Freiberg, was developed and then established formally at Schemnitz[3]. One Samuel Mikoviny had been employed at Schemnitz as instructor in mechanics and pumping from 1735. Following his premature death in 1750 classes lapsed until, in 1763, Nikolaus Jacquin (1727–1817) was appointed to a Chair of Chemistry there. Two years later a Chair of Mathematics and Mechanics was created. Jacquin, of French extraction, came from Leiden where he had studied medicine under Hermann Boerhaave. In the six years he held the Chair, before going to Vienna as Professor of Botany, he built up a most successful course that acquired an international reputation. In 1770 the Royal Hungarian Mining Academy at Schemnitz was formally created with an organized three-year course and three professors, the Chair of Mining Arts being held by Christoph Traugott Delius, who published his lectures under the title *Anleitung zu der Bergbaukunst* (1773), a book that was sufficiently highly regarded to be translated into French. Over these years the Academy was at the height of its fame; by the end of the century it had been overtaken by Freiberg.

From these cases we can draw a simple, general conclusion. Technical colleges, or technological universities, developed where the direct interest of the state was involved[4]: defence in the cases of all nations, a mining monopoly in the case of Austria–Hungary and Saxony, communications in the case of France. The exemplar here must be the famous Ecole Polytechnique, founded in 1794 when France recognized a need for a general scientific and engineering education for those who were to staff the technical branches of her armies and civil service. Ironically, Lavoisier, who had played his part in planning the Ecole des Travaux Publics, as the Ecole Polytechnique was originally called, was executed in the same year. The offence with which he was charged was that of being a private-enterprise tax-gatherer.

The half-century before the French Revolution was the time when technology became autonomous and publicly recognized. This is confirmed by the number of societies and institutions that were founded in France, Britain and elsewhere for the promotion of agriculture, the 'arts' and industry[5]. The different modes of technical advance were more clearly defined than ever before owing to the greatly improved means of publication and to the shortening of the time scale in which changes took place. Where previously centuries had passed between one kindred invention or improvement and the next in the series, now decades or less were sufficient. A good example is the invention of Samuel Crompton's spinning 'mule'. A hybrid between Arkwright's water-frame and Hargreaves' jenny, it appeared in 1779, just ten years after Arkwright's and Hargreaves' patents[6]. In the meantime Arkwright had invented his carding engine. On the other hand at least a hundred years elapsed between the invention of the original spinning wheel – the Great Wheel – and the improved Saxony Wheel that incorporated a flyer. It is not known by whom or where these medieval inventions were made. One thing all these textile machines – ancient and modern – have in common is that they satisfy Bacon's criterion for a certain type of invention: they incorporated no principles, materials or processes that would have puzzled Archimedes.

Textile machines reached a high point with the loom invented by J. M. Jacquard (1752–1834) in 1801. This remarkable machine was a complete solution to the problem of producing fabrics with exactly repeated patterns. The problem could be solved, simply and relatively cheaply, by printing; it could be more expensively but more precisely solved by using different coloured warp threads on a draw-loom and employing a 'draw-boy' to lift selected threads for each transit of the fly shuttle and to do this in an unvarying sequence so that the same pattern repeated itself. Jacquard's great contribution was to mechanize a known technique of using punched cards to encode a textile pattern (figure 8.1). Each card of a closed loop of strong cards carried an individual pattern of small holes; these holes represented the pattern to be woven. The holes were 'read' or detected by thin horizontal steel probes pressed against each card in turn by small springs. Each probe carried a small eye through which a vertical wire ran, the lower end of which carried a warp thread and the upper end a hook that could engage a horizontal rod. If a probe encountered and went through a hole, the eye and with it the vertical wire would be so displaced that the hook would catch the horizontal rod; if there was no hole the probe would not move and the hook would not catch the rod. Lifting the rod would therefore raise only those warp threads that corresponded to a hole in the card. In this way the progress of a linked procession of cards past the probes caused the pattern of holes to dictate which coloured threads were lifted and which were not, so that the desired pattern was woven. As the cards formed a closed loop, their continued rotation meant that the pattern was repeated without variation of any kind.

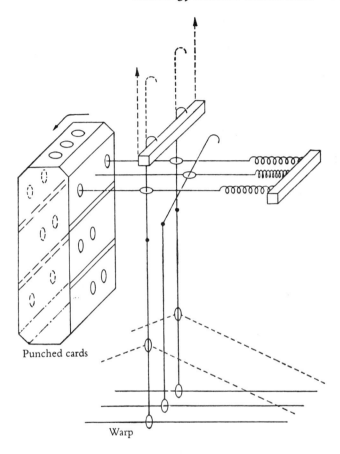

Punched cards

Warp

FIG 8.1   *The Jacquard action*

The Jacquard loom was immediately successful and it remained in use, substantially unmodified, up to the recent age of electronics. It would be fair to describe it as the most original, the most ingenious, of all inventions concerned with the textile industries[7]. Unlike the water-frame, the carding engine and the mule, the Jacquard loom had a strong and lasting influence outside the world of textiles.

The presence or absence of a hole in a punched card corresponded to 'on' or 'off', or, in binary notation, to 0 or 1. Jacquard's method of coding information to 'program' a machine and then 'reading' it by probes was the starting point of computer technology. It influenced the first inventor of the computer, Charles Babbage.

The Jacquard loom was by no means the only major French contribution to the revolutionary textile industry at this time. Sulphuric acid, the basic ingredient of so many industrial processes, was, as we have seen, commonly used in textile finishing. It was appropriate that the man responsible for bringing the lead chamber process to France should have been John Holker (1719–86), a Manchester textile worker who had associated himself with the wrong side in the British dynastic struggle of 1745 and had felt it advisable to leave the country. He settled in France where his knowledge and skills were appreciated so that he rose rapidly in, and in the service of, his adopted country. In 1769 and 1770 his son made extended tours in England and Scotland, studying various chemical works. As a result in about 1772 Holker senior began making sulphuric acid by the lead chamber process at his works in Rouen. Now burning sulphur produces two gases, sulphur dioxide and sulphur trioxide, but only the second is involved in the production of the acid. Lavoisier's classic researches led to the plausible belief that the function of the nitre was to expedite the combustion of the sulphur and so increase the proportion of acid produced. Insufficient oxygen coupled with the inevitable loss of gases when the chamber had to be opened for recharging accounted, it was thought, for the long-recognized inefficiency of both the *per campanum* and the lead chamber processes. However, the substitution of other additives in place of nitre was unsuccessful, as was the use of a current of air to increase the available oxygen.

Understanding of the true role of nitre came with the publication of the researches of two industrial chemists,

N. Clément and C.-B. Desormes, in the early years of the nineteenth century. They showed that a cyclic process takes place. Burning nitre generates nitrogen dioxide, which reacts with the sulphur dioxide to give sulphur trioxide and nitrogen oxide. The latter then reacts with the oxygen in the air to give nitrogen dioxide, which again reacts with the sulphur dioxide. This process goes on until either all the oxygen or all the sulphur dioxide is exhausted. Dr J. G. Smith has commented that this analysis is[8]:

> . . . a happy instance of the application of scientific knowledge and skills to the investigation, explanation and (in due course) amelioration of an industrial process. At the same time it illustrates the converse contribution which technological work could make to pure science, for the process they elucidated constituted the first well-characterized example of a catalytic reaction, and their theory was also in effect the first intermediate compound theory of catalysis.

Whatever the practical merits of this analysis, other improvements were required to enhance the efficiency of the process. They included combustion in a furnace outside the lead chamber, the application of a current of air (not, of course, as a substitute for the catalyst) and steam injection so that the process was, essentially, continuous. It was therefore, according to Dr Smith, the first continuous flow process in chemical industry. The consequence of these improvements was that by the second decade of the nineteenth century French production of sulphuric acid was markedly greater than British.

Bracketed with sulphuric acid as one of the two most widely used industrial chemicals was soda, or sodium carbonate, an essential ingredient in soap and glass making as well as a cleansing agent in the textile industries. Until the end of the eighteenth century soda had been derived from burning vegetation impregnated with sea salt, which in practice meant either seaweed (kelp) or the much favoured Spanish barilla. Water was used to leach out the soda from the resulting ash, after which the solution was evaporated to leave the soda crystals. It was, therefore, clear that salt, or sodium chloride, was closely related to soda and could be converted into it. Even before the mid-century it had been shown that soda could be derived from salt by a simple chemical technique. Sulphuric acid acted on salt to give sodium sulphate, which was then burned with charcoal; the resulting ash (sodium sulphide) was treated with vinegar to yield soda. But there was no question of setting up an industrial process. For one thing there was a crippling tax on salt; for another, sulphuric acid was far too expensive.

The first attempts to manufacture soda by such methods were associated with the names of Garbett, Roebuck, Watt and James Keir. Whatever successes they, and others engaged in the same kind of project, attained on the small scale they were unable to repeat them on the industrial scale. News of these attempts (with rumours of success) reached France, where Watt, for one, had many scientific contacts and correspondents. French attempts to transform what had been done successfully in the laboratory into a full-scale industrial process were no more successful. Besides technical difficulties the salt tax, as in Britain, was a serious discouragement.

The name most closely linked with the development of the soda industry is that of Nicholas Leblanc (1742–1806). The son of an employee of an ironworks, Leblanc went to Paris to study, first pharmacy and then surgery, before being appointed a surgeon in the entourage of the Duc d'Orléans, known as Philippe Egalité. He stated that he became interested in the soda problem in 1784 and, by 1789, had found the answer. In his process the first stage was the common one of allowing sulphuric acid to act on salt to give sodium sulphate and hydrochloric acid. In the second stage the sodium sulphate is burned with a mixture of charcoal and limestone, which yields a mixture of soda, calcium sulphide and carbon dioxide; the soda was then leached out by water. The essential innovation was the addition of limestone. It is not known what led him to take this step.

Leblanc was granted a patent in 1791 under the new system. At the same time he and his associate Dizé, financed by the Duc d'Orléans, were setting up a soda works near Paris. Although Leblanc had solved the production problems of the manufacture of soda and although the salt tax was abolished in 1791, the plant was not to be a success. The arrest and subsequent execution of the Duc d'Orléans in 1793 ended the supply of funds. The national crisis of that year when France was threatened by every nation in Europe meant that defence had total priority; all sulphur, charcoal, nitre and saltpetre had to be devoted to the manufacture of gunpowder. Whatever the reasons for the failure of the plant Leblanc was never to enjoy a reward for his work. The plant had been sequestered by the state; when it was returned to him he could not afford to run it. He committed suicide in 1806.

The success of the Leblanc process was realized in the first decades of the nineteenth century. It was a triumph, not of the application of science, but in the overcoming of problems of development. Readily available and abundant materials were used, but they had to be in the right proportions and at the right temperatures. Furthermore an entirely new plant had to be designed to deal with a wholly novel process. In short, the heavy chemicals industry began with the Leblanc process, the credit for which belongs to France with a significant contribution from Scotland. France and Scotland were, of course, the leading countries in 'pure' chemistry as well as in chemical industry[9].

The other profitable advance in the chemical industry at the end of the eighteenth century also took place in

France and, with the establishment of the sulphuric acid and soda processes, effectively completed the revolutionary stage of the textile industry. This was the introduction of chlorine bleaching. The Swedish chemist K. W. Scheele had discovered chlorine in 1773 and its bleaching properties were established by the French chemist C.-L. Berthollet (1748–1822), who immediately published his discovery (1785). Unfortunately chlorine gas is hardly suitable – or safe – for direct use in bleaching and is not very soluble in water, although even a dilute solution could be used as a bleaching liquor. There followed a series of researches, in which Berthollet played a prominent part, to find a way of combining chlorine so that its bleaching properties could be most efficiently used. One answer was to combine chlorine with a solution of potash (potassium carbonate). This formed the so-called *eau de Javel*, an effective and convenient bleaching liquor. A better recipe was the result of correspondence between Berthollet and Watt. Charles Macintosh, a partner of Charles Tennant, chemical manufacturer of Glasgow, found that a good bleaching liquor could be obtained by passing chlorine through lime in water. This was patented in Tennant's name in 1798. In the following year Macintosh invented bleaching powder: chlorine absorbed by dry slaked lime. This was easy to transport and easy to use by adding it to water. It, too, was patented in Tennant's name (1799).

Most of the French and Scottish pioneers of the chemical industry had been trained as medical men. After 1794 French chemists were increasingly trained at the Ecole Polytechnique and at other specialized institutions.

Progress in pneumatic chemistry, so typical of the later eighteenth century, received startling and truly dramatic practical application when, in June 1783, Joseph Montgolfier and his younger brother Etienne, paper makers of Annonay, in Languedoc, launched the first unmanned hot-air balloon[10]. The conception of such a machine and its satisfactory launch implied reasonable knowledge of the simple physics of gases as well as enterprise and courage. Before long, manned flights were undertaken to the delight of wildly excited crowds. In the following year, 1784, the first hydrogen balloon ascent was made and very shortly afterwards the parachute was invented and successfully tried out. (The hydrogen balloon was made possible by relatively cheap sulphuric acid.) Although by early 1785 a balloon had carried J. P. Blanchard and John Jeffries across the English Channel, the lighter-than-air craft were never to prove as useful, in peace or in war, as courageous pioneers had hoped (the parachute was another matter). And the enthusiasm of the crowds set beside the public indifference to the first Newcomen engines might seem to confirm the fickleness and unreliability of public opinion; or, for that matter, the foresight of the elderly Dr Johnson, who observed that a cure for asthma would have been of greater value. Such a conclusion would not be entirely fair: balloon flights surely confirmed in the minds of ordinary people the reality of technological progress and suggested, even to the least imaginative, the possibilities that lay in the future. At last humankind had achieved, if only tentatively, an ancient aspiration: the conquest of the air.

The invention of the lighter-than-air craft by the industrialists and technologists Joseph and Etienne Montgolfier was a consequence of the 'pure' science of the pneumatic chemists and 'physicists'. In the same way the Newcomen engine was the offspring of the scientific discovery of the atmosphere and Benjamin Franklin's lightning 'rod' was based on his researches on thunderclouds and his repulsive fluid theory of electricity.

Inventiveness is obviously the basis of technological advance. Eighteenth-century achievements showed how complex this can be. Watt's researches, based on the latest scientific knowledge, revealed the ineluctable limitations of the Newcomen engine but did not suggest an answer. His inventive abilities enabled him to envisage the separate condenser so that, consistent with his scientific knowledge, the engine could be radically improved; transformed, in fact, into a new type of engine. His inventive capacities were further confirmed by his parallel motion, sun-and-planet gear and the double-acting principle, although in these cases no scientific knowledge was involved. The contemporaneous advance in water power technology reveals the same basic principles, even though several men were involved and the time span was much longer. Systematic experiments revealed the basic flaws in Parent's original theory. Improved theory pointed the way to more efficient water-wheels.

It has been urged that Smeaton's work in effectively increasing the energy resources of Britain was an essential factor in the progress of the Industrial Revolution. Surviving early textile mills and the many more drawings and diagrams of ones that have vanished all testify to the widespread use of the Smeatonian industrial water-wheel. Unlike the numerous battles that have been so repeatedly described, this vital element in history has been neglected. Smeaton could well be described as the Great Improver. His method ensured the most efficient product or process using the materials available. It is probably true to say that Smeatonian method has been responsible for the steady improvements in automobiles, airplanes and television sets over the past forty years, even though the principles governing these machines are basically unchanged. Smeaton's abilities and limitations were summed up by John Farey[11]:

> Mr Smeaton must be mentioned as a most eminent example of sound judgement in mechanical combinations, without great invention or power of creating original ideas; his works and mode of reasoning are upon record . . . Mr Watt in addition to his great powers of original invention was not inferior to Mr Smeaton as a practical engineer . . .

There is some force in this judgement, although it is by no means the whole truth. Smeaton showed great originality in devising his experiments on the efficiencies of water-wheels and in accounting for the difference between the two main forms. Nevertheless, the Smeatonian method of parameter variation did not and, *in principle*, could not lead to a new invention; nor did it involve scientific research into the basic laws of the subject or system under investigation (although individual exceptions to the rule may be found). What it certainly did was formally to banish from the realm of technics the element of craft mystery, passed down from master to apprentice. With Smeaton, engineering in England became autonomous, with at least one common method for all problems and all circumstances. In other words, the peculiar merit of John Smeaton – and it can easily be missed – is that he demonstrated that his method was applicable over the whole range of what had been crafts. This common element establishes his claim to be considered the father of British civil engineering – and a grandfather of mechanical engineering – as distinctive and recognizable professions. Before the century ended the Smeatonian Society of Civil Engineers was founded in London. It was the forerunner of the Institution of Civil Engineers.

The remarkable improvements in agriculture and in the iron industry were consequences, no doubt, of faith in improvability, keen observation, intuition and hard work. Apparently little or no theory, or science, was involved and indeed did not become significant until the nineteenth century, although it is quite possible that further study will show that methods similar to parameter variation were used. Tull, in particular, seems to have used such a method. If this is so then we must probably look for the original source in the seventeenth century. In any case all these new methods and techniques must be seen against the background of a firm belief in, and expectation of, progress. If eighteenth-century *philosophes* could contemplate the perfectibility of humankind how much easier to believe in the perfectibility of technology! Indeed, the evident success of technology must have reinforced all aspects of the belief in progress.

Smeaton may have provided British engineering with a common method and the beginnings of a prestigious social institution. What he did not, and could not, provide was a common system of education and training. There were several different routes to the top of the tree in British engineering. The vocation of instrument maker was the one used by Watt and Smeaton; another, humbler, way was that of the millwright; a third was apprenticeship to a leading engineer such as Watt himself (much the same procedure was followed in the professions of law, medicine and surgery). Other routes were via coal-viewing, surveying and architecture.

An interesting and informative comparison with Smeaton's work is provided by the career of his younger contemporary Charles Augustin Coulomb (1736–1806)[12]. Born into a provincial family of some local standing, Coulomb studied in Paris before entering the Ecole du Génie at Mézières where his tutors included de Borda and Bossut. As an officer in the Corps he was sent to Martinique in the West Indies where he was responsible for constructing the fortifications. On his return to France eight years later he divided his time between military duties and writing memoirs on technology and physical science. One of these was his important memoir on friction.

His best-known technological memoir was his 'Essay on the Application of the Rules of Maximum and Minimum to Some Problems in Statics, Relevant to Architecture' (1773). He began the memoir by analysing the behaviour of a beam under load. Understanding that woods and metals are elastic while stone is not, he recognized that there must be a neutral layer some way between the top and bottom surfaces; he does not mention Parent of whose work he was, presumably, unaware. Coulomb then shows that the sum of tensions across the top part of the beam must equal the compressive forces across the lower part. Further, the vertical component of these forces must equal the load applied and the moment of bending force must equal the moment of resistance.

He then demonstrates that, taking account of cohesion and friction, a vertical pillar or column would fail along an inclined plane – rather like a sausage cut at an angle – when subjected to a critical load. In this demonstration, as in others, he necessarily used calculus. But the solution was not complete: he had ignored Leonhard Euler's theory of buckling and the strength of struts. Sliding along an inclined plane also characterized his study of a wall under the pressure of earth. A wedge-shaped mass of earth, pressing against an insufficiently strong wall, would push it over as it slid along an inclined plane. Coulomb was able to calculate the slope of the plane, knowing the cohesion and friction of the earth.

An arch can collapse in two ways, inwards or outwards, depending on the balance of forces: the resistance of the abutments and the weight of the voussoirs (figure 8.2). Coulomb systematically analysed the forces and moments acting in an arch and found the maximum and minimum

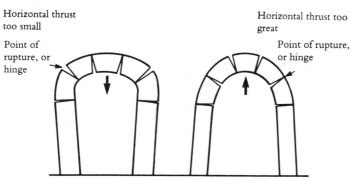

Horizontal thrust too small

Point of rupture, or hinge

Horizontal thrust too great

Point of rupture, or hinge

FIG 8.2 *Stability of arches*

values between which the horizontal thrust at the top of the arch must lie. If the thrust did not lie between these figures the arch would collapse. He was then able to specify the position of the point of rupture for a given arch.

Coulomb began as an engineer but in his later years made major contributions to physical science as well. He made extensive studies of elasticity and rediscovered Hooke's law. As a consequence he invented the torsion balance and with it established directly the inverse square law of force between electrically charged bodies and between magnetic poles. Up to that time researches in electricity had been largely in the rather casual hands of experimental philosophers; with Coulomb they became systematic, professional (the international name for the unit of electric charge is now the 'coulomb').

Coulomb's career illustrates three things. In the first place it shows the close connection between physical science and physical technology that had emerged early on in the professionalization of technology and of science. And it shows how easy it was to move from one of these to the other. Lastly, it shows how, in France, the formal institutions for technical education were being assembled well before the Revolution. In the last few decades of the eighteenth century the pace of technological innovation increased so rapidly that a historian of technology and science wholly ignorant of political and social changes (a scholar from another world?) would hardly have suspected that a revolution began in 1789 and ended with a military empire that subdued virtually the whole of mainland Europe.

The editors of *Annales de Chimie* had no doubts about the importance of science in the war effort and in industry. After the Revolution, after the 'Terror', they could claim that chemistry illuminates the practical arts:

Already France has ceased to depend on other nations for many useful products: the preparation of mineral acids, different salts, metallic oxides, dyestuffs and glass works so multiplied in France over fifteen years, are authentic proofs of the immeasurable advantages of chemistry. Without the guidance of this science could we have made quantities of saltpetre, powder and arms that were made in four years? Would we have had the copper, the iron, the steel, the potassium, soda, leather and many other precious materials that have helped us to defeat our enemies and maintain our existence? *Without chemistry, could we have perfected, as we have done, aerostatics?*

Accordingly, volumes 19 and 20 of the *Annales* were devoted to accounts of the services chemistry had rendered the nation during the critical years, 1793–97. The claim was debatable. Organization and developed practical skills had played a great part in the expansion of French chemical and metallurgical industries. Perhaps the most notable innovations related to the wars, over the whole period, were the development of the French sugar-beet industry to replace West Indian sugar, cut off by the British blockade, the discovery of methods of food preservation and the encouragement, by Napoleon, of the French madder-growing industry. Nevertheless, if we accept a broader definition of chemistry than the modern academic would concede, the claim had some substance and was to become more and more valid as the years passed.

These views and examples of the application of scientific discoveries might lead the modern reader too readily to project on the technology of the past the common assumptions of the present day. It can be too easily assumed that there were two distinct groups, scientists and technologists; and that the scientist formulates laws, proposes theories or discovers facts that can be used by the technologist. Seen in this light the technologist, whether chemist, metallurgist or engineer, merely applies scientific knowledge gained by others. This is a mistake. As far as England was concerned there is little evidence for a distinct, self-conscious, class of 'pure' scientists before the latter part of the nineteenth century. Newton was the exception that proves the rule; more typical of his time was Robert Hooke. The majority of men now classified as scientists made contributions to technology; many of those regarded as technologists contributed to science. As Shakespeare wrote, one man in his time plays many parts, although the parts do not necessarily fall into Shakespeare's temporal sequence. In France, further along the road towards the professionalization of science, a class of self-aware, publicly acknowledged, men of science is more readily identifiable and the same may be true, to a lesser extent, of Scotland. The formal separation of science and technics is due to educational and organizational factors. It is, as Dr Webster has pointed out, a distortion to interpret the history of technology and of science from the perspective of modern educational arrangements.

The main forms of technological method established in the eighteenth century include the first clear represent-ations of the essential development phase in which a 'laboratory' innovation or model is scaled up – and toughened up – to meet the requirements of productive industry. We know nothing of the development problems that confronted the inventors of the Great Wheel and the Saxony Wheel; we are ignorant about the first attempts to make a successful weight-driven clock, a mariner's compass, a cannon or a hand-gun. Although we know a little about Johan Gutenberg and his print shop at Mainz, we know nothing about the difficulties he faced, the problems he solved and the time it took to find the right type metal, to design his type mould and to select the right ingredients for his printing ink (if, indeed, he did this last). We can only speculate about, or discover by recon-struction, the problems that Newcomen overcame. When we come to James Watt and his condensing engine we have a tolerably clear picture of the ways in which he overcame the difficulties on the way and the factors that determined the design of the engines he and Boulton put on the market. The same is true of John Harrison with his chronometer and the pioneers of the soda process and chlorine bleaching, as well as of the beginnings of 'aero-statics'. Only innovations that are reliable, safe, meet a recognized need and are profitable can pass the test.

Historians, considering the emergence of an organized, self-aware and efficient technology and faced with the enormous commercial successes of Britain, have tended to overlook the achievements of organized science and technology in France. After all, France suffered enormously in the wars that ended in 1815; in comparison only one enemy musket was discharged (once) on British soil and the only casualty was a grandfather clock (Fishguard, 1797). To attempt to draw up national balance sheets would therefore be a futile exercise. It would, on balance, be fair to say that the industrialization of Europe after 1815 owes much to France as well as to Britain. All the forms of technology were clearly apparent in France before the end of the eighteenth century. The major features owed little to the requirements of war and revolution. And, in the case of Coulomb, we have perhaps the first important instance of a man who began as an engineer, the product of a great engineering school, using his engineering knowledge to effect a fundamental advance in a branch of science that was of apparently limited utility. Coulomb's invention and use of his torsion balance was the first step towards the quantification of the science of electricity.

Steam power and the use of iron as a structural material, both of which began in the eighteenth century, were to pose substantial problems for engineers in the centuries to come. Up to the end of the seventeenth century the use of traditional building materials – timber, masonry and bricks – was governed by long-acquired experience. There were rules by which men knew what they could and could not do with them. At the same time the power sources – wind, water, animal muscle – and machines used in production remained basically unchanged for centuries. Steam power and iron structures together with the revolutionary machines of Arkwright and his successors raised problems beyond the capacity of traditional knowledge and skills to solve. At the same time economic forces, in part stimulated by these developments, made solutions increasingly urgent.

With remarkable prescience Galileo had, as we saw, realized and explained the effects of scale in determining the behaviour of simple structures, natural and man-made. As no new materials were involved there was no immediate application for his analysis. With the introduction and widespread use of the new material – cast iron – the English craftsmen–engineers, ignorant of Polhem's work, learned the practice of scaling from Banks. And it was because of Banks that it became possible to answer the question: What form should the iron pillars and beams take to ensure that the building can bear the load of the machines it has to carry? Yet far more complex problems posed by varying loads under extreme conditions had to be faced by nineteenth-century engineers building railway bridges and large iron and steel ships.

Napoleonic Europe

The turmoil and the idealism of the French Revolution and the violence and brutality of the succeeding, Europe-wide, wars that lasted until 1815 altered the course of economic, political and social development of the Continent and must therefore have affected the course of technological development. It is, however, hard, if not impossible, to point to any specific major change that can be attributed to these upheavals. In France the metric system was established and the Ecole Polytechnique was founded, but these were events that could have been foreseen, or were even in train, before 1789. The moribund universities were abolished and a great national, all-embracing university established in 1808. The Académie Royale des Sciences and the various provincial scientific societies were abolished; they were replaced by the Institut National. In spite of these rather dubious changes the drive and creativity of French technology and science actually increased until, in the first decades of the nineteenth century, France could claim, in numbers and diversity, a scientific and technological community unmatched anywhere else and unsurpassed at any time.

Prominent among the engineers who achieved fame and power during the Revolution and the Napoleonic period was Lazare Carnot (1753–1823), the engineer officer and graduate of Mézières, who became known as the 'organizer of victory' for his success in summoning up raw conscripts and forming them into efficient armies. As an academically inclined engineer he had made an exhaustive study of the application of the *vis viva* principle to machines, generalizing the principle that the water, or driving agent, must enter the machine without shock or turbulent impact and leave without appreciable velocity. Lazare Carnot also made the first, tentative, contribution to the theory of dimensions.

Carnot's near-contemporary, Gaspard Monge (1746–1818), was not of a class exalted enough to be socially acceptable as an officer cadet at Mézières. He could, however, be appointed a student draughtsman in the fortifications office where his abilities were soon recognized. In 1795 he published his book *Géométrie descriptive*, a work that inaugurated the discipline of engineering drawing[1], at least in mainland Europe. In it Monge showed how objects could be represented, by projections of their defining or generating lines, on two planes at right angles to each other. The two planes would normally form a flat diagram, but the true lengths, shapes and angles of objects could be redetermined by rotation through 90°. This rational procedure enabled him to represent three-dimensional objects in two dimensions in such a way that they could be accurately re-created using the flat drawing. Such was the political and military power of France at the beginning of the nineteenth century that engineering drawing on Mongean principles was adopted all over mainland Europe. In Britain and America on the other hand the development of engineering drawing was more empirical. The Rev. William Farish (1759–1837), Jacksonian Professor of Experimental Philosophy at Cambridge University, popularized, through lectures, his development of isometrical perspective. This was an excellent way of showing the complete assembly of the components of a machine. We can assume that, when devising and making a new machine, the engineer would begin by sketching out his ideas; he would then prepare orthographic projections; finally he would rely on skilled craftsmen to interpret his detailed drawings and make the individual components. The assembly of the complete machine or engine would entail the work of skilled fitters to ensure that the individual components were within the right tolerances.

The quality of French technology was reflected in the high standard of books published over the period[2]. Among the most notable were Riche de Prony's *Nouvelle architecture hydraulique* (1790, 1796), admitted, even in Britain, to give the best contemporary account of the steam engine. Following the works of Lazare Carnot and Monge, Lanz and Betancourt's *Essai sur la composition des machines* (1808) set out a system of ten classes of mechanisms that could relate the different forms of circular and reciprocating motion, thereby inaugurating the science of kinematics; J. N. P. Hachette adopted this classification in his *Traité élémentaire des machines* (1811). Other works of the period were A. Guenyveau's *Essai sur la science des machines* (1810), C. L. M. Navier's revised version of Belidor's *Architecture hydraulique* (1819), and A. M. Heron de Villefosse's *De la richesse minerale* (1819), a successor to Gabriel Jars' book. Shortly afterwards came J. A. Borgnis' *Théorie de la mécanique usuelle* (1821), and G. J. Christian's *Traité de la mécanique industrielle* (1822-25).

It would be fair to say that the engineering textbook began in France. In Britain, where technology and industry were almost entirely matters of private enterprise and where State involvement – always excepting the Woolwich Academy – was minimal, such textbooks as there were amounted either to elementary treatises like those of Emerson, Banks or Olinthus Gregory or to feebly descriptive popular works. Even John Farey's encyclopedic *A Treatise on the Steam Engine* of 1827, while most useful as a source book for historians, belongs, in spirit, style and understanding, more to the 1770s than to the 1820s.

Another means of publicizing invention and innovation, this time to a far wider audience, was the first industrial exhibition, held in Paris in 1798. At the heart of the exhibition and supplemented by other industrial products were displays of the works of the three great establishments, formerly owned by the Crown, for the manufacture of tapestries, carpets and porcelain. The original idea was that exhibits should be disposed of by lottery and to that extent the exhibition resembled the great fairs whose origins go back to antiquity. It differed from a national fair, however, in that a jury, under the chemist J. A. C. Chaptal (1756–1832), awarded prizes for the best exhibits. The industrial exhibition of 1798 did not escape Napoleon's attention. Aided by Berthollet, Chaptal, Monge, Prony and others, Napoleon organized a second and much bigger industrial exhibition, held in the (once Royal) Louvre in 1801. Thereafter exhibitions were held at frequent, if irregular, intervals throughout the nineteenth century. Wars, nationalism, political fanaticism and rapidly mounting costs have meant that great exhibitions have been far fewer in the present century.

The consequences of the French Revolution, more particularly of the 'Terror', 1793–94, together with the subsequent conquests of Napoleon, led to a strong political reaction in Britain that was to have a lasting effect on the country's social, political and, quite possibly, technological development. At the same time the detached, cool distaste for religious 'enthusiasm', reaching as far as the scepticism of Black, Hume, Gibbon and Hutton that had characterized eighteenth-century thought, was replaced by the conventional religiosity of the earlier Victorian age. Economically, the wars boosted British industrial development, particularly the iron industry and industries that supplied armies and navies. They had, however, little lasting effect on technological innovation. Indeed innovations that were brought about as a direct consequence of the wars were discarded as soon as peace was finally established. The extensive visual telegraph system established in France by Claude Chappe ('ingénieur télégraphique') was copied in Britain when a telegraph system linking London to naval bases on the south coast was instituted. It was closed as soon as peace was declared, the relics being a few landmarks in the south of England that still bear the name 'Telegraph Hill' (cf. the various 'Beacons'). The novel missile 'system' known as the Congreve rocket was abandoned after the wars; there is a reference to it in the words of the American national anthem and a distant echo of it in the name of Stephenson's successful locomotive (see chapter 10). Montgolfier's balloons found a very limited military application but military requirements did not lead to development in lighter-than-air flight. At the height of the Napoleonic wars a Yorkshire land-owner, Sir George Cayley, was experimenting with heavier-than-air flying machines[3]. He had realized that the only solution was a fixed-wing airplane; to try to imitate the birds, as so many would-be aviators had done, was futile. Cayley saw that a hot-air engine, and not a steam engine, would be required to power his machine and, in an effort to reduce weight, he proposed using the lightest possible landing wheels with spokes in tension. There is no evidence that Cayley – or the British government – thought of his airplane and glider as possible weapons of war.

On the face of it the Portsmouth block-making machinery, due to Marc Isambard Brunel and General Sir Samuel Bentham, should have had considerable consequences for productive industry. Among the most numerous items required by the greatly increased number of British naval vessels during the Napoleonic wars were common pulley-blocks, which were standardized items for all ships. The machinery designed to meet the enormous demand for pulley-blocks followed the precedent set by Arkwright. Special machines were devised to carry out a series of single operations so that rough blocks of wood, fed in at one end of the process, emerged as fully formed pulley-blocks at the other end. Ingenious as it was, greatly as it increased productivity, it seems to have been a dead end and had little effect on the development of high-production machinery. This may have been a consequence of the fact that British naval and military bases were geographically – and perhaps socially – remote from the areas of intense industrial development in the midlands and the north. Additionally, iron was increasingly replacing wood in machine construction.

The steam engine was surely the most obvious candidate for adaptation for military and, particularly, for naval purposes. However, despite one or two suggestions, nothing seems to have been done about it. This is the more surprising as Kanefsky and Robey found that about a thousand steam engines were built in the last decade of the eighteenth century, as many as were built in the previous nine decades. Evidently the steam engine was rapidly becoming much more versatile and powerful.

The availability of better cast and wrought iron led to changes in the construction of steam engines as it had done in the cases of textile mills and textile machinery. In about 1795 the little-known firm of Aydon and Elwell, near Bradford, started making engines with elliptical cast iron beams. In part, at least, this owed something to John Banks who was associated with the firm. As, increasingly, all the components were made of iron or copper, so engines tended to become autonomous power units and no longer inseparably associated with and part of the buildings that housed them. This meant that engines could be designed, made and sold without regard to their individual destinations and the market for steam power was accordingly widened.

There had remained, up to the end of the eighteenth century, one major hindrance to the development of the steam engine. Extended by Act of Parliament in 1775, Watt's master patents still controlled virtually every form of practicable steam engine. And Watt was implacably opposed, all his life, to the use of high-pressure steam. Boiler explosions, due to bad design, faulty manufacture or more probably careless working, would assuredly discredit all steam engines in the eyes of the public. Watt had established a long time ago that a small increase in steam temperature resulted in a disproportionately great increase in pressure. An engine working with high-pressure and therefore high-temperature steam would be a hazardous affair, for a slight, accidental, rise in temperature in the furnace could cause a great, perhaps disastrous, increase in pressure.

With Watt's withdrawal from industry and with the end of his patent monopoly, bolder spirits were free to explore the possibilities of the high-pressure steam engine. High-pressure steam offered an obvious advantage: the higher the pressure, the more pounds per square inch on the piston and therefore the greater the power of the engine; alternatively, a smaller and more compact engine could be built to give the same power. Among the first to build high-pressure steam engines were Oliver Evans (1755–1819), in America, and Richard Trevithick[4] (1771–1833), in Britain. Their high-pressure steam engines were, during the nineteenth century, progressively to dominate the economies of western Europe and North America; and it might be supposed that their introduction reduced Watt's contribution to that of a minor advance between the suggestions of Leupold and the seventeenth-century inventors, on the one hand, and those of Trevithick and Evans on the other. This, however, would be a mistake based on the wisdom of hindsight. Apart from his kinematic inventions – the parallel motion and the sun-and-planet gear – Watt had established the form of the reciprocating engine and the double-acting principle; he had significantly raised the standard of engineering precision; most important of all, he had realized and demonstrated the importance of heat economy in the operation of the engine, while his invention of expansive operation was to prove truly fundamental, in science as well as in the operation of steam and other forms of heat engine. In a few words: Watt set the agenda for the development of all subsequent heat engines.

Trevithick, who had put up a number of column-of-water engines at Cornish mines, may have been prompted to consider the use of high-pressure steam by analogy with the high efficiency of the water engines. But his first notable use of high-pressure steam engines was to drive a road vehicle. He was not the first to do this; as early as 1769 Nicolas Cugnot (1725–1804) had built a steam-propelled road vehicle. It was premature. By Trevithick's time the technology of steam engines had improved to such a point that a mobile steam engine or 'locomotive engine' was a practicable proposition. The problem proved to be the unsuitability of contemporary roads for steam locomotives. An alternative application was offered by the numerous railroads or tram tracks serving mines and other industrial enterprises; these were particularly numerous in the north-east of England coalfield and in south Wales. In 1804 Trevithick built the first steam railroad locomotive at Samuel Homfray's ironworks at Pen y Darren in south Wales. The single horizontal cylinder was immersed in the boiler, presumably for compactness, safety and, possibly, to minimize heat losses. The piston drove the wheels by means of a piston rod and a train of gears; a fly wheel was fitted to ensure uniform motion. The locomotive hauled, at five miles per hour, five wagons carrying seventy men on the railroad that linked the ironworks to the Glamorgan canal about ten miles (sixteen kilometres) away. Unfortunately it repeatedly smashed the cast iron, L-section, track and the project was given up. Four years later Trevithick laid down a circular track near Euston, in London, and offered trips in a carriage hauled by a steam locomotive named, reputedly, *Catch Me Who Can*. Thereafter this very able but restless man lost interest in steam railroads and turned his attention to other things.

It was widely believed that the adhesion due to weight and friction between the wheels of a locomotive and the rails would not be sufficient to enable the locomotive to move a substantial load. Trevithick himself had roughened the surfaces of the driving wheels of his locomotive to ensure a better grip. And in 1811 John Blenkinsop took out a patent for a system in which the locomotive worked its way along the rails by means of a pinion, or cogged wheel, that engaged short, horizontal studs protruding from the side of one rail. The pinion was driven by a linkage from two vertical cylinders which, following Trevithick's example, were sunk into the boiler. Blenkinsop's engines started working on the Leeds to Middleton Colliery railroad in 1812, a distance of about six kilometres, and continued in useful service for a number of years. This was the first sustained demonstration of the utility of the steam locomotive. The engines themselves were built by Matthew Murray, a highly original and able Leeds engineer. There were various other attempts to overcome the supposed problem of inadequate frictional grip; they were uniformly unsuccessful. Evidently there was no conception of systematic experiments to clarify the nature of the problem – if any. It hardly mattered. Two years later, in 1814, George Stephenson built an engine that he named *Blucher*. It, too, had a pair of vertical cylinders sunk into the boiler but this time the drive was to the wheels and the frictional grip proved quite adequate.

In the meantime Trevithick had been building small stationary engines. In these engines, too, the cylinder was horizontal and inside the boiler. This was an unusual feature; horizontal cylinders were disapproved of as it was believed that the weight of the piston would result in excessive wear on the lower half; the wear in a vertical cylinder must be uniform all round. Whatever the merits and demerits of the different types, once the high-pressure steam engine was accepted one feature seemed to open up intriguing possibilities: the great rise in pressure due to a modest rise in temperature suggested (at least to some) that, with careful design and due precautions to prevent a boiler explosion, an enormous – almost unlimited – amount of work could be obtained from a limited amount of fuel. This was the first time that there appeared reasonable hopes of enormous power for little cost. The hopes were soon disappointed. Euphoria over the immediate prospects for very cheap power broke out on three subsequent occasions in the nineteenth century and at least twice in the present century; all such hopes have been disappointed so far.

Besides making possible the development of steam railroads, the high-pressure steam engine, compact and powerful, found widespread application in mills and factories. Nevertheless, for the first four decades of the nineteenth century the water-wheel continued to provide the major share of industrial power. It, too, was improved, although not as radically as the steam engine[5]. Iron water-wheels were built with ventilated buckets so that less water would be spilled. Massive spokes were replaced by thin wrought iron rods in tension, just as the spokes of a bicycle or perambulator wheel are in tension. The innovation made much lighter wheels possible with the drive taken from the rim of the wheel and not from the axle. Even more advanced, feedback arrangements were devised, so that if the machines in the mill were running too fast, a governor reduced the flow of water into the wheel; if they were running too slowly, the rate of flow was automatically increased. This was practically the limit of improvability of the water-wheel. The next advance was to be the hydraulic turbine due to Burdin and Fourneyron.

It would be wrong to suppose that ingenuity and inventiveness were limited to improvements in the reciprocating steam engine. On the contrary a wide variety of different forms were tried and the possibilities of working substances other than steam were explored. It was plausibly but wrongly believed that if a vapour with a much lower latent heat than that of steam could be used then the duty of the engine would be increased in proportion, for less heat would be needed to fill the cylinder with vapour than with steam. No economy was found when liquids with low latent heats of vaporization, such as alcohol, were used. Experiments were made on air engines and engines worked by different gases. Again, no advantage was found. The thermal expansion of solids and liquids was also proposed as a power source, but without success. Experience and experiment showed that nothing was superior to steam as a working substance, and no form was better than the reciprocating engine of Watt, Trevithick, Evans and the other leading engineers – or if it was, it was well beyond the engineering capabilities of the time. By about 1820 virtually all the possibilities of harnessing the expansive force of heat had been exhausted. From the strictly practical point of view no engine was found to be superior to the reciprocating steam engine.

Little contemporary science was involved in these many and varied experiments, certainly as far as Britain was concerned. One remarkable enterprise, while hardly scientific in the conventional sense, was certainly both systematic and co-operative. This was the publication of Joel Lean's *Monthly Engine Reporter* which, by its nature, diminished the subjective and personal element in assessing the performance of steam engines and introduced a measure of objectivity. The peculiar conditions of the Cornish mining area required that pumping engines worked as economically as possible. The nature of the community, isolated and at the same time relatively concentrated, made a degree of co-operation possible. It is said that the withdrawal of Boulton and Watt from Cornwall, following the lapse of Watt's patents, led to a serious decline in the efficiency of Cornish engines. Whatever the cause, however, the Cornish miners asked Joel Lean, a widely respected mine captain, to publish every month the details of the performance of as many engines as possible; in practice this meant the great majority of engines in the area. It was hoped that, as a result, the most efficient engines and practices could be identified and copied by the rest of the community. The *Engine Reporter* was published every month from 1811 until the decline of the industry in the 1880s.

One immediate consequence of the publication of the *Engine Reporter* was the confirmation of the much greater economy, or duty, of Arthur Woolf's engine. Woolf (1776–1837), a self-taught practical mechanic, patented, in 1804, a two-cylinder high-pressure steam engine. Steam, at high pressure, first expanded in a small cylinder after which it was admitted to a larger, low-pressure cylinder wherein it expanded down to near atmospheric pressure. (The high-pressure cylinder was smaller than the low-pressure one in order to ensure that the net force on the pistons should be about the same in both cases.) Woolf based his first design on a startlingly erroneous 'law' that he thought he had discovered. It took him several years to correct his error but when he had done so and installed his first engine at Wheal Abraham mine in 1814, the *Engine Reporter* confirmed that it returned a duty more than twice the maximum considered possible by Watt himself. From then on Cornish engines, the most efficient in the world, were, increasingly, high-pressure engines. There were those who, like John Farey, denied the superiority of the high-pressure engine, but common experience and in particular the Lean reports confirmed it beyond question. When peace returned in 1815, Woolf's partner Humphrey Edwards went to France to build Woolf-type engines there. He was successful; and it was realized in France, as in Britain, that the high-pressure engine was superior in economy to the low-pressure, Watt-type engine. Although the Woolf design was soon superseded, for many years high-pressure compound engines were referred to in France as Woolf-type engines.

The last major innovation that came from the Soho

foundry was due to Watt's assistant and later partner, John Southern. This was the indicator, invented about 1796 but not made public until 1824, and then in a very discreet fashion. A small cylinder is connected by a pipe to the working cylinder. A spring-loaded piston in the small cylinder is attached to a pen, under which is a board with a sheet of paper pinned to it. The board is connected to the working piston of the engine. As the board moves to and fro the pen traces out a closed loop, representing the variation in pressure with the movement of the working piston. As the product of pressure, or force, multiplied by distance moved is the work done, the area of the loop will be proportional to the work done for one cycle of the engine. In this way, by taking an indicator diagram, as it was called, and multiplying by the number of cycles performed, the total work done by the engine in a given time could easily be computed. The indicator might not be essential for calculating the work done by a non-expansive engine, where steam pressure multiplied by length of stroke would give (at least approximately) the work done for one cycle; but where expansive operation was used it was essential.

The compact steam engine, made almost entirely of iron, required manufacturing techniques very different from those that had been needed to build the Newcomen or even the first Watt engines. At the same time the continuing rapid expansion of the textile industry made the extensive use of cast and wrought iron in textile machinery economically and technologically advantageous. Frames, pulleys, shafts, gear wheels and pinions could be cast in large numbers, cheaply and accurately. Machines made largely of iron could stand greater stresses, were fire-proof, less bulky, more accurate and lasted longer than wooden machines. The power loom, introduced at the turn of the century, was substantially an iron machine. The extensive use of iron in making advanced machinery was to have consequences far outside the steam engine and textile industries. Parenthetically, it was significant that the first two large buildings to be lit by coal gas light were Boulton and Watt's Soho foundry and the Phillips and Lee twist mill in Salford. Gas lighting had been pioneered by Boulton and Watt's employee and representative in Cornwall, William Murdock, although priority for the idea belongs to the Frenchman Philippe Lebon and the German F. A. Winzer.

In short, textiles and steam engines combined to stimulate new technologies. One of the most important of these was the design and manufacture of machine tools, or machines to make machines[6]. Primitive machine tools, such as the pole lathe, had existed for a long time. The simple boring engine appeared when, in the Renaissance period, it was used, as we saw, to bore out cast bronze and then cast iron cannon. In these machines a hardened cutting tool was mounted on the end of a long mandrel that advanced along the bore of the cannon as it rotated, cutting away the metal; the power source was a water-wheel. These were inaccurate machines although a considerable improvement was made when John Wilkinson built his famous boring engine in 1775; it provided James Watt with his most accurate cylinders. 'Mathematical instruments', such as telescopes and sextants, were commonly of brass, a soft metal, and were made in large numbers by the skilled instrument makers of the eighteenth century using small lathes, screw cutting lathes and drills. There were also rose engines, used to produce ornamental patterns on soft metal. But the industrial machine tool working on hard iron or steel was very much a product of the demands of the textile and the steam engine industries.

Joseph Bramah (1748–1814), was an ingenious mechanic and fertile inventor who, among other things, devised a thief-proof lock (important in a lawless age), the pump for raising beer from barrels in a cellar, and the hydraulic press. More important than Bramah was his young foreman Henry Maudslay (1771–1831), who is credited with having devised the machine tools for the manufacture of Bramah locks. Maudslay, who soon set up in business on his own, can fairly be described as the father of British machine tool technology. He has been credited with the invention of the slide rest but this is incorrect; a German manuscript of *c.* 1480 represents a slide rest and there is a drawing of one, used on a rose engine, in the *Encyclopédie* (1771). What Maudslay certainly did was to incorporate the slide rest in lathes for turning iron. And, although screw cutting lathes for soft metals had been used by eighteenth-century instrument makers (and much earlier for the ornamental turning of wood), Maudslay's lathe of 1797 with its extremely accurate lead screw was an industrial machine tool. The spindles and dead centres of screw cutting lathes had to be accurately aligned, the lathes themselves absolutely rigid and therefore made wholly of iron or steel and the surfaces over which the slide rests and cross-feeds moved had to be true, perfect planes. Maudslay therefore imposed the highest possible standards of accuracy and precision in his works. Each one of his craftsmen, engaged on making lathes, had to have, beside him on the bench, a perfectly flat, standard plane with which to check his work. These standard planes were made by the lengthy and painstaking process of rubbing three planes together, the surface of each one being lightly coated with red lead. Places where the coating was rubbed off were protuberances, to be carefully filed down and the process repeated until three true planes were achieved.

As in the case of Boulton and Watt, Maudslay's reputation attracted talented young men to work – and learn – under him. One such was Joseph Clement (1779–1844), the son of a handloom weaver. Clement devised a self-regulating lathe; he standardized his screw threads and, it is said, made a planing machine, but did not patent it. So high was Clement's reputation that Charles Babbage employed him to make the components for his (never to be completed) difference engine, a precursor of the computer. Among the last of Maudslay's protégés was the famous Joseph Whitworth, an improver rather than an original inventor. But the most original of Maudslay's apprentices was Richard Roberts[7] (1789–1864). His inventions included the back-geared lathe, the pillar drill, the radial drill and a punching machine controlled by a Jacquard mechanism and therefore the first digitally controlled machine tool. Roberts had certainly invented a planing machine by 1817 (James Fox, of Derby, was yet another claimant to have invented a planing machine). With the planing machine it became possible to make accurate guide rods for the pistons of steam engines so that Watt's parallel motion became obsolete. Roberts also perfected Crompton's mule by making it self-acting and he introduced, although he did not invent, the first industrial electromagnet and the stroboscope.

Roberts was also associated with the further application of the differential[8], which had had an intermittent life, being forgotten and then reinvented through the centuries, mainly by clockmakers. In 1822 an American mechanic, Asa Arnold, used differentials to control the speeds of flyers and spindles in a roving frame, so that as the rovings built up on the bobbins their weight caused the applied power to shift to the flyers. The spindles therefore slowed down so that the rovings did not build up too rapidly. This invention spread from America to Britain and the rest of Europe. In 1828 a Frenchman, Onesiphore Pecqueur, applied the differential to a road vehicle and, four years later, Roberts was granted a patent for the use of the differential on the powered wheels of road locomotives. Steam- or horse-powered road transport was soon superseded, or so it seemed, by the railroads and the application of the differential to powered road vehicles had to await the automobile at the end of the century. Roberts turned his hand, successfully, to the building of railroad

locomotives and his firm of Sharp, Roberts became world-famous.

Other pioneers of machine tool technology were associated with the manufacture of steam engines; among them were Henry Maudslay, Matthew Murray and William Fairbairn; the last two, together with Whitworth, also had close links with the textile industries. On balance it seems that the latter connection was the more important. The machine tool industry in early nineteenth-century Britain was located in Glasgow, Leeds, Bradford, Manchester and London, all, with the exception of London, textile cities. The textile industry was the only industry that demanded large numbers of identical and complex machines substantially made of iron. On the other hand, Birmingham – the home of Boulton and Watt – Liverpool and the Cornish mining area, famous for its steam pumping engines, were not centres of machine tool technology. The rationale behind this seems to be that, although the statistics are not available, the number of individual textile machines must have greatly exceeded the number of individual steam engines. Richard Roberts, in particular, had a high reputation as a manufacturer of large batches of identical textile machines that were reliable and of high quality. He had, in other words, achieved batch production and was well on the way to mass production. The difference between the lines of identical textile machines in a Lancashire cotton mill of the early nineteenth century and the lines of identical automobiles in a Detroit or a Japanese car plant is more one of degree than of kind.

Contemporary American technology was conditioned by the requirements demanded and the opportunities offered by a vast country. In addition, the social order was different from that in Britain and other European countries. Land was cheap and freely available; there was no apprentice system, which may have been a disadvantage, but there were no guild restrictions. In one respect, at least,

America followed the British pattern. The rivers and streams along the Fall Line in the north-east of the country provided ample power to drive the textile mills of New England, which were modelled on those of Britain. And these mills were, as in Britain, the source of the earliest machine-making firms.

The first noteworthy American technologist was Eli Whitney (1765–1825). The son of a farmer and a Yale graduate, he went to Georgia to be a teacher. There he was struck by the time-consuming practice of extracting the seeds from the cotton bolls by hand. By 1793 he had invented and built his cotton gin ('gin' = engine), a combination of a combing device to remove the seeds and rollers to collect the fibres. It has been said that the cotton gin had the effect of prolonging the institution of slavery by increasing the prosperity of the cotton states; but it would hardly be fair to blame Whitney for that. Having patented his cotton gin Whitney set about manufacturing it and this led him to devise and make appropriate machine tools. His skill at making machine tools was recognized and the government commissioned him to manufacture 10 000 muskets, the individual parts of which were to be so accurately made that they could be freely interchanged between different weapons. That is, any component could be replaced from stock without filing or fitting of any sort.

Interchangeability in this field was a new idea although it is claimed that it had – most improbably – been achieved by a French gunsmith. As we saw, Gutenberg had achieved interchangeability in printing technology many years before Whitney contemplated it for gun making. Indeed, the term is strictly relative; everything depends on the acceptable level of tolerance. The average household today has a multitude of items with interchangeable components: for example, fluorescent tubes, light bulbs and power plugs, washers, floats and other components of the water system, all are easily replaced from a hardware store when worn out, damaged or lost. On the other hand, the more precise the item – for example, a gun – the more difficult it is to achieve interchangeability. The advantage of interchangeability – if it could be achieved – in a country like the United States, where skilled men were relatively widely dispersed, was that parts could be easily replaced by unskilled men without having to travel long distances to the nearest armoury or workshop. In practice, however, complete interchangeability was not to be achieved for a very long time; some fitting had to be carried out, although the skills of the gunsmith were not always required. Whitney did not achieve interchangeability but he gave currency to, or publicized, the idea.

A New Form of Electricity

In spite of the hopes of the editors of *Annales de Chimie*, science had not played a leading part in these technological advances in Britain and America or even in France itself. And yet it was a particularly fruitful age for science, nowhere more than in France. At the head of many distinguished physicists stood P. S. de Laplace; and, while there was no one to replace Lavoisier, undeniably Berthollet, Proust, Fourcroy and Gay-Lussac were chemists of the first rank. The first of two seminal advances that were to set the pattern for nineteenth-century science came, however, not from France but from Italy.

In 1791 Luigi Galvani, a physiologist, described how frogs' legs, suspended from a metal lattice by metal hooks, jerked convulsively if the hooks were moved and the legs touched the lattice. He ascribed the phenomenon to 'animal electricity'. A more satisfactory explanation was given by Alessandro Volta (1745–1827), Professor of Physics at the University of Pavia. Volta showed that the phenomenon was due to the fact that the hooks were made of a different metal from the lattice. In confirmation he demonstrated that when two different metals touch and are separated one acquires a positive charge of electricity, the other a negative charge. Metals are, he remarked, movers of electricity and he was able, by careful measurement, to arrange the metals in an electromotive series according to their capacities to acquire positive or negative charges by contact. Volta did more than discover and analyse a new phenomenon; he developed it to make a vitally important invention. Two discs of different metals were put in contact and separated from a similar pair of discs by a non-metallic substance such as cloth or cardboard soaked in brine. A pile of metallic couples, each couple separated from the next by non-metallic materials, would, he found, give an appreciable and continuous electrical effect. The function of the non-metallic separator was, of course, to prevent the cancelling effect that two similar metals in contact with the opposite sides of one dissimilar metal would have on each other. Volta, in short, had invented the electric battery. The electricity it gave was described as galvanic electricity and for a long time it was doubted whether it was the same as the electricity generated by friction.

The voltaic pile was announced in 1800. Very large and powerful piles were built, with little delay, in London and in Paris. They were first used to make discoveries in electrochemistry. In the same year, 1800, the pile was used to 'electrolyse' water into its constituent gases, oxygen and hydrogen. And, a few years later, the young Humphry Davy discovered, with the aid of the pile, a whole series of new Lavoisierian elements. He showed, too, how the pile, connected to two pieces of charcoal that almost touched each other, could be made to give a brilliant light, an arc light. Of course it was far too expensive to be of practical use.

The use of static electricity as a means of signalling had been proposed in the middle of the eighteenth century but it never got beyond the stage of an entertaining parlour trick. The pile, however, suggested better prospects. In 1809, S. T. Sommering (1755–1830), Professor of Anatomy at Kassel, demonstrated his electric telegraph. The stimulus to his invention had been the speed with which Napoleon, relying on the French visual telegraph system, had moved to drive the Austrians out of Bavaria. Sommering's telegraph used 27 insulated wires bound into a cable, each one of 26 terminated, at the receiving end, in a gold point at the bottom of a glass vessel full of acidulated water. At the transmitter the wires were connected to 26 terminals each bearing a letter of the alphabet. The remaining wire was for the return. On connecting a pile to a terminal, gas bubbles, due to electrolysis, appeared at the corresponding gold pin at the receiver. Although Sommering demonstrated his telegraph over a distance of over 600 metres it was too expensive, slow and limited in range to be of use.

It was not until 1820 that the pile led to a discovery in the realm of physics. The Danish physicist Hans Christian Oersted (1777–1851) found that if a wire were connected to the two ends, or poles, of a pile the electrical effect (it would be premature to call it a *current*) deflected the needle of a magnetic compass placed under the wire. This was a wholly unprecedented experience. Everybody knew of central forces – gravitation, the mutual attraction or repulsion of magnetic poles or of electrically charged bodies – but a *deflecting* force was something entirely new. What, it may be asked, would Newton have said to that?

The Royal Institution, London, at which Davy worked, had been founded in 1799 by the energetic American loyalist (to the British Crown), Benjamin Thompson, later Graf von Rumford, as a science museum and technical college. However, so successful was Davy as a man of science and as a popular lecturer, so skilled was he in the social graces, that the Royal Institution rapidly became a research centre and a place where popular scientific lectures were delivered to fashionable and appreciative audiences. Before long Davy, having identified the composition, interests and habits of his audiences, was giving lectures on agricultural chemistry. The gentlemen-improvers who had done so much for British agriculture were quite willing to learn from the science of chemistry.

The second seminal advance was made, improbably enough, in the parvenu industrial town of Manchester. The Literary and Philosophical Society of Manchester, founded in 1781, differed from the Royal Institution. It was a club for men interested in science and learning generally and it had no public function. Nevertheless it was to play much the same role in John Dalton's life that the Royal Institution did in Davy's. Dalton[9] (1766–1844) was, like Davy, of humble origin. He was by vocation and interest a meteorologist and a teacher of mathematics before (and indeed after) he was a chemist. He joined the 'Lit & Phil' in 1793 and it was in its rooms that he worked out the atomic theory on which all subsequent chemistry and much physics was to be built. As a meteorologist he had puzzled over the problem: Why do the two main gases of the atmosphere – oxygen and nitrogen – remain mixed at all levels? They should separate out, like oil and water, with the heavier gas, oxygen, at the bottom. (This question, it should be remembered, could hardly have been asked before Lavoisier had formulated his theory of chemical elements.) A chain of reasoning, clear enough in outline but still obscure in some details, led him to the concept of relative atomic weight. Realizing, or perhaps being told of, the importance of this concept he put aside his meteorological work and, from 1803 onwards, devoted himself to working out the chemical implications of his theory. After this was done he returned to meteorology. No one could have predicted that the next major advance in chemistry following Lavoisier would have come about through an interest in meteorology on the part of a teacher of mathematics in an unlikely place such as Manchester. It is most probable that Dalton was greatly helped by his friends and fellow-members of the 'Lit & Phil', Thomas Henry and his son William, chemical manufacturers of Manchester. The Henrys, according to Drs W. V. and K. R. Farrar and Mr E. L. Scott, may well have been the first in Britain to operate the Leblanc process, which was later to become such a feature of the chemical industries of the Manchester area.

For most people in England, however, the most conspicuous technological advances were the new and improved roads, the bridges and the canals that the new breed of civil engineers, the disciples of Smeaton, were building. They were there for all to see. British roads had improved slowly under the numerous Turnpike Acts and then, from the end of the eighteenth century onward, improved rapidly with the work of John Metcalf and J. L. Macadam. They built roads that were well ballasted, surfaced and drained and as a result the journey time between the main towns and cities was sharply reduced. More important than Metcalf and Macadam was Thomas Telford (1757–1834), for he was a civil engineer with very wide interests and an innovative spirit. A Scotsman, he effectively began his career when he was appointed County Surveyor in the English county of Shropshire. In 1805 he was faced with the problem of carrying the Ellesmere canal across the steep valley of the river Dee at Pontcysylltau. He solved the problem by constructing a trough of cast iron plates about three hundred metres long and supported on masonry arches about forty metres high. The towpath was also made of iron plates, supported by iron struts. Telford had the benefit, in carrying out this work, of close proximity to the leading ironworks of the day: Coalbrookdale, Ketley and Bersham (John Wilkinson).

Telford, in fact, was the first to realize the structural possibilities of iron. The problems of building iron-framed mills and warehouses were comparatively simple; bridges presented a wholly different set of challenges. In 1814

Telford was commissioned to rebuild the politically important London to Holyhead road, the shortest route between London and Dublin. In the course of this work he built the elegant Waterloo bridge (1815) at Betws-y-Coed. This was one of his first bridges to make full use of the qualities of iron as a structural material. No theory could establish the optimum form for the bridge. Telford, like Smeaton confronted with the best form for the Eddystone lighthouse, had to rely on his intuition. Unlike Ironbridge, the Waterloo bridge was slim, making full use of the strength of iron; the arch forming a segment of a circle of considerable diameter and the deck being comparatively flat. The Holyhead road project also involved the construction of a large harbour at Holyhead and carrying the road over the tidal Menai Straits. His suspension bridge over the Straits, completed in 1826, is a magnificent achievement[10]; still in use, it now carries motor traffic. It, too, is about three hundred metres long and over thirty metres above the surface of the sea. This time the main structural material had to be wrought iron. The links of the chains supporting the roadway were flat plates, bolted together. For these reasons Professor Billington has described Telford as the father of structural engineering. The evaluation is authoritative.

By the time Telford was engaged on the Holyhead road he had become the most distinguished civil engineer in Britain. Smeaton's informal society had not developed as might have been expected (cf. the Lunar Society) and the need for a formal institution was beginning to be felt. In 1818 the Institution of Civil Engineers was founded and two years later Telford was invited to become its first President. For many years the Institution included what would now be called mechanical engineering within its range of interests. In recent years a new town has been named after Telford, a unique distinction, at least in Britain.

An alternative technique to Telford's use of chains for

the Menai bridge was pioneered by Marc Séguin (1786–1875), an able and versatile engineer who was a great-nephew of Joseph and Etienne Montgolfier. In 1822 he had built a small foot bridge in which suspension was by wire ropes and not by chains. Professor Gillispie points out that Séguin never explained how he hit on the distinctive idea of substituting wire rope in place of chains. He did not invent wire rope; most probably it originated in the German mining industry. One clear advantage was that all the metal used would be in tension. With chains the horizontal components, or bolts, add to the weight without directly supporting the load. Wire ropes have progressively replaced chains for all suspension bridges.

Civil engineers in Britain were trained as apprentices to established masters. There were, in Telford's day, no institutions comparable to those of France and other European countries. In 1799 the Berlin Bauakademie was founded; it was followed by the Prague Polytechnic in 1805 and the Vienna Polytechnic in 1815. But the most remarkable educational foundation was Wilhelm von Humboldt's University of Berlin, established in 1810 as a gesture of national defiance following Napoleon's defeat of Prussia at the battle of Jena. The widely proclaimed ideals of the French Revolution had been enthusiastically accepted, particularly by the young, all over Europe. The subsequent invasion by Napoleon's armies had aroused other emotions, notably strong feelings of affronted nationalism. These were the two forces that were to create a unified Germany well before the end of the century. The revival of German science and technology can be taken to date from von Humboldt's gesture. Before long Germany could reclaim her leading place, first of all in science, then in technology.

Roads, Railroads
and a New Philosophy of Power

From the middle of the eighteenth century to the first
decade or so of the nineteenth century the British textile
industry was one of the most important strategic industries
on record, for it initiated the Industrial Revolution. Its rapid
growth caused it to push service and related industries and
technologies to the limits of their capacities, its insistent
demands stimulating invention in the related technologies.
Structural engineering, power generation and
transmission, machine tools, gas lighting, chemical
industry and transport all advanced under pressures from
the textile industry. The first modern elevator or lift was
the teagle, a by-product of structural engineering made
possible by multi-storey textile mills with power laid on.
The later progress of the industry itself was less influential.
Textiles had much to tell the world of the advantages of
organization, of mechanization, of mass production and
of marketing. These were all in the field of industrial
management. The basic principles governing the Jacquard
loom were, much later, to find application in a quite
different field. Apart from the introduction of the
differential this was the most widely influential innovation
in textile technology.

The mining industry, which in medieval and
Renaissance Germany had been an exemplary strategic
industry and technology, continued to benefit other
industries when it fostered the birth and development of
steam power. And, at the beginning of the nineteenth

century, it offered another benefaction when the railroad, so long in gestation, and the steam locomotive transcended the boundaries of the mining industry. British roads, evolved and designed for horse-drawn traffic, were unsuitable for heavy and relatively fast steam-propelled vehicles. Accordingly, after the technical, social and legal problems of public railroads had been solved, rail locomotives could begin to challenge the domination of the horse in land transport and dispose of the brief competition of the road locomotive. The canal builders were, in many ways, trail-blazers for the railroad.

Cugnot's road locomotive may have been a more feasible proposition on the fine, straight roads of France. But for many years progress in land transport was limited to horse-drawn vehicles. The scope for improvement was, initially, wide. There were regular coach services on the roads of France before they started in Britain. In August 1784, a regular coach service between Bath and London was instituted to carry mail together with a strictly limited number of fare-paying passengers. Receipts from the fares helped to keep down the cost of carrying the mail. Passengers travelling inside the coach paid more than passengers travelling in the open, on top; in both cases the fare charged was so much per mile. Before long the country was served by an extensive network of mail coaches, well organized and running strictly to timetables. As the roads improved so the speeds of the coaches increased, until by the third decade of the nineteenth century an average speed of nearly ten miles an hour (sixteen kilometres an hour) over long distances was maintained on the most important routes. Although the mail coach is now an object of vicarious nostalgia, a subject for greetings cards, in its day it represented – and was widely accepted as – a thoroughly modern technical advance and a clear proof of social progress. Jane Austen's young ladies could travel by coach safely and unaccompanied; fifty years earlier such a thing would have been unthinkable. However, the limitation of this particular advance in communications technology was ineluctable. An average speed of sixteen kilometres an hour was close to the absolute maximum possible, given the strength, endurance and speed of the most suitable horses. Over the centuries some horses had been bred to run as fast as possible, others to carry or pull the greatest loads and yet others to carry a rider as far as possible in all weathers. The limitations of the species were well known. The stage coach system represented the final perfection of a technology.

On the other hand, the crude steam locomotives designed and built by colliery mechanics and millwrights in the north-east of England represented a technological innovation whose enormous future potential could be grasped by the far-sighted. William Hedley, who built the famous *Puffing Billy* locomotive, Timothy Hackworth and George Stephenson (1781–1848) with his son Robert (1803–59) transformed Trevithick's idea of a steam locomotive running on iron L-plates into the railroad system that was to change the world. The first railroad to carry passengers was the Surrey Iron Railway, opened in 1803; the carriages were drawn by horses. In September 1825 the Stockton and Darlington Railway was opened. It was about 28 kilometres long, passengers as well as freight were carried and locomotives were used as well as horses. The engineer responsible was George Stephenson and the first of two locomotives he designed and built for it – *Locomotion* – has been preserved. It had two vertical cylinders mounted in the boiler, and an elaborate linkage, similar to Watt's parallel motion, connected the piston rods to the driving wheels. To increase the rate of transfer of heat and hence the generation of steam, the chimney was connected to the firebox by a large tube that passed through the boiler. Exhausted steam was blown up the chimney to draw hot air through the tube and to improve combustion in the firebox. Two further points of interest about the Stockton and Darlington Railway are that wrought iron as well as cast iron rails were used and that the gauge Stephenson adopted for it – 4 feet 8½ inches – was to become the standard gauge all over the world. At the same time that the Stockton and Darlington Railway was being planned and built a scheme was under discussion for a railroad between Liverpool, a rapidly growing and important port, and Manchester. Although the distance was only about fifty kilometres the roads and the canal linking the port with the textile mills in and around Manchester were unable to cope with demand, such had been the growth of the cotton industry. After much discussion and a great deal of opposition from interested parties, Parliament approved the proposal for a Liverpool and Manchester Railway. The question then arose: What means of haulage should be used? Horse haulage was seen to be inadequate. Initially, the favoured means was cable haulage, powered by stationary steam engines along the track. Cable haulage had its merits. It survives to the present day in the famous San Francisco system of cable-hauled street-cars: a clear demonstration of the versatility and reliability of the system, particularly in very hilly places. The land between Liverpool and Manchester is, however, flat so that the distinctive advantage of cable haulage had no force. The success of the Stockton and Darlington Railway and the reputations of Stephenson and his gifted son Robert were enough for the proprietors to decide in favour of locomotives.

The proprietors were nothing if not fair. The project had aroused enormous interest and the only way to decide the issue was to hold a competition. The Rainhill (near Liverpool) trials of October 1829 were between four steam locomotives and one oddity. The locomotives were Braithwaite and Ericsson's *Novelty*, Hackworth's *Sanspareil*, the Stephensons' *Rocket* and Burstall's *Perseverance*. The oddity was a horse-propelled contraption of interest to connoisseurs of eccentricity.

Rocket, designed and built by George and Robert

Stephenson, was the winner. It was the only entrant to meet and, indeed, to exceed the stated requirements. The issue was decided: the Liverpool and Manchester Railway was to be worked by steam locomotives. *Rocket's* boiler had 25 copper tubes running from the firebox to the chimney; this greatly increased the surface area exposed to the heat of the fire and hence increased the rate of generation of steam. As with *Locomotion* a steam blast was incorporated. Outwardly, at least, the most distinctive feature was the positioning of the two cylinders. These were on either side of the boiler and at an angle of about 45° to the vertical. The piston rods were coupled to connecting rods that drove the two large front wheels. Vibration of the piston rods was prevented by the use of parallel guide rods with cross-heads. This was a far simpler and more effective arrangement than previous linkages and no doubt helped to account for *Rocket's* phenomenally high speed (about 50 km/h). What is clear is that the Stephensons must have used the new machine tool – Robert's planing machine – to make the guide rods. Guide rods with sliding cross-head remained a feature of all subsequent steam locomotives.

The first two locomotives to reach France were imported in 1828 by Marc Séguin where they were to be used on the Lyon–St Etienne Railway[1]. Three years earlier, in 1825, he had founded a company – Séguin, Dayme, Montgolfier & Cie – to operate steam boats on the Rhone. In order to buy the most suitable steam engines he toured England where he met, among others, Maudslay, Babbage, Brunel and Stephenson. Séguin was most impressed by the new Stockton and Darlington Railway. Experience with the steam boats, back home in France, showed that the boilers could not deliver enough steam to develop the full power of the engines. This, he knew, could be remedied by increasing the area of hot metal exposed to boiler water and, he realized, the best way to do this was by carrying the hot air from the firebox through metal tubes traversing the boiler. In order to ensure an adequate flow of hot air through the tubes he used fans to force air into the furnace. The invention was tried out on a locomotive he built for the Lyon–St Etienne Railway before trying it on a marine engine.

Séguin's invention of the fire-tube boiler, patented in February 1828, seems to have pre-dated Stephenson's invention by several months. That Séguin used a forced draught to ensure a flow of hot air through the tubes instead of Stephenson's exhaust steam blast tends to confirm that the inventions were made independently; although quite possibly both were consequences of discussions of a problem of mutual interest that Séguin and Stephenson may have had in Newcastle. Séguin's forced draught was, much later in the century, adopted as common practice in marine engines when compound engines working with high-pressure steam became commonplace; it was never used in locomotives where the exhaust steam blast remained in use until the end of the steam era.

It is not known how many witnessed the first, clumsy working of the Newcomen engine or what their reactions were. It was quite different with the opening of the Liverpool and Manchester Railway on 15 September 1830. The great national hero and much-disliked politician, the Duke of Wellington, attended, as did Mr Huskisson, the Foreign Secretary. There was an impressive assembly including many members of the nobility with their ladies, several bishops, Members of Parliament and people of distinction from all walks of life, among them Miss Fanny Kemble, the popular actress, and Mr Charles Babbage, Lucasian Professor of Mathematics at Cambridge; there were also huge crowds of ordinary folk. The fuss was entirely justified. Unlike the stage coach the railroad had enormous potential and this must have been sensed by the majority of those present.

The Liverpool and Manchester Railway was the model

for all subsequent railroads. All passenger and freight traffic was hauled by steam locomotives. There were two tracks: one for east-bound, the other for west-bound trains. Railway stations were built for the convenience of passengers. (The Manchester terminus of the line now forms part of the Manchester Museum of Science and Industry as does the fine 1830 goods shed on the opposite side of the tracks to the station.) Passenger trains ran to a timetable and there were three classes: first, second and third, in diminishing order of comfort and price. Fares were charged at so much per mile. This, too, with local and national variations, was to be the pattern for railroad development everywhere.

The Stephensons must have the major share of the credit. As we have seen, the individual components of the railroad, considered as a system, had been assembled over the previous fifty years or so. The canal builders had mastered the techniques of drilling tunnels, building up embankments and digging out cuttings. They had established the legal precedents for compulsory purchase of land and they had learned how to muster and manage large bodies of skilled and unskilled men. The millwrights, blacksmiths, mechanics, enginewrights and coal-viewers of the mining industry had invented and developed the locomotive. The stage coach system had popularized the idea of public passenger transport at so much per mile and the discipline of the timetable. These components had been progressively refined. What was required was the ability to put them all together and weld them into the public railroad so very different from the little mine trackways with their rough, primitive steam locomotives, 'iron horses'. The Stephensons came to realize that the railroad could extend to working people a luxury that only the affluent had been able to afford. Their vision of a rail network to provide ordinary folk with a cheap, fast, safe and comfortable means of travelling wherever they wanted to go amounted, when combined with their practical abilities to bring it about, to the achievements of genius.

Robert Stephenson, whose education included attending Edinburgh University, was a better trained and, indeed, a better engineer than his father. Over the years 1831–32 he systematically improved – perfected – the components of the steam locomotive, just as Smeaton had perfected the Newcomen engine, with the result that *Planet* (see plates), when completed for the Liverpool and Manchester Railway in 1831, was the prototype from which all subsequent steam locomotives descended.

The triumph of the steam-powered railroad from 1831 onwards was rapid and complete. The advent of the steam ship was far more hesitant, although experimental vessels had been built before Trevithick's Pen y Darren locomotive. There had been an experimental steam ship in France even before the Revolution and, in 1801, William Symington's *Charlotte Dundas*, a small steam tug fitted with a paddle wheel, was tried out on the Clyde–Forth Canal. It was said that the wash from the paddle wheel damaged the sides of the canal so there was no further development. However, Robert Fulton (1765–1815), who had witnessed the trials of the *Charlotte Dundas*, took the idea to the United States where he built the steam boat *Clermont*, powered by a Boulton and Watt engine. Plying between New York and the state capital, Albany, a distance of about 220 kilometres, the vessel was a commercial success and was soon followed by others. The first commercially successful vessel in Europe was Henry Bell's *Comet* of 1812; it operated on the Clyde.

As the steam engine requires a continuous supply of, preferably, fresh water, it is hardly surprising that the first successful steam boats operated on big rivers or fresh-water lakes; and this, in turn, meant that North America with its great river systems and huge lakes was to provide the widest scope for the development and utilization of the early steam boat. Travel by steam boat was relatively fast, comfortable and safe. And fuel, in the form of wood, was freely available on the river banks or lake sides. By 1824 a steam boat had reached St Anthony's Falls, as far as possible up the Mississippi, and the site of the city of Minneapolis.

The difficulty of applying the steam engine to sea-going ships was simply that of the supply of fresh water. This might not have been a serious problem on important cross-Channel routes such as Dover–Calais and Holyhead–Dublin. Watt, we remember, had originally specified a surface condenser for his steam engine but had given it up, together with the steam jacket, to save the extra cost. On a sea-going steam ship a surface condenser, cooled by a flow of sea water, would enable the boiler water to be continuously recycled. The development of such a condenser for sea-going ships proved difficult. Apart from the stresses that a marine engine was bound to undergo in rough weather, the problems of corrosion and differential expansion of the metals used in the condenser were difficult to resolve. Not until the middle of the century was a reliable surface condenser developed for marine engines. In the meantime sea water had to be used in the boilers which, consequently, had to be flushed out regularly to prevent excessive deposits of salt. This was clearly inefficient and shortened the life of the boilers. Apart from ferry steamers there was an incentive to develop steam ships for the intensive trans-Atlantic trade. Even on the routes to North and South America steam ships were barely economical and when they were introduced a subsidy was paid in return for carrying the mails. Elsewhere, the advantages of the sailing ship were plain enough. The wind was free. Ports world-wide could provide everything needed for a sailing ship, including skilled seamen waiting for berths. How many ports could provide the engineering services of London, Glasgow, New York, Boston, Cherbourg or Le Havre? A sailing ship, dismasted in heavy weather, could make port under a jury rig; what would happen to a steam ship whose engines broke down in mid-ocean? No wonder all steam ships continued to carry sails until the last decades of the century.

A New Technology and a New Science

The first three decades of the nineteenth century were a time when, beginning in France, science was organized into something like the structure we know today[2]. Textbooks on physics included, in addition to mechanics, light, sound, electricity and magnetism and heat, the last being taken over from chemistry as ways were found of expressing heat phenomena mathematically. This owed something to the Scotsman John Leslie and more to the Frenchman J. B. J. Fourier. The new physics was both narrower and deeper than the old, post-Aristotelian, physics, which was concerned with the study of movement and change and therefore included biology and kindred sciences. In Britain, of course, natural philosophy, experimental philosophy and chemistry continued to be recognized as the main branches of physical science. Men of science were commonly referred to as philosophers and it was not until 1839 that the word 'scientist' was coined; by analogy, it was said, with the word 'artist'. Many decades were to pass before the word 'scientist' was commonly accepted.

As we saw, the development of the steam engine had benefited greatly by the fortuitous relationship between James Watt, the *quondam* instrument maker, and Joseph Black. While the study of heat remained, at least in Britain, the province of the chemist, another such partnership was effectively ruled out. The steam engine in the country of steam engines (Britain) was regarded as a steam *pressure* engine and any thought given to the theory of the machine was concentrated on saving, optimizing and applying steam pressure. The example of the Black–Watt relationship was forgotten. It was not until 1828 that Captain Samuel Grose rediscovered the importance of preventing heat losses. He found that if he insulated – 'clothed' as he put it – the steam pipes, valves and cylinder of the engine at Wheal Towan mine in Cornwall the average duty was raised to '80 000 foot pounds per bushel' of coal, three times the best that Watt thought was possible (26 000 foot pounds per bushel). The performance, publicized in the *Monthly Engine Reporter*, was noted by other engineers and the benefits of heat conservation recognized. Nevertheless this did not imply any theoretical understanding of the basic processes of the steam engine such as Watt had possibly glimpsed.

When peace returned to Europe in 1815 French observers were deeply impressed on discovering the remarkable progress made by British industry. Although Boulton and Watt engines had been imported into France before the Revolution they had been confined to a few manufacturing establishments, remote from public attention and interest. The long years of revolution and war had concealed the rapid development of the steam engine from the attention of French and other Continental engineers. The greater, then, was the impact of the radical new technology after 1815. This was made quite apparent by the success of the Woolf engine in France. As Professor Fox has remarked, it must have been something of a blow to the pride of the well trained and able French engineers to find that such progress had been made by scientifically uneducated British craftsmen. They did not, however, allow any chagrin to overcome their intellectual curiosity. In fact, Cornish engineers were to complain, with some justification, that their engines were better understood in Paris than they were in London.

The most important of these French observers, although not recognized as such for many years, was N. L. S. Carnot (1794–1832). The son of Lazare Carnot, Sadi Carnot[3] was educated at the Ecole Polytechnique and in due course became an officer in the Génie. Opportunities for promotion in the times of peace and demobilization after 1815 were slim, more particularly for the son of an exiled regicide. Young Carnot was, therefore, posted far away from Paris to frontier districts where he could cause no mischief; and he was allocated routine, repetitive jobs such as resurveying bridges. In due course, bored with a dead-end post, he went on half-pay and returned to Paris where he occupied himself studying economics and, of course, the steam engine. In 1824 he produced his short book of just over a hundred pages, *Reflections on the Motive Power of Fire.*

Seen from the French perspective the steam engine had undergone a sudden increase in efficiency and versatility that, Sadi Carnot reasoned, demanded explanation. There was, he argued, a complete theory of water power; a theory that made it possible to calculate the work available from any flow of water and that showed conclusively how to develop that work with the maximum efficiency. Apart from isolated, *ad hoc*, investigations, such as measurements of the increase in the pressure of saturated steam with the temperature, there was no comprehensive, complete theory of steam power similar to that of water power. It was not even known whether the power of steam was finite (here, he was plainly referring to the hopes that the rapid increase in pressure with temperature presaged virtually unlimited power for a finite amount of heat, or fuel).

Several things operated in favour of Carnot's investigation. Like the young Watt, he was not engaged in the steam engine business; he was therefore free of occupational prejudices and able to take a detached, outsider's view. As a well trained engineer he was familiar with the theories and practices of hydraulic power technology. He had immediate access, through his friends Clément and Desormes, to their work on heat and on the expansion of gases. Lastly, it was fortunate, paradoxical as it seems, that he accepted the conventional theory that heat was due to a tenuous fluid called caloric. These factors, taken together, enabled him to frame a theory of heat power on the limited analogy of water power.

Heat, he observes, taking a cosmological view, is the grand moving agent of the world. It powers the great wind systems and the ocean currents. The universal requirement is that wherever there exists a temperature difference there is the possibility of generating motive power. We have learned how to harness this power; thanks, he generously admits, to the genius of British engineers like Watt, Trevithick and Woolf. Now, likening a temperature difference to a head of water, he argues that the greater the temperature difference – hence the word 'fire' in the title of his book – the greater the motive power for the same quantity of heat: water falling twenty metres will do twice the work of the same quantity of water falling ten metres. The analogy here was between water and caloric (by fitting in an air reservoir the expansive principle had been applied to water pressure engines: this could only strengthen the analogy). Support for Carnot's inference came from the experience that high-pressure steam engines were more efficient than low-pressure ones and high-pressure steam was hotter than low-pressure steam.

The conditions for the maximum efficiency of a water power machine – a water-wheel or a column-of-water engine – are that the water should enter without shock and leave without velocity. It should fall out exhausted, as it were, having given up all its effort to the machine. Similarly, with any heat engine – using steam, air or any other 'working substance' – the heat or caloric should enter at the same temperature as the working substance and leave at the same temperature as the condenser, or exhaust. If there is a difference in temperature at either of these stages it follows that an opportunity to generate motive power has been missed. At this point we should note that Carnot, unlike his contemporaries who thought of steam engines as pressure engines, was putting forward the idea of the *heat* engine as a general concept covering all engines that worked by the agency of heat.

A perfect heat engine would work like this: The steam or air in the cylinder would be at furnace temperature and fully compressed. It would then expand, pushing the piston and absorbing heat from the source, or furnace, through the base of the cylinder so that the working substance remains at the same temperature. At a certain distance the supply of heat is cut off and the working substance continues to drive the piston out, but now, deprived of the supply of heat, the temperature falls[4]; and continues to fall until the working substance reaches the temperature of the condenser or exhaust. At no point is there any flow of heat between the cylinder walls and the working substance (which postulates a cylinder made of a rather unusual material). At the low temperature, the working substance is compressed so that it gives out heat through the base of the cylinder, remaining at the same temperature until, at a predetermined point, the base of the cylinder is insulated. Compression continues so that the temperature of the working substance must rise until it and the pressure are the same as at the beginning.

All that has happened in this *cycle* is that a certain amount of heat has flowed from a high temperature to a low temperature and work has been done. No heat has leaked away, uselessly; nothing has been heated up or cooled down, expanded or contracted. All is as it was at the beginning. An engine working like this will be the most efficient possible. This can be proved quite simply, for an engine working in this way would be reversible. It could be driven backwards and *all* the heat restored to the source with no other change apart from the performance of work. No engine can be more efficient; for if a better one could be made, then, by using it to drive the reversible one backwards, heat could be accumulated in the source with no other net change. And the accumulation of heat in the source would make a perpetual motion machine possible. A parallel argument shows that a water engine better than a perfectly reversible one is impossible.

From this argument Carnot deduced that no working substance can be superior to any other; for if one could be found then perpetual motion, or the unending performance of work without the consumption of fuel of any kind, would be possible. Further, he drew certain conclusions that were important in the physics of gases, but these do not concern us here. In conclusion he offered brilliant insights into the operation of all heat engines. He pointed out that the excessive rise in pressure of steam as the temperature increased was a practical disadvantage, for it meant that it would be impossible to work with steam at the same temperature as the furnace (about 1000°C). Instead, he suggested that the air engine offered better prospects for the really efficient exploitation of heat power: the pressure of air at 1000°C would not be disastrously high. He was able to reach, by a curiously erroneous argument, the correct conclusion that the same amount of heat 'falling' the same amount through a perfect engine would do less work the higher the ambient temperature; that is, for example, less work between 100 and 99°C than between 1 and 0°C. From the fact that the work done by a perfect heat engine depends only on the temperature drop and is independent of the working substance, he hints that it might provide the basis for an absolute scale of temperature, one independent of the properties of mercury, or alcohol, or air.

Carnot had drawn on the well proven practice of expansive operation and on the work of his friends Clément and Desormes to set out his ideal cycle. His conclusion that the superiority of high-pressure operation was due to the fact that high pressure meant high temperature was, essentially, a circular argument as it was not derived from independent considerations. There was, in reality, little that Carnot could tell practical engineers.

They had already concluded that no working substances were superior to steam, they knew the practical drawbacks of the air engine and the better informed of them were aware of the superiority of the high-pressure steam engines. There were three sound reasons why this last should be so: high-pressure engines tended to be newer, they were more compact so frictional losses were reduced and, thirdly, the higher the pressure the more effective expansive operation becomes. Even his contributions to the physics of gases could be – and were – deduced on other grounds. Carnot himself seems to have lost faith in his work shortly after it was published. He became convinced that the caloric theory of heat was wrong and that therefore the whole basis of his argument must be rejected; perhaps he thought himself a failure. He published nothing more and died, tragically, in a lunatic asylum. His must have been a sad life; and yet events were to show that nothing more fruitful was published in the 1820s – a very creative decade – than his *Reflections on the Motive Power of Fire*.

It is ironic that, by 1816, an obscure Scottish clergyman, the Rev. Robert Stirling, had already patented an air engine that was certainly reversible and could, in principle, be made perfect by Carnot's criteria[5]. A long cylinder, closed at one end, was fitted with two pistons, the outer one being the working piston; the inside, bulky, one was the 'displacer' piston. The closed half of the cylinder was within the furnace, the outside half was kept cool by the surrounding air or by water. Heated air drove the working piston out, rotating a fly wheel, while at the same time a mechanical linkage caused the displacer piston to move in the opposite direction. This pushed the hot air outward through narrow strips of metal running along the inside to the cylinder (the diameter of the displacer piston was less than that of the inside of the cylinder). The air gave up its heat to the metal strips (called the heat exchanger, or regenerator) until – in theory, at least – it was at the same temperature as the outside air. At that point the displacer had reached the closed end of the cylinder, the pressure was low and uniform, so that the fly wheel could carry the displacer out again, the working piston could go back towards the closed end, the air, passing back through the heat exchanger, could pick up some of the heat it had deposited there and the next cycle could begin.

The Stirling engine was before its time. Metallurgy was not sufficiently advanced to deal with the extreme conditions to which the cylinder and other components were subjected, nor were suitable lubricants available. Nevertheless it was worked successfully, for a time. Some 35 years later William Thomson[6] (Lord Kelvin) drove one in reverse to act as a refrigerator and freeze water. As for Stirling, he, like Carnot, had his disappointment. As Professor Daub points out, it may well be that his devoted labours in his parish during a cholera epidemic prevented him from continuing with the development of his engine. He had, at least, a vicarious triumph, for his son Patrick became one of the best known of British locomotive engineers.

We do not know how Stirling came to invent his engine. And we have no evidence of any connection between Carnot and Stirling. Neither man, it must be stressed, was a member of the community of steam engineers. Both, particularly Carnot, were concerned with the operation of *heat engines* and not just *steam engines*. And both died in obscurity.

F. M. G. de Pambour took a different approach[7]. Apart from numerous papers he published two books, one on the steam engine, the other on steam railways, that were respectfully received in Britain, where they were granted the accolade of translation into English. De Pambour, who had carried out experiments on the Liverpool and Manchester locomotives, argued that the steam pressure in the boiler was no guide to the work done by the locomotive. Steam pressure in the cylinders must, if the locomotive is travelling at a uniform speed, exactly equal the load it is overcoming. If the pressure is greater the locomotive must accelerate; if it is less it must slow down. The argument here is exactly the one that was familiar to hydraulic engineers since Parent's time. The problem for de Pambour was to calculate steam pressure under different conditions. And it would be fair to say that for him a complete theory of the steam engine would be one that gave steam pressure at every point and at any time in the working of the engine. This, engineers of the time must have thought, was the right mixture of theory and practice; the correct approach to the problems of the steam engine. Put like this the implausibility, to the contemporary mind, of Carnot's idea of a perfect engine, following an impossible routine in some Platonic heaven, was obvious.

Nevertheless, Carnot's theory of the heat engine, derived from power technology, had provided the scientific study of heat with what was eventually recognized as a fresh and powerful impetus when, in other respects, the science had become static, perfect within the limits of Fourier's analytical theory. In other words, power technology was to lead to the development of a major new science, that of thermodynamics.

The Perfection of Water Power

A vertical column of water represents so much energy that can be converted into useful work by two methods. It can be used to drive a machine like the column-of-water engine – the water pressure engine as it was usually called in Britain – or, rushing out in the form of a jet, its *vis viva*, or kinetic energy, can be used to drive an impact, or impulse, water-wheel. By the beginning of the century it was generally recognized that the second method was inefficient as power was wasted in generating turbulence; but it had the potential merit that, if such waste could be eliminated, it could readily drive (relatively) high-speed machinery without the use of expensive and energy-wasting gears.

In 1824, the year of Carnot's *Reflexions sur la puissance motrice du feu*, Prony and Girard reported favourably to the Académie Royale des Sciences on a memoir of C. Burdin entitled 'Hydraulic Turbines; or High Speed Rotative Machines'. Burdin, a Professor of Mechanics at the St Etienne Mining School, analysed the conditions under which the maximum work can be obtained from a rapid flow of water parallel to the axis of a machine with suitably curved blades. The idea of using curved blades, so that the water could enter with minimal shock, had occurred to de Borda; Burdin had developed the idea and had given a name to such machines. He had built a prototype at St Etienne and claimed to have got promising results from it but had not yet got enough data. If the blades of Burdin's machine were so curved that the water left in a direction exactly opposite to the direction of rotation, then the outflowing water must strike the leading side of the following blades; if, on the other hand, the blades were arranged to avoid this by directing the stream downwards the water would leave with appreciable velocity and the machine would be less efficient. In 1827 Benoit Fourneyron (1802–67), resolved this dilemma by arranging for the water to flow outwards from the machine. He was able to build and run the first admittedly successful turbine[8] although it was some time before engineering workshops could build high-speed turbines for public use. The Fourneyron outward flow turbine worked partly by the pressure of the water on the, suitably curved, blades and partly by using the reaction resulting from the acceleration of the water moving along the blades. Ideally, the velocity of the water as it left the runner, or wheel, should have been equal and opposite to that of the wheel; in practice it never was; it was always slightly greater so that some energy ran to waste. Prony and Girard were sufficiently impressed to observe that: 'Trained engineers like [Burdin], in favourable circumstances, have the duty to apply the principles of rational mechanics to the improvement of industrial processes, and to prove the advantages of theory by useful applications which they can make in practice.' At much the same time J. V. Poncelet described an undershot water-wheel in which the blades were curved, pointing upstream; he claimed an efficiency of 67 per cent.

Shortly before Fourneyron had built his first turbine Riche de Prony (1755–1839) described his friction brake, or dynamometer (1826). This extremely simple device measured the power output of rotative steam engines by balancing the work it did overcoming the friction of the brake against a weight at the end of a long lever arm. It was not even necessary to know the force of friction to calculate the power output at a given speed. The Prony dynamometer could be used to measure the power output of turbines, water-wheels, and all other rotative hydraulic engines. Watt's indicator could not, of course, do this. It is most improbable that de Prony knew of the indicator; its existence, as we saw, was kept secret until 1824.

Almost simultaneously, two fundamental advances in power technology, one theoretical and the other practical, had been made in France. At the same time, and again in France, a simple and efficient way of measuring power output had been devised. Unlike von Guericke and Papin's ideas, the hydraulic turbine was not taken up in Britain for many years: cheap coal was more plentiful than great falls of water. In America it was another story.

The Beginnings of Electrotechnology

Oersted's discovery was extended, particularly in France, as rapidly as the invention of the pile had been. Among the many discoveries made was that of D. F. J. Arago (1786–1853), who, in the same year (1820), showed that soft-iron filings became temporarily magnetized if placed near a wire carrying a current, while a steel needle became permanently magnetized if put inside a 'solenoid' – a long wire wound round and round to form a tube or cylinder – carrying a current. This suggested that magnetism might be due to electricity associated with the atoms of the steel needle. In 1825, William Sturgeon (1783–1850), an old soldier who had educated himself while serving in Wellington's army, invented the soft-iron electromagnet, based on Arago's experiment. The recognition that a solenoid with a soft-iron core could provide a more powerful magnetic attraction than any permanent or natural magnet, or lodestone, and that the attraction would vary immediately with the energizing current, becoming reversed if the connections to the poles of the battery were reversed and vanishing the moment the current was stopped, constituted an extremely fruitful and original invention.

Sturgeon invented the electromagnet as a teaching aid, to help him in lectures on the electromagnetic phenomena discovered by Oersted, Arago, Faraday and others. The development of the electromagnet was carried on by Gerard Moll, a Dutchman, and Joseph Henry, an American. They improved the design and Henry introduced the now universal practice of insulating the wires themselves. Between them they increased the lifting power of electromagnets to 750 pounds and then to over 2000 pounds. This meant that a small teaching aid, a laboratory instrument, was effectively brought within the range of practical or engineering application. Although neither Henry nor Moll made extravagant claims, one is inevitably reminded of von Guericke's immensely suggestive proposals for harnessing the pressure of the atmosphere (chapter 5). One immediate consequence of the electromagnet was the invention of the first electric motors (or electromagnetic engines as they were called). In this machine the magnetic attraction between a fixed electromagnet and an electromagnet mounted on a spindle caused the latter to revolve. Half-way round each revolution the current flowing in one electromagnet had to be reversed by a commutator, otherwise the machine would stop dead. It is no exaggeration to say that all subsequent electrotechnology has been based, at least in part, on the electromagnet of Sturgeon, Henry and Moll.

The close analogies between electricity and magnetism suggested to several investigators that voltaic electricity could, in some way, induce electricity in a nearby 'circuit', or 'arc' as it was sometimes called. (Although there was still doubt in some quarters about voltaic electricity, we can, from now on, refer to an electric current.) In 1831 Michael Faraday[10] (1791–1867), at the Royal Institution in London, discovered electromagnetic induction. In a brilliantly conceived series of experiments he found that when a voltaic current started to flow in a wire it produced a brief transient current in a nearby circuit; when the voltaic current was stopped there was a brief current in the opposite direction in the nearby circuit. Systematically, Faraday explored this elusive phenomenon that could, on the face of it, have been due to a host of accidental factors of the sort familiar to all experimenters, today as in the past. Faraday determined the true principle of induction: a changing current induced a changing current; a steady current had no effect. He went on to show that rapidly moving a permanent magnet near a circuit produces exactly the same effect. It was not, therefore, the current *per se* that had caused the induced current in his first experiment; it was the changing magnetic effect due to the changing voltaic current. This, following Oersted, marked the very beginning of 'field theory'. It was all very un-Newtonian.

Faraday's discovery of electromagnetic induction led almost immediately to Hippolyte Pixii's invention of the magneto-electric machine, or magneto for short, by which an electric current could be generated mechanically. In its original form a horseshoe magnet was spun rapidly close to a coil of wire. Two improvements soon made were to spin the coil rather than the magnet and to incorporate a commutator to convert the alternating current from the magneto into a direct current.

The inventions of the magneto and the motor laid the foundations for the development of electric power technology and much else as well. Sixty years, however, were to elapse before electric power began to be socially and economically important. The reasons for this long delay will be discussed below. For the present it is sufficient to note that the magneto was a relatively feeble affair. It was believed, plausibly enough, to work by converting magnetism into electricity. The strength of the best permanent magnets was small, while there seemed to be no limit to the power obtainable from voltaic batteries and no perceived limit to their capacity for improvement. The electric motor together with the voltaic cell seemed to offer prospects of power that might well rival and even displace the steam engine. The actual course of development of electrotechnology was, however, to be one that few would have predicted.

Oersted's simple apparatus constituted an instrument for the detection of an electric current. The sensitivity of the instrument was greatly increased when S. C. Schweigger, in 1822, substituted a coil of wire of many turns round the magnetic needle for Oersted's single wire. Schweigger's 'multiplier' was soon called the 'galvanometer'. And this instrument together with Sturgeon's electromagnet brought the electric telegraph, so long on the register of the possible, much closer to realization. In 1832 Baron Schilling had a telegraph system using galvanometers and in the following year J. C. F. Gauss (1777–1855), and W. E. Weber (1804–91), used much the same system to link the University of Gottingen to the Magnetic Observatory that operated as part of a great international network of similar observatories. The line was about one kilometre long and communication was by means of a code of signals. This was the first practical electric telegraph. A further advance came in 1838 when C. A. von Steinheil described the principle of the earth return, which meant that only one wire was needed to activate a distant instrument. Steinheil had his own telegraph line from his house in Munich to the Observatory at Bogenhausen, a distance of about ten kilometres.

The development of an extensive telegraph network could hardly depend on the requirements of observatories and universities. There were only two agencies that could spread telegraph lines over a whole country. One was military authority; the other was the growing network of railroads. These were the only two agencies to whom instantaneous communication would be invaluable and, at the same time, who could afford to finance it. As with the British canal system in the eighteenth century, the great success of the Liverpool and Manchester Railway was followed, without appreciable delay, by the construction of other railroads: the London and Birmingham, the Manchester and Birmingham, the Great Western towards Bath and Bristol and masterminded by Isambard Kingdom Brunel (1806–59). The rapid spread of railroads heightened the demand for the fastest means of communication along the lines. In 1836 W. F. Cooke (1806–79), travelling in Germany, saw Schilling's telegraph and took the idea back with him to London. There, in collaboration with Charles Wheatstone (1802–75), Professor of Natural Philosophy at King's College, London, he designed a new type of telegraph for railroad use. Their first demonstration telegraph used five wires to activate five electromagnets in such a way that one pair of five compass needles, over the electromagnets, could be made to point to each one of twenty letters on a vertical board, the letters being arranged in two triangles, one pointing up and one pointing down with the five needles along the common base line. The system was tried successfully on the Great

Western line in 1838. In 1842 it was extended to Slough, nearly thirty kilometres west of London. It aroused intense public interest when, a few years later, a young man travelled out to Slough where he murdered his girlfriend. Having a return ticket, he travelled back by the same route but he had been seen; his description was telegraphed to London and he was arrested at the London terminus, Paddington.

The telegraph system was steadily improved and the method invented by Samuel Morse (1791 – 1872) replaced the slower and clumsier methods previously used. The telegraph had far wider applications than serving railroads and helping in the arrest of murderers[11]. As the railroads and their associated telegraphs spread, so timekeeping was made uniform everywhere. Before the railroads, time had been fixed locally and differences from Greenwich Mean Time of a few minutes were of little account. With the telegraph, 'railway time' became the standard in Britain and in all other countries. And a difference of a few minutes could mean the difference between catching and missing the train.

The Public Face of Technology: Artistry and Intelligence

The formation of the Zollverein, or Customs Union, in 1833 was a major advance towards the unification of Germany. Another, less public, step in the forward march of Germany had been taken eight years earlier, in 1825, when Wilhelm von Humboldt, Prussian Minister of Education and twin brother of the explorer and naturalist Alexander von Humboldt, authorized the appointment of Justus Liebig (1803–73) to the Chair of Chemistry at the small University of Giessen. Liebig was a brilliant chemist. He was also an inspiring teacher; not so much in the didactic sense as in his capacity to inspire in his students a love for, and an understanding of, scientific research. Such was his reputation that before long students from Britain, the United States and all over mainland Europe were attracted to his laboratory in increasing numbers[1]. Although others have some claims in the matter, this inspired appointment can be taken as the beginning of the revival of German science, which, by the end of the century, was to dominate the world. German contributions to the early development of the electric telegraph showed that the long period of quiescence was ending; Liebig's international reputation as a chemist and a teacher of chemists confirmed this. From Renaissance times onward famous teachers had attracted students to the medical schools of Europe; Fabricius, Boerhaave and Black were among the best known. At Giessen Liebig created a school in organic chemistry; it was the first of the modern international research schools.

The medieval university, as we saw, never envisaged the possibility of discovering new knowledge. In the seventeenth century a few universities were associated, for short periods, with brilliantly original scholars. The great industrial movement of the eighteenth century passed the universities by, those of Scotland being exceptions. The universities of England were by then quiescent, having largely become finishing schools for wealthy young men who also subscribed to the doctrines of the Church of England. French universities were moribund and so were those of Germany, with the exception of the new University of Göttingen. With von Humboldt's idea and more particularly with Liebig's teaching and research the university, as it is understood today, was born.

In the meantime, in Britain, the spread of the railroad networks and the rapid development of a public telegraph system brought technology to the attention of a far wider public than the inhabitants of the industrial areas of the midlands and the north of England, lowland Scotland and south Wales. It amounted to more than a mere increase in numbers. Matthew Boulton may have appreciated the world-wide potential of James Watt's steam engine, but he was exceptional in his understanding. Generally, informed and philosophically inclined people in the eighteenth century had welcomed advances in the arts (applied as well as fine) and had taken them to be concomitants in the progress of humankind. These were not the only indications of progress equally important were advances in politics, law, manners, the comforts and conveniences of life and, not least, religious toleration. The ultimately successful campaign for the abolition of slavery was a surer index of progress, and more cogent for most people, than the latest Watt or Trevithick engine.

Montgolfier's balloons and the Liverpool and Manchester Railway, however, could not be ignored, even by the least imaginative. The most hidebound conservative could not deny that they were achievements that no

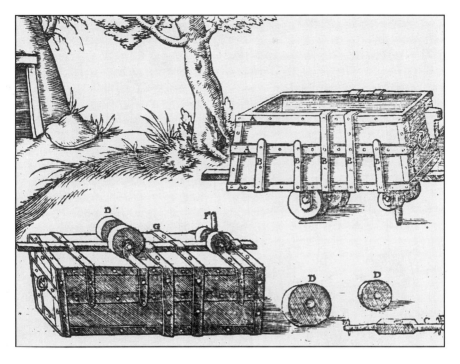

1. The development of the railroad, I. 'The Dog', from
Agricola, *De re metallica* (Basle, 1556). The illustration shows
the large and small wooden wheels (D) and (D), the iron
axle (C) with small iron pins (E) and (E), and the big blunt
pin (F) that guides the truck along the wooden track, or
rails. A lamp is fitted to the right-hand end of the truck (top
figure). (Reproduced by courtesy of the Director and
University Librarian, the John Rylands University Library of
Manchester).

2. Details of silk throwing machine from Zonca, *Nuovo teatro di machine*. The S-shaped flyer, to the left, imparts a twist to the silk threads as it feeds them on to the bobbin A. (From the National Paper Collection in the Manchester Museum of Science and Industry. Photo by Ms Jean Horsfall).

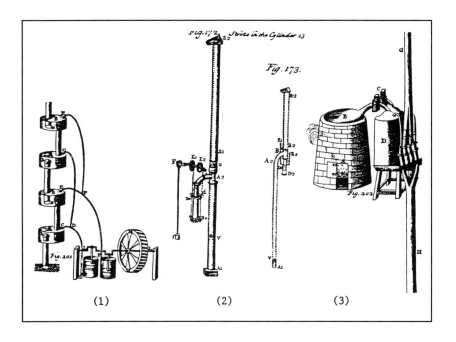

(1)   (2)   (3)

3. Papin's double-acting pump (1). As the piston in the left-hand cylinder rises so the air is rarefied in the first and third vessels and air pressure drives water up the first and third pipes. At the same time the piston in the right-hand cylinder is descending and therefore increasing the air pressure in the second and fourth vessels. Non-return valves ensure that the water can only move upwards. Both pistons are driven by the water-wheel on the right, (from *Philosophical Transactions*, 1685).

The problem with all such pumps was the leakage of air (or water) between the piston and the sides of the barrel. To obviate this Joshua Haskins replaced the piston with a bulky plunger surrounded by mercury. An improved version of the machine was described by J. T. Desaguliers in *Philosophical Transactions* in 1722. Leupold referred to it as the 'English pump'. Two forms of the Haskins-Desaguliers machine are shown in (2).

For comparison, Savery's engine is shown in (3), from *Philosophical Transactions*, 1698).

4. Arkwright's water frame. (Manchester Museum of Science and Industry. Photo by Ms Jean Horsfall).

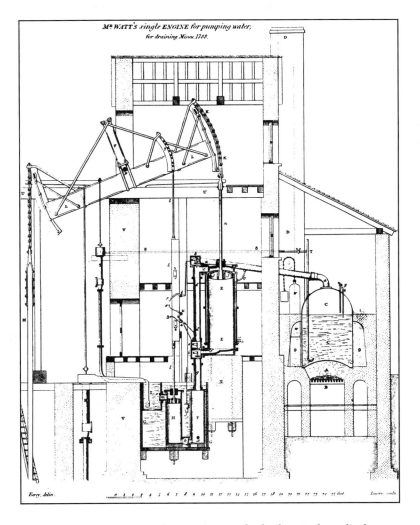

5. James Watt's Pumping Engine. C: the boiler; E: the cylinder; F: the condenser, with jet aperture, X, at the base; H: the pump to remove air and water from the condenser; 1–1: the plug rod that operates the valves and also the pump, H. The great beam, L–L, and the arch heads are strengthened with wrought-iron tie-rods. (John Farey, *The Steam Engine*, 1827).

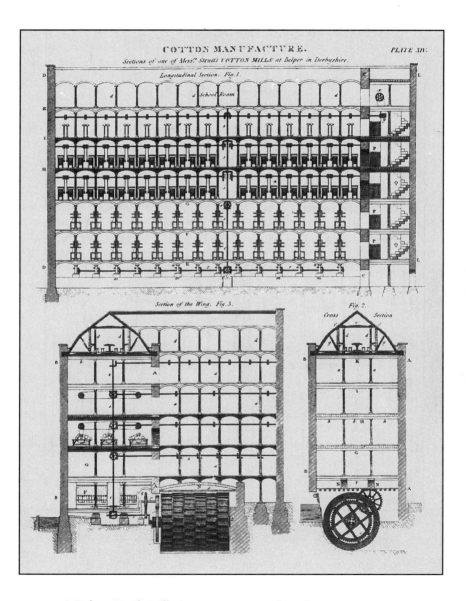

6. Belper North Mill. (From A. Rees, *Cyclopaedia*, 1819).

7. The development of the railroad, II. *Planet*. This is a working replica
of the original Stephenson locomotive of 1831. (Manchester Museum
of Science and Industry).

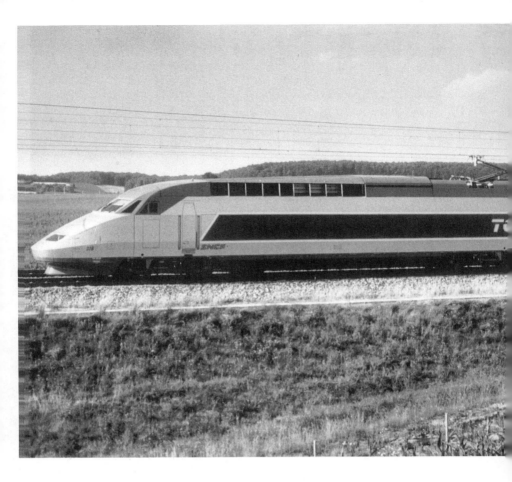

8. The development of the railroad, III. The high-speed train of French Railways (SNCF), the TGV. This remarkable train can reach a speed of 300 mph (nearly 500 kmh). (By courtesy of French Railways. Photo by M. P. Olivain).

previous age could rival. Greece and Rome at the heights of their power and civilization could show nothing like them. And with the ascendancy of technology the engineer became a far more public figure than he had been before and his tangible achievements received far wider public recognition. Fashionable intellectuals affected, as always, to despise the latest achievements of the engineers. Inevitably, a few generations later, no one doubted that these same achievements were indispensable for a civilized way of life and essential to the proper functioning of society.

As engineers assumed a more public role so the more extrovert of them showed a greater awareness of sound publicity. As we saw, this had been foreshadowed in the eighteenth century, but the increasing pace of innovation, the radical improvements in communications and inventions like photography greatly enhanced the scope for publicity – personal as well as technological. In Britain, the great master of engineering publicity was the Anglo-Frenchman, Isambard Kingdom Brunel, creator of the Great Western Railway, bridge builder, revolutionary naval architect and versatile, sometimes misguided, inventor[2]. Brunel's positive achievements and his successful propaganda on behalf of bold engineering enterprise outweighed his less happy enterprises such as the atmospheric railway and the seven-foot railroad gauge. He is immortalized in the fine photograph that shows him standing in front of the massive chains of that huge ship, the *Great Eastern*, and for the confident, even arrogant, way in which his name is displayed on the portico of the Royal Albert bridge across the river Tamar, his last major achievement: 'I. K. BRUNEL, ENGINEER'.

An equally impressive but more important bridge was the one that was built across the Menai Straits to carry the London to Holyhead railway. The political, administrative and economic importance of this route was noted in connection with the building of Telford's impressive suspension bridge. It was doubtful if a suspension bridge could carry a railroad: the vibrations set up by a train could well wreck the bridge – and the train. There had been disturbing precedents. A short suspension bridge near Manchester had collapsed due to the vibrations caused by a small squad of soldiers as they marched across. The men had not broken step and, although the links were of wrought iron, one of them had failed completely.

The problem confronting the designers of a railway bridge across the Menai Straits was this: the British Admiralty insisted that any such bridge must be sufficiently high to allow the tallest mast of the biggest warship to pass underneath at the highest of spring tides; and, furthermore, the width between the arches must be sufficient to allow free passage to the ships with the widest yard-arms in the Navy. Telford's solution to the same problem had been his fine suspension bridge but Robert Stephenson, the engineer in charge, doubted if a true suspension bridge could be made to carry trains, far heavier and faster than the biggest road vehicle. In 1830 Samuel Brown (1776–1852) had built a suspension bridge to carry a railway over the river Tees in the north of England, but it was unsuccessful. It was said that when a locomotive towed coal wagons across, a 'two-foot wave like that on a carpet' preceded the locomotive. No more railroad suspension bridges were built in Britain.

The only solution acceptable to Stephenson was a huge iron tube supported, if necessary, by chains: a kind of semi-suspension bridge. A tube is, weight for weight of material, more rigid than a solid rod, as Galileo had realized. Such an unusual bridge would represent an extreme extrapolation from the known and the tested. It would predicate extensive practical research on an enormous scale as well as advanced mathematical study of the stresses to be expected in such a novel structure[3].

Fortunately the resources, human and industrial, were available. One of Stephenson's two collaborators was

William Fairbairn (1789–1877), who had an engineering works in Manchester and another in Millwall, near London. Besides great engineering ability he had considerable scientific talent and collaborated with William Hopkins, of Cambridge University, in studying the rigidity of the earth. The third member of the triumvirate was Eaton Hodgkinson (1789–1861), an able mathematician who had been tutored by John Dalton. Hodgkinson had been interested in the failure of the Manchester suspension bridge and, in 1830, had written a seminal paper on the design of cast iron beams with comments on the use of wrought iron. Fairbairn was engaged on the construction of iron-framed mills and, by the mid-1830s, was, according to R. S. Fitzgerald, the best-known mill builder in the country. Using Hodgkinson's theory he had brought the cast-iron-framed building to the limit of economical and safe use, a limit that was set by imperfections in foundry and casting techniques.

Cast iron would, in any case, have been totally unsuitable for the proposed tubular bridge over the Menai Straits. The theories worked out by French engineers in the first half of the century were known in Britain but the size of the proposed bridge made a straightforward application of any theory extremely hazardous. In addition, the theories had only been tested on cast iron structures. The relatively easy decision having been made that wrought iron was the only feasible material, the next step was to decide the form of the tube. There were three possibilities to consider; it might be of circular, elliptic or rectangular cross-section. Experimental tubes of the three forms were made and extensive tests were carried out using Fairbairn's apparatus for the examination of beams. The areas of cross-section were varied, as were the thickness of plate and the length of span. In each case the tube was supported at each end and a load on the middle increased until at last it failed. This technique, of parameter variation, was similar to that used by Smeaton, but the aim was somewhat different. The tests, Professors Rosenberg and Vincenti point out, were exploratory; the object was to identify the key problems. The common cause of failure was, Fairbairn and Hodgkinson found, buckling of the plates at the top of the tube. This was a new and surprising discovery. The plates on top were under compression, those beneath under an equal tension (this was the opposite of the case of Galileo's beam, which was loaded at the unsupported end).

Although the material was strong enough to stand the tension, the *structure* could not stand the equal compression. It seemed that a structure would have to be found that could resist the buckling of the top plates. For simplicity's sake the decision was taken to concentrate on a tube of rectangular cross-section; although the other two forms offered certain theoretical advantages, there were certain practical difficulties. Furthermore the flat top plates of the rectangular section concentrated the zone of buckling. To prevent the top buckling under compression Fairbairn substituted, in place of the flat plates, two small parallel tubes. By using a cellular structure on top, instead of flat plates, Fairbairn found that he had almost doubled the load the tube could carry before buckling. He then constructed a new model tube, about 24 metres long, at his Millwall works and carried out extensive tests on it, the tube being repaired after each failure. The experiments showed that the best structure would have eight parallel tubes in place of the top plates. Ideally, the structure should be such that both top and bottom should fail at the same load (the two other circumstances would indicate either excessive material on one side or inadequate strength on the other). Since the bottom plates would have had to be massive to equal the strength of the top they were replaced by six parallel tubes. In the meantime Hodgkinson had worked out how to brace the side plates, or webs, against premature buckling. The data for the tubular bridge were now complete. A smaller prototype bridge was built to carry the railroad over the Conwy river on the Welsh mainland. It consisted of two tubes, one for each track, each 130 metres long.

Three masonry towers were built: one, 70 metres high, sprang from Britannia Rock, in the middle of the Straits, and the other two, slightly shorter, were close to the opposing shores. Four iron tubes, each 150 metres long, were assembled on staging by the shore before being put on pontoons and floated down the Straits to the towers. Powerful hydraulic presses, mounted on the towers, raised each tube. It was a slow business as the supporting masonry had to be built up underneath as each tube was lifted. Four tubes, each 70 metres long, spanned the gaps between the abutments on each shore and the two shorter towers. Finally the separate tubes were joined together to form one long tube 500 metres long. This reduced and equalized the strains along the whole length. The bridge was opened to rail traffic in March 1850.

By a combination of systematic investigations, trial and error, mathematical theory, corner cutting and application of existing knowledge, the engineers had arrived at the optimum design for the tubular bridge. Supporting chains were found to be unnecessary. The bold endeavour to build something that was far beyond contemporary experience was a great success; so successful, in fact, that the bridge proved capable of carrying the last and heaviest steam locomotives to run on British railroads, far heavier than the puny machines of 1850. The bridge was at once an object of great public interest. Queen Victoria went out of her way to visit it in 1852 and was shown round by Stephenson. Two years later a less exalted visitor, the inveterate traveller George Borrow[4], dismissed it, patronizingly after the manner of the literary gentleman, '. . . a wonderful structure, no doubt, but anything but graceful . . .'. ('I despise railroads', he wrote later, 'and those who travel by them.') Samuel Smiles passed a far sounder judgement when he concluded that, 'The

Britannia Bridge is one of the most remarkable monuments of the enterprise and skill of the present century.' From another point of view it could be said that, at last, Galileo's theory of the strength of tubes and cylinders had reached practical fulfilment.

The Britannia tubular bridge was the source of the last significant contributions to the common pool of knowledge from the strategic technology of railroad construction. Much was learned about the properties of thin metal cylinders and about the techniques of riveting. This new knowledge was applied in a wide variety of fields. Joseph Whitworth, for example, used the new technique of cellular construction to replace the heavy (and often needlessly ornamented) frames of machine tools by light and strong hollow ones. The thin but strong metal tube became a feature of engineering structures down to the present day. And, although the Britannia bridge was built of wrought iron, the advances in knowledge that accrued from it enabled the best use to be made of the steel that was increasingly to be produced by the Bessemer and then the Siemens–Martin processes.

The two Jacquard-controlled punching machines that Richard Roberts designed and made to deal with two million rivet holes which the bridge required represented a dead end; digitally controlled machine tools had to await the development of electronics. The large hydraulic presses made to lift the tubes found another use when they were acquired to help launch the largest ship in the world.

Progress in Shipbuilding

The original specification for the *Great Eastern* (or, as she was first called, the *Leviathan*) was for a ship that would be able to steam to Australia and back without refuelling. (The Suez Canal was not opened until 1869; in any case it was originally too shallow to take a vessel with the draught of the *Great Eastern*.) The conception of this huge ship – six times the displacement of the largest ship of the time – was due to Brunel. It was to be about 230 metres long, 26 metres beam and to have two sets of engines, one driving a propeller, the other driving paddle wheels. It was also to have five masts and be equipped with sails. We do not know whether the idea came entirely from his own bold imagination and willingness to tackle huge new enterprises or whether it was stimulated by the realization, following the success of the Britannia bridge, that technology could now leap several stages. Brunel's associate, John Scott Russell (1808–82), was a gifted naval architect who, from 1835 onward, had worked out his theory of the wave-line form of hull design. He had been the first to recognize that the generation of waves and turbulence by a ship represents a waste of power. No doubt, he was extending the established principles of water power engineering to ships. But the recognition of this particular waste would hardly have been possible before the advent of the steam ship. A hull, he argued, designed according to wave-line theory, would be able to reach the maximum possible speed for its length without wasting undue power generating waves. It was, incidentally, through the study, by naval architects, of wave generation by ships that the difference between group velocity and phase velocity was first recognized.

After working at a shipyard in Scotland Scott Russell moved south, to Millwall, on the Thames, where he bought a yard that had belonged to Fairbairn. It was therefore natural that Brunel should give him a contract to build the great iron ship. And there he built the hull (of wave-line form), paddle engines, propeller shaft and rigging of the huge ship. The propeller engine was built by James Watt and Company of Soho, Birmingham. In view of its enormous length the ship was, with great difficulty, launched sideways into the Thames; Scott Russell being responsible for the launch. The influence of the Britannia bridge was most evident in the double hull.

Up to the water line the vessel had an inner and an outer hull, the space between being cellular. A ship as long as the *Great Eastern* would routinely have its bows lifted, in the long ocean swell, by a large wave as its stern was being lifted by the proximate wave. This would mean that much of the weight of the ship had to be carried by the hull which would therefore be under severe strain. The cellular double hull greatly increased the strength of the hull.

The tragi-comedy of the *Great Eastern* has often been told[5]. Construction began in 1852 but it was not until January 1858 that the ship was safely launched. She made a few successful crossings of the Atlantic but rolled excessively in bad weather. She was also an unlucky ship, suffering many mishaps and accidents. The most useful job she did (see chapter 13) was entirely different from the purpose for which she was designed. She ended her days as an advertisement for a Liverpool department store and then as a home for vaudeville and circus acts. And yet, as Professor Emmerson remarks, 'The *Great Eastern* caught the imagination as no other large ship has ever done, or perhaps may ever do again. She epitomised the industrial and entrepreneurial spirit of her times . . .' More, she foreshadowed the fast, steel ships that from the 1890s to the coming of cheap air travel dominated passenger traffic across the Atlantic and Pacific Oceans. These 'ocean greyhounds' were believed to represent the ultimate in luxurious, fast and safe travel until the *Titanic* disaster of 1912 shook public confidence in the omnicompetence of modern technology.

It is easy enough to spot the shortcomings of the ship. Not enough was known about the behaviour of a very large steam ship in a sea-way for her designers to foresee that she would roll excessively. She was not fitted with bilge keels and, of course, had no stabilisers. Furthermore, she was under-powered. The huge, low-pressure engines may have been masterpieces of engineering but it was already old-fashioned engineering. Sea water was used in her boilers and the steam pressure in her cylinders was less than the pressure of the sea on the bottom of her hull. By the time of her maiden voyage other ships were at sea with efficient high-pressure compound expansive engines equipped with surface condensers to recycle fresh water to their boilers. One minor detail, however, should not be forgotten. Her huge rudder would have been very difficult to handle in heavy weather. Accordingly, Macfarlane Gray, an able naval architect and engineer, invented the first steam steering gear and fitted it in the *Great Eastern* in 1867. It was so effective that one man could steer the ship in the worst possible weather. Macfarlane Gray's invention was the first in a series that now includes the powered steering of airliners and the power-assisted steering of the private car.

William Froude (1810–79) was the fourth son of a wealthy Church of England clergyman. At Oxford University he was tutored in mathematics by his elder brother, Hurrell Froude, who was a leading member of the Oxford group to which the Rev. J. H. (later Cardinal) Newman belonged. After Oxford, where he gained a first-class degree in mathematics, Froude worked with Brunel on railway construction for a time before retiring to Devonshire to the life of a gentleman, with leisure to pursue his interest in problems of the design of ships' hulls. He was consulted by Brunel over the *Great Eastern* and investigated the factors determining the rolling of ships.

In the mid-1860s began a series of classic experiments on hull form for which he built six model hulls made of wax. Two were three-feet long, two six-feet long and two twelve-feet long. The pairs were called 'Raven' and 'Swan'; the three 'Ravens' had wave-line hulls, the three 'Swans' had more traditional hulls. In each experiment a 'Raven' and a 'Swan' were towed from the ends of long booms projecting horizontally from a steam launch; in this way the waves from the launch did not affect the passage of the models through the water. From these experiments

Froude deduced laws of resistance in terms of the displacement of the models and their speeds through the water. The experiments also provided a test (unfavourable) of the wave-line theory.

In 1871 Froude completed a large test-tank[6], equipped with a dynamometer and a track for towing models, near his home at Paignton in Devonshire. He had a small grant from the British Admiralty and gave his own services free. There had been some tank experiments on model hulls in France, England and Denmark in the late eighteenth and early nineteenth centuries, but the complex motion of a sailing ship through the water made the results of little practical value. Froude's test-tank, in the age of the steamship, was the forerunner of similar, government- or industry-supported tanks and therefore of systematic design of ships' hulls.

The construction of large marine engines, much bigger than those required for mills or railroad locomotives, required new and bigger machine tools: in particular, bigger lathes and boring machines. Bigger marine engines also brought about the invention of an entirely new metal forming machine. This was the steam hammer. The largest paddle wheel and propeller shafts that could be forged had been limited by the capacity and the flexibility of the traditional tilt hammers. James Nasmyth (1808–90), a Manchester machine tool maker and a pupil of Maudslay, invented his steam hammer in 1839, specifically to forge the propeller shaft for Brunel's steam ship, *Britannia*. A large cylinder fitted with a piston was mounted vertically, the piston rod being connected to the hammer head. Steam pressure raised the hammer head, which was then released to fall on the workpiece below; alternatively steam pressure could be applied above the piston so that the hammer head struck the workpiece with additional force. Nasmyth's steam hammer was not only massive and powerful, it was capable of very delicate control. In a famous demonstration the hammer head was brought down so that it touched but did not break a hen's egg placed on the anvil. The Victorians were suitably impressed.

Civil Engineering and Agriculture

The excellence of French civil engineering in the eighteenth century and the subsequent conquests of Napoleon led to the extension of French practices over Europe. One enduring indication of this is the custom of driving on the right-hand side of the road, extended to Austria, which had escaped French occupation, by Hitler in 1938. Sweden fell into line only after 1945. The railroad systems of mainland Europe follow French, rather than British, practice. A factor here may have been the immense strategic importance of railroads on the mainland; an importance that, for obvious reasons, they never had in Britain.

The United States also borrowed from France, its strong ally in the War of Independence; reminders of the relationship are the long-established decimal coinage and the foundation of West Point Military Academy, inspired by the Ecole Polytechnique. But American problems were not those of a long-established country like France, or indeed of any other European country with the possible exception of Russia, whose political and social systems were radically different.

Not long after the opening of the Britannia tubular bridge a successful railroad suspension bridge was built in the United States. The engineer responsible, John A. Roebling (1806–69) was born in Muhlhausen, in Thuringia, in Germany. After training at the Bauakademie, in Berlin, he was employed by the government for some years before deciding to emigrate to the United States. His first engineering contract was to build a suspension wooden aqueduct to carry a canal over a river. In place of chains he used brass wire ropes. It is said by David B. Steinman and Sara Ruth Watson[7] that he had remembered that wire ropes were used in the mines of Freiburg, in Saxony (he does not appear to have known of Séguin's work). After his success with the suspension aqueduct he tackled the great challenge of a railroad bridge across the Niagara river. Charles Ellet had already suggested a rail suspension bridge and had built a bridge to carry foot passengers, but had soon turned to other projects, leaving the field to Roebling, who commenced his bridge in 1851 and completed it four years later. It carried a single railroad track and underneath that a road-way, the decks being firmly braced to be as rigid as possible to prevent dangerous oscillations.

Roebling's Niagara bridge was a success; apart from the great design skill, a contributory factor had been the wire ropes he used. These were of wrought iron and were not twisted as had been the practice, the individual wires being bound together by loops of soft iron. After the Niagara bridge – which remains in service although strengthened and supported by arches – Roebling suggested that a suspension bridge be built over the East River, New York. Unfortunately financial crises and the great Civil War delayed the start of the bridge. Roebling himself was never to see the completion of the bridge; he lost his life as a result of an accident on the site. The great Brooklyn bridge was completed in 1883 under the supervision of Roebling's son, Colonel Washington A. Roebling, himself an invalid as a result of an industrial accident. Brooklyn bridge, too, had an important novel feature: it was made of steel.

From Roebling's time onward many, probably most, of the great bridges of the world were built in the United States. American engineers faced challenges on a more daunting scale than any European engineer was called on to face (at least, in his homeland). The distances were greater, the rivers were wider, the terrain more varied, the mountains as high and the mountain ranges much more extensive than in Europe. In addition, the range of climate was more extreme. The skills and knowledge of American civil engineers together with the abilities of the locomotive designers made possible a railroad network that bound the individual territories into states that were to form one nation.

Topography, geology and natural resources did much to determine the course of American technology in the first half of the nineteenth century. The enterprise and inventiveness of the people together with the social order did the rest. America's problems were those of a free people engaged on the rapid development of vast territories. Accordingly, agriculture and its requirements predominated. Skilled agricultural labour being in short supply, there was strong pressure for labour-saving equipment of all kinds. Barbed wire was a good example of American response to an acute need. Here was a cheap artificial hedge that could be put up quickly by unskilled labour to enclose the biggest fields. Enormous forests meant that timber was much more freely used than it could be in Europe (visitors from Europe may be surprised by the common use of timber in house building in the United States, the land of the soaring skyscraper). Woodworking machinery was therefore highly developed, one particular triumph being a profiling lathe to form the stocks of guns. The gun was another essential agricultural implement. In a newly settled land farms had to be protected against marauding animals and human enemies. Users, who were often far from gunsmiths and workshops, wanted guns that were reliable, accurate and easy to use. Although, as in Britain, the early machine tool industry in the United States had strong links with textile mills (in New England), it was the manufacture of hand-guns that gave the greater and most sustained impetus towards the goal of machine tools that could be used to manufacture guns with interchangeable parts. In the first decades of the nineteenth century the invention of the percussion cap made the old flintlock redundant and, with the cartridge case, opened the way to much improved hand-guns. Simeon North

(1765–1852) has been credited with inventing the milling machine in which a wheel fitted with hardened steel teeth performs the same job as a hand-held file but with far greater precision and speed. John H. Hall developed the milling machine so that it could be operated by an unskilled man or a boy. Hall, according to Dr Merrit Roe Smith, played an important, if not the principal, role in the development of machine-made interchangeable parts[8]. Large orders from the US Army, a customer who was prepared to pay, not only for quantity but also for quality, made possible the development of high-precision and labour-saving machine tools at the government armouries at Harper's Ferry and Springfield. Interchangeability was obviously a desideratum as far as the Army was concerned; for the private customer, who would normally buy one or, at most, two guns, it was a needless luxury.

In 1836 Samuel Colt (1814–62), patented his six-chamber revolver. This, rather like Watt's original steam engine, demanded new standards of workshop precision. The action of the revolver was such that the mechanism had to carry out the sequence of movements that the fingers and hand of a rifleman perform in placing a cartridge in the breach. Accuracy of alignment was essential, the movements of the components had to be precise. Colt's revolver was by no means the first repeating firearm but it was certainly the most successful and the basic pattern is still used today. To meet the great demand for the Colt revolver and similar weapons the turret lathe was invented in the United States in about 1845. In this machine tool a vertical octagonal turret is mounted on the lathe bed in place of the slide rest. Eight cutting tools are bolted to the faces of the turret, which can be rotated so that such operations as cutting, drilling, boring, reaming or tapping can be carried out rapidly in the correct order. Once a skilled operative has set the tools and the stops, any unskilled worker can carry out the prescribed operations easily and repeatedly. The increase in productivity achieved by the turret lathe and the larger capstan lathe was enormous. In spite of these advances, Colt had not achieved interchangeability by 1851.

Less well known than the Colt revolver but arguably no less important was the reaping machine invented by Cyrus H. McCormick (1809–84) in about 1840. There had been previous attempts in Britain and elsewhere to make a reaping machine but without success. In Britain and European countries the relatively small fields and the ready availability of skilled reapers reduced the need for such a machine. The high crop yield on British farms reduced the need still further. In South Australia, on the other hand, circumstances similar to those in the United States – shortage of skilled labour, relatively low crop yield and large fields – led to the invention and use of a successful harvesting machine. In 1843 John Ridley (1806–87), built his first 'Stripper', a simple but highly efficient machine designed to be pushed by a pair of horses[9]. A long horizontal 'comb', fixed at a height just below the ripe grain, caught the heads of grain between the teeth and a rapidly revolving paddle wheel, or 'beater', driven by a belt from the wheels, knocked the grain off and swept it back into a storage compartment; chaff and dust, being lighter, was carried up and out of a chimney. The Ridley 'Stripper' collected and threshed the grain but left the stalks standing, uncut. It was, Dr Jones points out, very economical in that little grain was lost and much labour saved. Moreover, it was particularly suited to the very dry climate of South Australia and many thousands were subsequently built. In McCormick's reaper, horses pulled the machine with the row of horizontal knives on one side. A paddle, driven from the wheels, bent the stalks over the knives and the cut stalks were carried to the back of the machine by a moving belt. The principles of the McCormick reaper are incorporated in the modern combine harvester but the arrangement of the paddle in front and the inclusion of threshing in the same process as harvesting are those of Ridley's 'Stripper'.

Although less revolutionary, British farming continued to progress. Farm machinery was increasingly made entirely of iron. The steam 'traction engine' was introduced and enabled steam ploughing to be practised on the big, flat fields in the eastern half of the country. The traction engine would be placed at the end of the field and, by a system of wire ropes, stakes and pulleys, haul a plough to and fro across the field. At about the same time the threshing machine was introduced. A highly efficient, labour-saving machine, it could be driven by the same steam tractor that towed it from farm to farm. British farming, in fact, was reaching a peak of efficiency. There was, however, one conspicuous failure. In 1845 the potato harvest in Ireland failed following an outbreak of the potato blight. The potato – 'Ireland's debilitating root' as the political journalist William Cobbett presciently, if rather inaccurately, described it – was the staple crop and the consequence of failure was mass starvation. The occasional failure of crops and resulting starvation had been the lot of all rural communities from time immemorial. It is, however, hard for the modern observer to understand how the old pattern of rural starvation could have been allowed to recur in part of the wealthiest country in the world, in the age of the steam ship and the railroad. The explanation must lie with administrative failure coupled with the rigorous application of the doctrine of *laissez faire*: on no account should public authority or the state feed the people, for that would discourage healthy independence. The dead, however, were not expected to bury themselves.

Some Social Considerations

As if in compensation, the miseries of the new industrial communities were being exposed by Friedrich Engels in his famous *Condition of the English Working Class in 1844*.

Engels was living in Manchester at the time, working in a branch of the family business. He had an excellent opportunity to study the living conditions in a city that could fairly be described as the intellectual and technological capital of the Industrial Revolution. The conditions that he described were truly dreadful. Against that it must be remembered that the areas he studied were very restricted and that Manchester itself was still a relatively small town, being barely a couple of kilometres across in any direction. Furthermore the causes were not necessarily endemic. They were simply precipitate growth coupled with inadequate administrative arrangements. Vast numbers of hopeful immigrants were pouring in, seeking jobs and a standard of living that were quite unattainable in the rural areas of England, Scotland, Wales and Ireland from which they came. Politically and administratively, Manchester could not cope with the influx; the town had no more than the administrative and political resources of a village during the years of rapid growth.

Surprise is often expressed at the relative ease with which people from rural areas took to the discipline of work in the great textile mills. The seasons impose a discipline on rural folk but, that apart, the custom of working regular hours was unfamiliar and, it is inferred, most unwelcome to the countryman, who could – and often would – take time off when he felt like it, confident in the knowledge that the job would be waiting for him when he felt like tackling it. But 'leisure preference', as the habit is called, was surely a male attribute. Women have never been able to take such a relaxed attitude to work. The discipline of the family was hardly less – perhaps more – demanding than the discipline of the mill. It is not surprising that many mill workers were women; many of the others were children who could also be easily conditioned to regular hours of work.

If Engels was the eloquent critic of the effects of

industrialization on the lives of people there were a number of other writers who had set out to understand and explain it. In 1832 Charles Babbage (1792–1871) turned his versatile and enquiring mind to a study of the new, technological society that had emerged in Britain. His book, *On the Economy of Machines and Manufactures*, was his second general study on the theme of science, technology and society; the first had been his *Reflections on the Decline of Science in England* (1830), a work that had influenced the formation of the British Association, in imitation of the Deutsches Naturforschers Versammlung (1822). The *Economy of Machines* is a remarkably perceptive book in which Babbage shows a high degree of technological understanding. He looks forward to the division of labour in science and to the emergence of systematic applied science. He had read Lean's monthly reports and was well aware of the benefits of pooling technological information. He even prepared a remarkably modern-looking questionnaire that an enquirer could use to assess the technological efficiency of any particular factory. But perhaps his most interesting observation related to obsolescence:

> Machinery for producing any commodity in great demand seldom actually wears out; new improvements by which the same operations can be executed either more quickly or better, generally superseding it long before that period arrives; indeed to make such an improved machine profitable, it is usually reckoned that in five years it ought to have paid for itself and in ten to be superseded by a better.
>
> 'A certain manufacturer', says one of the witnesses before a committee of the House of Commons, 'who left Manchester seven years ago, would be driven out of the market by the men who are now living in it, provided his knowledge had

not kept pace with those who have been, during that time, constantly profiting by the progressive improvements that have taken place in that period.'

The rate of innovation in the industrial areas of Britain was evidently very rapid indeed. It is clear, too, that a class of machinery makers had grown up that was responsible for innovation and thereby put pressure on machine users. They did not wait for a demand from the market; they invented, developed and made machines that their customers could not afford to be without. Some light was thrown on this by Andrew Ure (1778–1857), in his book *The Philosophy of Manufactures* (1835). Ure's book is less discursive than Babbage's and his concern was almost entirely with the textile industries. Nevertheless, as the textile industries were associated with a wide range of other industries and technologies, the book was not so specialized as it might appear. The frontispiece of Ure's book represents the weaving shed of a Stockport cotton mill. The artist obviously exercised his licence to the full and greatly exaggerated the scale of the building. Nevertheless, the rows of identical power looms stretching interminably into the distance, like a vast mechanical army on parade, clearly represented something new in manufacture: the mass production of complex machines, similar to the spinning machines in Strutt's Belper North Mill. Ure makes the point clearly:

Indeed the concentration of mechanical talent and activity in the districts of Manchester and Leeds is indescribable by the pen and must be studied confidentially behind the scenes in order to be duly understood or appreciated.

The following anecdote will illustrate this position. A manufacturer at Stockport . . . being about to mount two hundred power looms in his mill fancied he might save a pound sterling in the price of each, by having them made by a neighbouring machine maker instead of obtaining them from Messrs Sharp, Roberts in Manchester, the principal constructors of power looms. [He] surreptitiously procured iron patterns cast from one of the looms of that company which, in its perfect state, cost no more than £9 15s. His two hundred looms were accordingly constructed at Stockport, supposed to be facsimiles of those made in Manchester and they were set to work. Hardly a day passed without one part or another breaking down in so much that the crank or tappet wheel had to be replaced three times, in almost every loom, in the course of twelve months . . .

In the end the foolish and greedy manufacturer had to go to Sharp, Roberts and order the looms he had been unable to make for himself. The specialized machine makers, capable of manufacturing in large batches, or of mass producing, surpassed the individual who wanted to make his own machines. Specialized and high-precision machine tools enabled them to do this. Of Sharp, Roberts, Ure remarks:

Where many counterparts or similar pieces enter into spinning apparatus they are all made so perfectly identical in form and size by the self-acting tools, such as the planing and key-groove cutting machines, that any of them will at once fit into the position of its fellows in the general frame.

It seems that in Manchester, in the first four decades of the nineteenth century, specialized machine tools had been developed that made possible the manufacture of interchangeable parts for textile machinery.

Ure has had a bad press. He was notoriously cantankerous and quarrelsome and his eulogy of the delights of factory life, particularly for children, together with his attacks on trade unions, made him the butt of bitter attacks by Marx and Engels. While his social views are wide open to criticism, his technological insights are very much to the point. Babbage, on the other hand, is now well regarded. It is said that in his *Economy of Machines* he pioneered the technique of operational research, and it is certainly true that his understanding of the principles of the computer were far in advance of the times and were not to be realized in practice until the rapid advance in electronics over the last five decades.

A practicable, if simpler and humbler, form of computer was proposed in an attempt to devise a means of applying the principle of Watt's indicator to railway locomotives. Watt's indicator was perfectly satisfactory when applied to engines that were running at a uniform speed and therefore recording more or less exactly the same diagram – doing more or less the same work – all the time. If, on the other hand, the pressure of steam and the speed varied – as they certainly did on all locomotives – a multitude of different diagrams would result and it would be impossible to measure the work done. In 1822 James White, a Manchester engineer, described an efficient differential dynamometer, superior in some ways to the Prony brake, that could easily be applied to mill engines. However, like Watt's indicator, it could not integrate automatically: that is, it could not sum up the work done if the speed and the load varied. In 1841 the British Association, aware of the problem, set up a committee of three – the Rev. Henry Moseley, Eaton Hodgkinson and J. Enys – to devise the best form of locomotive indicator[10].

In fact, the problem had already been solved, in principle, by General Morin and Captain Poncelet in 1838. They had invented an apparatus for measuring the work done by horses in towing heavy loads, such as guns, on roads that were anything but level. At any given moment the work done by the animals, or the power exerted, would vary. What was required was some device that would instantly integrate the continuously changing force

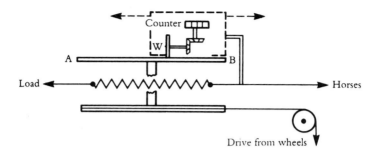

FIG 11.1 *Morin's* compteur

exerted by the horses and the distance covered. The dynamometer they devised, which they called the *compteur*, is shown in figure 11.1.

When the horses pull – doing work – the spring is stretched and the little wheel, W, is pulled from its neutral or rest position in the middle of the horizontal disc, AB. The disc is driven by the wheels of the limber or gun-carriage and as it rotates it turns the little wheel which, in turn, drives the counter. The greater the pull the more the revolutions of the little wheel; similarly the faster the disc rotates, the more the revolutions of the little wheel[11]. In short, the number of revolutions of the little wheel, as shown on the counter, is proportional to the product of the force exerted and the distance covered; or, in other words, to the work done. An obvious mechanical defect of this apparatus is that the spring drags the little wheel across the disc at right angles to its direction of rotation.

Moseley, Hodgkinson and Enys devised a more complex machine. In their apparatus steam pressure – the driving force – from a cylinder displaces the 'integrating wheel', as they called it, along a cone that is driven by the cord attached to the piston. The cone's rotation is to and fro and its speed in these opposing directions is necessarily proportional to the speed of the locomotive. As in Morin's apparatus, the integrating wheel is connected to the counter. However, for a variety of reasons, this, rather complex, apparatus was not adopted by the railway companies.

Babbage, too, was interested in the problem of measuring the work done by locomotives and claimed to have invented the dynamometer car (a version of Morin's apparatus). These efforts by Morin, Poncelet, Babbage, Moseley and his colleagues, on the very margins of contemporary science and technology, no doubt seemingly unimportant at the time, represented something most unusual; unprecedented, in fact. For the first time inventors were designing machines that could do mental work. Babbage, of course, was clearly aware of this. In his book, *The Exposition of 1851* (London: 1851), he wrote:

It is not a bad definition of *man* to describe him as a *tool-making animal*. His earliest contrivances to support uncivilised life were tools of the simplest and rudest construction. His latest achievements in the substitution of machinery, not merely for the skill of the human hand, but for the relief of the human intellect, are founded on the use of tools of a still higher order.

PART III

Power Without Wheels

Progress at the Flood

Although faith in human progress was at its height in the middle of the nineteenth century, certain, apparently well-founded, hopes had been disappointed. On the debit side of technology, electricity had, apart from telegraphy, failed to show the dramatic progress that could reasonably have been expected after Faraday's discoveries and the inventions of the electromagnet, the magneto and the electric motor. This was particularly so, since, in 1835, M. H. Jacobi had pointed out that, if imperfections like air resistance, friction and what is now called back-EMF could be eliminated there was no reason why a well-designed electric motor should not, once started, go on accelerating until it reached an infinite speed; and an infinite speed suggests infinite power[1]. Jacobi was a professor with an impeccable academic reputation and his paper was read with great interest in Europe and the United States.

Only the scientifically illiterate could believe that imperfections such as friction could be completely eliminated, but it was reasonable to suppose that when these had been reduced as far as possible a motor could be built that would deliver immense power at little cost. The vision of enormous power from a finite source – a battery of voltaic cells – by means of a simple, clean and compact motor (far superior to the clumsy, dirty and bulky steam engine!) aroused a veritable euphoria among experimentalists. Electric motors were built to drive boats (on the Neva and the Hudson), to work machines and, in 1841, to propel the first electric locomotive to run, battery-powered, on a standard-gauge railroad. Unfortunately, no matter how ingenious, how versatile the experimentalists, nobody succeeded in making a motor, driven by voltaic cells, or batteries, that could compete with the steadily improving, ever more versatile, steam engine.

The widespread euphoria about the immediate prospects of very cheap power, aroused by the invention of the electric motor, was the second such outbreak of enthusiasm in the course of the nineteenth century. As we saw, the first had been when the high-pressure steam engine was introduced. The third followed the appearance of the Ericsson 'caloric engine'. John Ericsson (1803–89), the Swedish–American inventor, was an able engineer whose elegant little locomotive *Novelty* had been the most serious rival to Stephenson's *Rocket* in the Rainhill trials (see chapter 10). The idea, popularly associated with his 'caloric engine', is easy to understand, plausible and quite erroneous. A volume of air, enclosed in a cylinder fitted with a piston, is heated so that it expands and does work, pushing the piston out. The heat is then abstracted from the air and stored in a heat store or regenerator, consisting of strips of metal; in this respect the 'caloric engine' resembled, but was otherwise inferior to, Stirling's engine. Deprived of heat, the cooled air contracts and the piston returns to the starting point. The next cycle begins with the heat being withdrawn from the regenerator and restored to the air, which therefore expands, doing work. In a few words, all that is required is that a certain amount of heat be moved, to and fro, between a heat store and a 'working substance' such as air.

A vessel named the *Ericsson* was built and fitted with 'caloric engines'. It was hoped – by some – that it would be able to cross the Atlantic using no greater source of heat than was required to compensate for unavoidable losses through conduction, convection and radiation. No such economy – verging on 'perpetual motion' – was or could be achieved. In fact the 'caloric engines' were a failure.

They were taken out and replaced by an ordinary steam engine. The significance of this, well publicized, venture is that it confirms the separation, or culture gap, between some leading conventional engineers and the scientific study of heat, a culture gap that was made apparent at meetings of the Institution of Civil Engineers in London in 1850 and 1852–53.

International Exhibitions and International Comparisons

It is, or at least it used to be, a widely held belief that architects in the nineteenth century failed to realize the potential of the new materials and the deeper understanding that technology had provided. Instead, it is, or was, claimed that they reverted to the styles of the distant past and relied substantially on the traditional materials of stone, brick and timber. Public taste changed early in the present century; contempt was then expressed for nineteenth-century affectations; faith in contemporary styles, using new materials, was asserted. Only in one area were nineteenth-century builders allowed to have shown the right spirit. It was the area in which the engineers had necessarily played the leading part. The greatest railroad stations of Europe and the United States could be admitted as admirable in their way; for the properties of iron and then of steel had substantially determined the forms of their structures. Another building of a revolutionary kind that owed little to aesthetic traditions was the great 'Crystal Palace', erected in London's Hyde Park in 1851. The creation of a landscape gardener, Joseph Paxton, it demonstrated what could be done by a judicious combination of glass and iron. Once again it had taken an outsider to show how new materials and techniques could be used in an old, indeed ancient, craft. Although intended as a temporary structure to house an exhibition, the Crystal Palace was so popular that it was later re-erected about twelve kilometres south of London where it stood until accidentally destroyed by fire(!) in 1936. Interestingly enough, the most suitable outline form for the Crystal Palace was found to be that of a medieval cathedral, with nave and transept, although there was no attempt to make it look like a cathedral. There were no spires, flying buttresses or gargoyles in cast iron.

The Great Exhibition of 1851, for which the Crystal Palace was built, was an enterprise of the Society of Arts, the President of which was Prince Albert, the enlightened Consort to Queen Victoria. Although the idea of an international exhibition originated in France, in 1848, the Crystal Palace Exhibition of 1851 was to be the first in which all civilized and some uncivilized nations were invited to participate. To modern eyes the most striking feature of the Exhibition was the decorative and applied art on display. On the evidence of the illustrated catalogues[2] public taste was at a low level. Ornamentation was carried to architectural extremes; steam and other engines had Ionic or Corinthian columns in cast iron or carried other needless decorative ironwork. Elaborate furniture, including a large sofa with arms and legs made of real rhinoceros horns and chairs made of stags' horns with the sharp points protruding at shin height, was supplemented by curious ornaments like the orchestra of stuffed kittens playing miniature instruments and wearing little crinolines. These and similar objects won prizes or were commended. It took forty and more years for public taste to recover from the low point of 1851. No nation, it seemed, escaped the collapse of taste in the decorative arts. It would be difficult to believe, on the evidence of the Great Exhibition alone, that the mid-century was also a period when great literature was written and great music composed. Critics of mid-nineteenth-century taste should, however, always remember that their judgements are based on evidence provided by the period in question and largely as a result of the Great Exhibition.

The main purpose of the Exhibition was, however, the

display of industrial achievements. The exhibits, taken as a whole, give some impression of the 'state of the art' in the various technologies; for example, the magneto was classified as hardly more than a toy. In another direction, as it were, the displays offered some rough idea of the progress made by different nations. In this respect it commemorated Britain's industrial success and marked the topmost pinnacle of Britain's leadership. Thereafter other nations were to take the lead, although no one nation was ever clearly in the van. The more far-sighted among British observers were uneasily aware of growing competition, although in what sectors and from what nations in particular were questions on which it was very difficult to agree. A poll of informed opinion would probably have picked out France, the old rival, as the nation whose competition was most to be feared. Certainly the French display was large and comprehensive. Judged by numbers of Council Medals awarded, France was easily the second nation after Britain (52 to 78); and, in the general class of 'Manufactures', France won 20 medals to Britain's 18.

Germany, while showing evidence of some interesting developments, particularly in chemical and metal products, was still an assemblage of three small kingdoms and a number of minor states. Belgium was plainly highly industrialized. The United States, on the other hand, gave the impression that it was primarily an agricultural country. McCormick's Virginia grain reaper was an admired, prize-winning, exhibit as was the Colt revolver. Robbins and Lawrence rifles with interchangeable parts were also on display. Charles Goodyear, of New Haven, Connecticut, exhibited some novel products made of rubber, including lightweight boats. Ericsson showed a number of his machines, including the 'caloric engine'; it aroused a good deal of interest. Less familiar names associated with very familiar articles were those of S. C. Blodget, of New York, who demonstrated his sewing machine in action, while C. Morey, of Boston, claimed that one of his sewing machines could do the work of four seamstresses and that they were widely used in ready-made clothing establishments (in themselves a most important innovation). For the rest, saddles, mineral and vegetable products, woodworking machinery and small-town craft manufactures such as examples of the book-binder's and printer's art formed the numerical majority of the United States display. Among novelties were Dennington's floating church for the particular benefit of sailors and the now forgotten inventions of Henry Pinkus. In short, remarkable understanding or a degree of prevision would have been required to forecast, on the basis of 1851, the tremendous industrial and technological power that was to develop in the United States over the next hundred years. As for Australia, that young colony had little to display apart from minerals, exotic trees and plants. The Ridley 'Stripper' was not exhibited. How it would have appealed to British and other farmers must remain unknown. The fact that it had no blades that needed to be resharpened was surely an advantage; that it left the stalks unharvested was no less a disadvantage for European farmers.

Immediately after the Exhibition closed the Society of Arts initiated a series of lectures[3] on what could be learned from it. The Rev. Dr Whewell, Master of Trinity College, Cambridge, gave the first lecture. It was, inevitably, couched in very general terms, but he did make two interesting observations. 'Art', he remarked, 'was the mother of science; the comely and vigorous mother of a daughter of far loftier and serener beauty.' Less pompously and with considerable insight, he pointed out that in chemical industry science was now the leader; art – technology – was dependent on and learning from science. Dr Lyon Playfair, the chemist and an early student and disciple of Liebig, pleaded for recognition of the importance of technological and scientific education. This was supported, with respect to a different technology, by

Captain Washington, RN, who praised the American
designers and builders of clipper ships, liners [sic] and
yachts. He commended the celebrated yacht *America*, first
winner of the world-famous cup, named after her. The
America had a wave-line hull, he said, and had been
scientifically designed purely for racing. In the lecture on
agriculture, the McCormick reaper was described and
warmly praised but there was no mention of the Ridley
'Stripper'. Under the heading of 'Philosophical
Instruments' the only mechanical generator of electricity
to be described was a curious, long-forgotten, machine by
one Westmoreland. The lecturer had some hopes of it as
a new source of power. No electric motor was mentioned.
There was no mention of the Colt revolver or of the
Robbins and Lawrence rifles.

Whatever impressions the general public took from the
Great Exhibition, British engineers like Joseph
Whitworth[4], concerned with machine tools and guns,
were much impressed by the Colt revolver and the Robbins
and Lawrence rifle with interchangeable parts.
Comparison with the small arms used by British forces was
inevitable. The British army of the nineteenth century has,
for too long, had a bad press. Since 1918 innumerable
writers have dismissed the officers as aristocratic imbeciles
and the common soldiers as brutally oppressed cannon
fodder. But this is a caricature. A small, entirely volunteer,
body like the British army could not have policed, with
minimal force, something like a quarter of the world's
population unless it had been efficient. The officers of the
technical arms – artillery and engineers – were
particularly intelligent and forward looking. Their
contributions to science and technology over the
nineteenth century would have done credit to a
contemporary university. Like the US army the British
army fought many, colonial-type, minor wars in the course
of the nineteenth century. If therefore, other things being
equal, the US army was equipped with modern and

efficient small arms while the British infantryman had to rely on 'Brown Bess', the musket used at Waterloo and long before then, the responsibility must have lain primarily with British politicians in their anxiety to keep public expenditure as low as possible.

Britain, the last major country to impose conscription (1916), had never raised a large national army so the demand for standardized small arms was limited. The manufacture of muskets and shotguns in Britain was in the hands of the craftsmen of Birmingham, the 'city of a thousand trades'. The individualistic Birmingham craftsmen relied on the manufacture of sporting guns to earn a living. Manual skill, not advanced machine tools, was the practice, even perhaps the boast, of the Birmingham gun trade; it was accordingly quite unable to meet, cheaply and rapidly, a sudden rush of orders from a government facing a national emergency. The Great Exhibition, if nothing else, must have convinced the British army that there were methods of producing better guns more quickly and more cheaply. And the place to find out about these methods was the United States. In 1853 two artillery officers and the Superintendent of the Woolwich Arsenal – the government factory for the production of field and siege guns – were sent to the United States to study American methods and to buy American machines.

The small group was very well received. They were shown over the major armouries, both government and private, and they inspected a large number of other manufacturing plants as well. They commented on the enterprise and intelligence of the American mechanic and they remarked that the endemic shortage of skilled labour meant that labour-saving machinery was at a premium. The workforce offered no resistance to labour-saving machinery; rather the contrary. This was in contrast to the situation in Britain where there was an abundance of labour, skilled and unskilled, and where there was little incentive on the part of employers and every disincentive on the part of employees for the installation of such machinery. The essence of the 'American system', apart from these economic and social factors, was the use of highly specialized machine tools. A particular product would be broken down into a sequence of operations, each one of which could be carried out by a special machine tool. And many, perhaps all, of these tools could be operated by unskilled men. What, Professor Hounshell points out, the British visitors did not know was that achieving interchangeable manufacture had been very expensive; only the state could meet the cost. Here, the visitors might have recalled the earlier occasion when the needs of war and the resources of the state had led to the development of the Portsmouth block-making machinery. It is not too much to claim that the whole process had begun when Arkwright had recognized the advantages of breaking down complex operations into simple ones, each one of which could be mechanized. And those massive assemblies of advanced textile machines in early Cheshire, Derbyshire and Lancashire mills indicate a no less advanced production technology, the details of which have been lost.

In a triumphant demonstration for the British visitors the components of ten muskets, made between 1843 and 1853, were mixed up; satisfactory muskets were then assembled from these components, selected at random. We should, however, remember that, as Professor Rae has noted, the German goldsmith, Johan Gutenberg, had achieved practically perfect interchangeability with his type mould four hundred years before the introduction of the 'American system'. It was the type mould and not the printing press that was at the heart of what was the revolution in production engineering of the fifteenth century, the production being of books.

The British government decided to build an entirely new factory for the manufacture of small arms for the army. American methods as well as the American high-production machines bought by the three visitors were to be used. The site of the new factory was to be at Enfield, to the north-east of London, where there had been a small government armoury. Construction began in 1855, too late for the factory to supply new guns to the army for the war with Russia that broke out in 1854. Consequently the British army, unlike the allied French army, went to war equipped with antique muskets. In the end, however, Enfield was a complete success; the rifles used by British armies in both World Wars bore the name 'Enfield'. In response to the challenge, a number of Birmingham gunsmiths came together and formed a private company to compete with the government factory; they called the firm 'Birmingham Small Arms' (BSA). The beginning of the manufacture of machine tools in the Birmingham area can be dated from this episode.

The outbreak of the war with Russia – the Crimean war – in which Britain had France and Turkey as allies, resulted in a slight setback in confidence in progress, which had been at a peak at the time of the Great Exhibition. The war is remembered in Britain for the Charge of the Light Brigade, a popular poem on the subject, and the story of the Thin Red Line. It is remembered in Britain and the world for the astonishing achievement of Florence Nightingale (1820–1910) – quite clearly an outsider to army medical services – in reforming the craft of nursing, military and civilian, and establishing it as an eminently respectable profession. No worthy poem was written about her. Another new technology played a part in the Crimean war; for the first time in a major war newspaper correspondents and the telegraph played a part in keeping a critical public well informed of the conditions under which the war was fought and its disasters were duly reported. It was also a war in which the camera recorded accurately, without the exaggerations of the poet and the imposed heroism of the artist, the scenes of fighting and the casualties after battle.

The war had other technological and socio-technical consequences. Lyon Playfair, a strong advocate of government support for science, made the interesting suggestion that shells filled with poison gas might be fired into Russian ships; the proposal was not accepted. British soldiers, returning from the Crimea, brought back the cigarette habit, learned from their allies, the Turks. The habit eventually gave rise to a new industry. They had also learned to grow beards as a protection against the Russian winter; the fashion was adopted at home, to the detriment of the Sheffield steel industry.

One substantial technological advance that had its origin, albeit indirectly, in the Crimean war was the invention of the Bessemer steel-making process. Although without any formal training, Henry Bessemer (1813–1898) was a prolific and versatile inventor. The outbreak of the Crimean war led him to take an interest in the design of heavy guns. Steel seemed to be the ideal metal as it had the strength of, but was lighter than, wrought iron. Unfortunately the craft techniques then in use for the production of steel meant that it was far too expensive. Bessemer therefore sought a method of producing steel cheaply and in great quantities. The puddling process removed carbon from iron but could hardly be used for steel making as the carbon content of the metal (between 0.1 and 1 per cent) would keep it liquid in the furnace. After many experiments Bessemer found the answer[5]. He described it to a meeting of the British Association in 1856, making the claim, which must have seemed absurd, that he made steel without using fuel. The Bessemer 'converter' is a large, drum-shaped vessel with an open top and an air pipe at the bottom. It is lined with a suitable refractory material and can be tilted through 90° to be filled with molten pig iron. The converter is then turned to the vertical position and a blast of air forced through the molten metal. The carbon burns more rapidly than in the puddling process and the heat generated keeps the metal molten.

When the spectacular display of flames and sparks dies down to the degree that indicates that the right carbon content has been reached, the molten steel can be poured out of the converter.

A few years before Bessemer invented his converter William Kelly, a Kentucky craftsman, discovered that an air blast through molten pig iron generated enough heat, by burning off the carbon, to keep the metal liquid. He was working this process before Bessemer delivered his paper in 1856, and in 1857 was granted a patent for it. But he was not able to take it any further and fame and fortune passed to Bessemer.

The Bessemer converter enormously increased the world production of steel and transformed industries that had previously relied on heavy wrought iron. There were, however, initial problems. Bessemer had used Swedish iron, which was non-phosphoric. When he tried to use British pig iron, which was phosphoric, the process failed. Another difficulty was caused by blow holes appearing in the steel. This was cured by the addition of manganese. It was not until 1875 that Sidney Gilchrist Thomas and Percy Gilchrist, who were cousins and neither of whom was engaged in the iron and steel industry, discovered that by adding limestone to the refractory material phosphoric pig iron could be used in a Bessemer converter.

Although Bessemer's invention was not based on theory, apart, that is, from knowledge of the role of carbon in steel making, his invention, with its wide-ranging technological and economic consequences, led to the establishment of metallurgy as a study on the border between science and technology. By varying the carbon content and with different additives such as manganese, tungsten, chromium, vanadium, etc., different types and qualities of steel can be obtained. The ancient craft was now acquiring a systematic or scientific basis.

As the century progressed the Bessemer process was steadily overtaken by the Siemens–Martin, open-hearth process. This depended on the prior invention of the regenerative furnace by Friedrich and Wilhelm (Sir William) Siemens. The original idea was that of a furnace in which waste heat preheated the gaseous fuel and air entering the furnace (cf. the Stirling regenerator). This made it possible to obtain far higher furnace temperatures with greater economy. The brothers Pierre and Emile Martin, in France, were the first to use the Siemens regenerative furnace specifically to produce steel. In their open-hearth process, a suitable mixture of pig iron (high in carbon content) and steel or wrought iron (both low in carbon), together with iron ore, as an oxidizing agent, is heated in a high-capacity open-hearth furnace. The process enables much larger quantities of steel to be smelted than the Bessemer process could handle and it allows closer control of the quality. A final stage in the steel-making process that came much later was the introduction of the electric furnace. The smelting in this process is done by heating due to heavy electric currents. This is not a primary process; it is used for the refinement of Bessemer or open-hearth steel.

Water Power in the United States

Another ancient craft that came close to technological perfection by the middle of the century was that of water power. Running water was still the major source of power for American mills, forges and factories. Steam engines were used largely for transportation, by river boat or railroad. American millwrights were craftsmen concerned with small water-wheel installations in relatively isolated farms or small communities. However, as in Britain, there were also large textile mills and, although the New England states were well endowed with rivers and streams, as a textile mill prospered and expanded so the demand for greater power and more efficient machines made itself felt.

It was easy enough to design a water-wheel which roughly approximated to the conditions that the water entered without shock and left without velocity. The buckets had to be big enough to hold all the water that poured in and the wheel should turn slowly so that the water left without appreciable velocity. The column-of-water engine could, more nearly, approach the ideal and, in addition, could utilize a far greater head of water than any water-wheel but, like the water-wheel, it had its limitations. The high-speed turbine was the nineteenth-century answer to virtually all problems. Its successful development depended on additional factors: the capacity to manufacture large iron machines with greater precision than the eighteenth century could attain and with better instrumentation and theory than the eighteenth-century millwrights and engineers possessed. In particular there was the problem of ensuring that the water entered the buckets tangentially, with minimum turbulence, which meant matching the velocity and direction of the water to the velocity of the wheel; and there was the highly complex problem of determining the optimum shape for the buckets. Losses due to turbulence, a minor factor that could be ignored in the case of the overshot water-wheel, were a serious matter in the case of high-speed turbines. The problems of the turbine were, therefore, both theoretical and experimental.

Professor Layton[6] has called attention to the work of two New England millwrights, Arthur and Zebulon Parker, in the third and fourth decades of the century. The tradition of the millwright-improver was one that America had in common with Britain. The Parker brothers were disciples of Oliver Evans who, in turn, had been a disciple of John Smeaton. The criteria for deriving the most power from a river or stream were therefore familiar to them. Their science might have been elementary, basic, by Parisian standards, but it was enough for their purposes. They designed and built a fast-running reaction turbine;

and, by a fortunate accident – this is reminiscent of Triewald's account of the discovery of jet condensation – were able considerably to improve the efficiency of the machine. More generally, the achievements of the millwright-improvers recall the work of the Cornish enginewright-improvers who, guided by experience, intuition, a modest amount of science and benefiting from the interchange of ideas with their fellow-workers, were able to take the pumping engine so far along the road to the practical thermodynamic limit for such engines.

In 1843 Elwood Morris, who was associated with the Franklin Institute, took the Fourneyron turbine and the Prony dynamometer to the United States. This turbine was, as pointed out, an outward flow reaction turbine[7]; it was therefore one of three types, the other two being axial flow and inward flow. It normally ran full of water and immersed in the tail-race. The residual velocity of the water leaving the turbine meant a waste of useful energy. This could be recovered by the 'diffuser' invented in about 1844 by Uriah Boyden, a consulting engineer. Increasing the diameter of a water pipe increases the pressure, as it reduces the velocity of flow. The *vis viva*, or kinetic energy, is reduced as the diameter is increased. Boyden flared the outlet of the turbine under the water in the tail-race until the pressure of the water flowing out was practically the same as that of the surrounding water. This meant that the energy previously lost, due to the water flowing out with appreciable velocity, was restored to the turbine and little of the head or fall of water was wasted.

In 1838 Samuel Howd, of New York, had patented the first inward flow turbine. With this type vanes guided the driving water inwards to the blades, or buckets of the wheel, or runner. Less energy was lost through turbulence in the machine. Later James Thomson, younger brother of William Thomson, later Lord Kelvin, designed a form of inward flow turbine in which the guiding vanes were adjustable so that the angle at which the water entered the runner could be varied to suit the load on and speed of the turbine. Thomson's machine is often called a vortex turbine.

A year or two after the introduction of the Fourneyron turbine Boyden designed and installed an improved version at one of the textile mills at Lowell, on the Merrimack river. The Boyden turbine proved to be remarkably efficient, but it was very expensive. Its particular interest seems to have been that it was followed by an exemplary series of experiments carried out on both inward and outward flow turbines by James B. Francis (1815–92), consulting engineer to all the mills in Lowell. These experiments, published in 1855 under the title *Lowell Hydraulic Experiments*, made extensive use of the Prony dynamometer. Francis, well aware of the complex effects that destroyed the simplicities of the received theory of the turbine, accepted that systematic experiment was the only way to achieve the optimum design. Layton makes the important point that Francis, like Boyden, 'had an affinity for an Anglo-Saxon empirical tradition and their attitudes may have reflected national styles in science'. Nevertheless, they borrowed what was necessary from theory so that their work was a synthesis of the best practices of the engineer and the millwright. The impulse turbine works by using a nozzle to convert the potential energy of a relatively static column of water into the *vis viva*, or kinetic, form. A fast-flowing jet of water is directed in such a way as to strike the blades of the runner tangentially to minimize shock. The simplest form of impulse turbine is the Pelton wheel, invented by Lester Pelton when working in the California goldfields. The wheel, in this machine, carries a number of hemispherical cups. A smooth jet of water is directed at the edge of the cup, the curvature of which then guides the water round so that it emerges on the other side of the cup, travelling in the opposite direction with a velocity practically equal and opposite to that of the wheel. In this way the energy of the jet of water is given up to the wheel. The smooth jet of water is formed in the same way that a garden hose can produce a smooth jet instead of the spray that is usually required.

Beginning with Parent's erroneous but very fruitful paper of 1704 a succession of millwrights and engineers, mainly French and American, had brought the technology of water power to a high pitch of efficiency by the middle of the nineteenth century. By that time the potential energy of any given fall could be fully exploited, the best machine specified for any purpose and power developed at high speed or low speed as required.

A New Form of Chemical Industry

A third antique craft was, during the 1850s, to begin an even more radical change than those effected in steel making and water power. By 1851 the dyestuffs industry, a vital ancillary to the mechanized and revolutionary textile manufactures, was a highly efficient commercial, although craft-based, business[8]. The staple dyes were red and blue, obtained from the madder root and indigo respectively. In addition there was a wide range of other natural dyes, such as woad, weld, fustic, cochineal, logwood, etc., that came from all over the world. The dyestuffs trade was organized on the basis of factoring, rather like much of the modern coffee, tea or tobacco trades. Using recipes passed on from master to apprentice and kept secret within individual firms, master dyers could produce a wide range of excellent and reasonably fast colours. However, in 1856 a discovery was made that was completely to change the industry and end the rule of the factors and the skilled dyer.

In 1845, under the auspices of the Prince Consort and Sir James Clark, the Queen's physician, the Royal College of Chemistry was established in London. Its purpose was to provide a research education in chemistry on German lines (British university education at that time was still almost entirely didactic and research had no place in the syllabuses). The head of the little college was A. W. Hofmann, nominated by Liebig, and teaching was on the Giessen model. In 1856 a student at the College, W. H. Perkin (1838–1907), was given a problem by Hofmann: he was to attempt the synthesis of quinine. The method could not have succeeded but in the course of his researches using aniline derived from coal tar Perkin found a black residue from which he extracted a promising purple substance that had the properties of a dye. Realizing its significance, encouraged by a favourable report from a firm of dyers and backed by his father and brother, young Perkin launched out as a manufacturer of the new dyestuff. The moment was propitious. Mansfield, another student of Hofmann's, had just discovered a method of separating benzene from coal tar; Zinin, a student of Liebig's, had found out how to reduce nitrobenzene to aniline and Béchamp had improved the process. Owing to the thriving gas industry, coal tar was freely available and, thanks to gas lighting, new, bright colours were very popular; finally, by a fortunate coincidence, purple, or mauve, was a fashionable colour.

A number of scientifically compounded dyes had been discovered and successfully marketed before 1856. For example, in 1822 Hartmann, of Munster, introduced a widely used bronze-coloured dye, derived from manganese; in 1840 a chromic oxide dye was marketed and it, too, was very popular. Perhaps the most interesting of these dyes was *murexide*, a derivative of purpuric acid, which Liebig and Wohler suggested could be used as a pink dye for silk. The proposal was not taken up, however, until 1851 when Saac, in France, succeeded in using its related compound, alloxan, to dye silk and wool and to print calicos. At about the same time a factory was opened in Manchester for the production of murexide.

These dyes had, however, been individual substances

and none had led to further development. Perkin's 'mauveine', on the contrary, was to be the first of a long series of aniline dyes. Three years later, Emile Verguin discovered another aniline dye that he named 'magenta', after the scene of one of Napoleon III's victories. After that organic chemists discovered more and more synthetic dyes. Notably, in 1862 Peter Griess discovered the first of the azo dyes that opened up wide prospects of further dyes, and in 1868 Graebe and Liebermann announced, one day ahead of Perkin, the synthesis of alizarine from anthracene. Alizarine was the colouring element in madder and the discovery destroyed the French madder-growing industry. As for Perkin, he made a large fortune and retired to a life of leisurely research at the advanced age of 38.

There were a number of interesting circumstances surrounding Perkin's discovery and his exploitation of it. He was fortunate in that the market for his 'mauveine', the huge British textile industry, had an impressive record of willingness to innovate and had already shown that it readily accepted new dyestuffs provided they met the technical requirements and satisfied the customers. He was also fortunate that his evident gifts as an entrepreneur were supported by his talents as a chemical engineer: he and his partners had to design and make the plant to manufacture the new dye. And, finally, he was fortunate in that although he was only eighteen his father and brother were willing to back him in his enterprise. Perkin, whose father was a boat-builder, had no prior connection whatsoever with the technology or the industry that he revolutionized.

But the significance of Perkin's discovery and its exploitation goes beyond the success story of a Victorian innovator. In the first place, although the individual episode was set in Britain, the scientific context was German. The Royal College of Chemistry was effectively an overseas branch of Liebig's laboratory. The style and content of teaching and research were German; it was fortuitous that the particular discovery was made in London. In the second place, the discovery of 'mauveine' was due, unambiguously, to scientific research both in the methodological and in the institutional senses. Finally, the direct link between the aniline dye industry and the scientific research on which it was based was to lead, in the following decades, to the first industrial research laboratories; understandably, these were to be in Germany. And the names of modern industrial concerns such as Badische Anilin and Agfa as well as the Swiss CIBA-Geigy all testify to the importance of the discovery of 1856. With Perkin, in short, a new mode of technological innovation appears.

It is, at first sight, surprising that medicine, among the oldest of crafts and historically closely related to chemistry, had shown least evidence of progress towards systematization or of the influence of science. The biological sciences, to which medicine is allied, are more complex, or basically more obscure, than the physical sciences. Furthermore, while a prototype engine or machine may well fail, the consequences of failure do not usually endanger the engineer in charge or his sponsors. Medical pioneers almost invariably faced danger; often acute danger. Physical risks and the absence of scientific knowledge did not discourage Edward Jenner who, at the end of the eighteenth century, had introduced the practice of vaccination against smallpox in the face of strong opposition from the fellow-members of his profession. Jenner's innovation was based, as he himself states, on the folk wisdom of country people. Although Humphry Davy had, in 1799, suggested the use of nitrous oxide as an anaesthetic for minor operations, it was only from 1844 onwards that William T. G. Morton (1819–68), a New England dentist, began experimenting with sulphuric ether as an anaesthetic (the word had not then been coined) for the extraction of teeth. In this work he was associated with Horace Wells, a fellow-dentist, and John C. Warren, a

Boston surgeon, who carried out the first operation (removal of a tumour) under anaesthetic. Morton, like Jenner before him, met with great opposition and had to make considerable financial sacrifices to carry out his work. There were such problems as the determination of the most effective substance, the best method of application and the correct dosage.

The transmission of diseases, in the absence of the germ theory, was a mystery. The problem could be denied by arguing that all diseases were *sui generis*, which was unconvincing and an admission of defeat; or, more commonly, that they were spread by a mysterious *miasma*, which concealed ignorance under a word; more plausibly, with the rise of scientific chemistry and the atomic theory, the miasma was thought to be due to a host of tiny chemical particles in the air. The last interpretation became popular in the early years of the nineteenth century. In some cases, however, the problem proved solvable before the establishment and acceptance of the germ theory. In 1843 Oliver Wendell Holmes (1809–94) demonstrated, on the basis of a large number of case studies, that puerperal fever (childbed fever, septicaemia) 'is so far contagious as to be frequently carried from patient to patient by physicians and nurses'[9]. Four years later the Hungarian obstetrician Ignaz P. Semmelweis (1818–65), reached the same conclusion, based this time on his experience at the Vienna maternity hospital. He noticed that cases of puerperal fever were far more numerous in the wards served by medical students who frequented the *post mortem* rooms than in the wards served by nurses who never entered the *post mortem* rooms. Semmelweis required all students to wash their hands and their instruments in a disinfecting liquid; the result was the virtual elimination of puerperal fever in wards where it had previously raged.

Cholera reached Europe in 1830 and during the remainder of the century spread throughout that continent and the United States. The chemical, or particle, theory, then popular, implied that its spread could be checked in hospitals by heat treatment of all things that had been in contact with a cholera patient. However the real method of propagation was discovered as a result of the outbreak in London in 1854. John Snow (1813–58), the pioneer of anaesthetics in Britain and a practising physician, marked on a large-scale map the houses where cholera patients lived. He found, in the area he was studying, a cluster of such houses around a source of water supply: the Broad Street pump. Cholera, he concluded, was water-borne and the solution to the problem was a supply of pure water, uncontaminated by contact with sources of cholera. Although thirty years were to elapse before Koch identified the cholera bacillus, Snow's discovery gave a great boost to civil engineering by hastening the development of public water works and sewerage.

The examples of puerperal fever and cholera confirm what we have noticed before that a craft, or a technology, can make substantial advances in advance of a satisfactory scientific theory. The successes of Semmelweis and Snow can fairly be described as examples of precocious technology. They also exemplify the truth of the Victorian proverb: cleanliness is next to godliness.

The progress of science and technology appears, to many scholars as well as to the general reader, to consist in a series of dramatic advances interspersed with periods of relatively slow evolutionary improvement. This view is substantially acceptable but its interpretation involves a certain danger. It would be quite incorrect to suppose that at any given time the future can be foreseen and the course of evolutionary improvement generally acknowledged. Still less can the next dramatic advance be predicted. What can surely be asserted is that, as the ragged front lines of science and technology advance, so the consolidating armies of technicians and craftsmen follow on at a more or less uniform pace. Although the great majority of technicians and craftsmen may have little or no contact with the front lines of science and technology, they will have at their command more scientific and technological knowledge than their predecessors had before them. Such knowledge will have been mediated – digested, filtered, interpreted – by their apprentice masters and teachers. This means that the area of empirical invention, defined as including those things whose principles would have been readily understood by Archimedes, must be restricted. While Archimedes would certainly not have understood the nature and process of mauveine manufacture, would he have understood the Parkers' turbine? The answer must be, only to a degree. The 'how' of it – a water-powered machine – he would understand; the 'why' would escape him. He could know nothing of the residue of eighteenth-century theory plus experimental knowledge involved in the construction of the machine. Such 'elementary' textbooks as those by John Banks and Oliver Evans have their place in the history of technology.

The new production technique exemplified in the manufacture of guns in the United States, with its antecedents in the Portsmouth block-making machines and Arkwright's textile processes, was an aspect of the division of labour. Adam Smith's conception of the division of labour was limited to strictly manual operations and, for that reason, represented a dead end. Further progress would have been impossible without genetic and surgical engineering (bionic men and women). The improvability of machines, as Marx observed, is without known limits and the only restrictions can be due to the nature of the raw material being used. When therefore the much-derided Andrew Ure talked of the automatic factory rendering Smith's doctrine of the division of labour untenable, he had a good point. The advances in specialized machines in the mid-nineteenth century were a major step in that direction. Beginning with Arkwright's mills, the skills of the individual craftsman working on his own were being displaced by systems of production that brought with them a new logic of their own.

The middle period of the nineteenth century, the time of the Great International Exhibition, which was a landmark in the history of technology, was an immensely fruitful era when the ancient arts of dyeing and iron and steel making began to be put on a scientific or systematic basis. There were also a number of isolated innovations in the ancient craft of medicine. It was also the time when there were clear signs of the scientific and technological revival of Germany and the beginning of the rapid technological growth of the United States. The Exhibition could almost be taken as symbolic of Britain handing over the torch to other countries in what was the relay race of technology. It certainly signified the completion of the Industrial Revolution in Britain. Economic historians have disputed whether there was an industrial 'revolution' or whether what happened was merely an acceleration in a trend that had been developing in British society for many years prior to the middle of the eighteenth century. Leaving aside the pedantic quibble as to whether the word 'revolution', used in this context, merely implies the return to an initial state, it is surely applicable to profound and irreversible changes. An over-critical, not to say carping, interpretation would dissolve all changes – even the French 'Revolution' – into a rapid acceleration of a developing trend. This, we suspect, would not be acceptable to most scholars. Gibbon could refer to the 'long revolution' in Roman law between the time of the Twelve Tables and the code of Justinian. The historian of technology can have no doubts. The tangible remains of eighteenth-century industry – at Cromford, Coalbrookdale and elsewhere – are testimonies to a quite remarkable qualitative and quantitative change. The unprecedented development of power technology, the transformation of the textile industries, the invention and growth of the steam railroad, the steam ship, the electric telegraph and the establishment of chemical industry between the mid-eighteenth century and 1851 allow of no other description than a technological revolution comparable to, and exceeding in importance (at least as far as the great majority of humankind is concerned), the Scientific Revolution of the seventeenth century.

13

Three Decades of Innovation

Indisputably, in popular opinion and in economic fact, the single most important technological agent over the greater part of the nineteenth century was the ever-triumphant steam engine. Indeed, if the popular biographies of Samuel Smiles and the science fiction of Jules Verne are reasonable guides, there was probably more interest in, and approval of, technology in the middle of the nineteenth century than at any time before or since then. After botany, the most popular and one of the most useful of sciences was geology. It was a thoroughly respectable, healthy, outdoor science that anyone could pursue and the rewards – intriguing collections of fossils – were freely available to all. Moreover, by the middle of the nineteenth century, geology was to occupy the centre stage of public debate on issues understood to be of profound importance: the nature of humankind and the evidence, if any, for design and purpose in the world.

The career of Charles Darwin (1809–82) must surely be more widely known than that of any other scientist, and his book, *The Origin of Species* (1859), has had a wider and more immediate impact than any other scientific work. To what extent this was due to a skilful publicity campaign waged by T. H. Huxley may be debated. What is undeniable is that Darwin came to scientific maturity when conditions favoured a grand synthesis.

A. G. Werner's 'Neptunian' theory of the origins of all rock formations had proved unsustainable. The bleak uniformitarian geology of James Hutton, i.e. that we must assume no agency ever acted in the formation of the surface of the earth that we do not see operating at the present time, was scientifically cautious, although its implications for the religious were uncomfortable. The works of William Smith, the canal engineer, and Georges Cuvier established the relationship between fossil forms and strata, together with the fact that great species had come into existence, developed and then become extinct. These ideas were expounded in the three volumes of Charles Lyell's *Principles of Geology* (1830–33), a work that may be said to have done for geology what Hermann Boerhaave's *Elementa chemiae* did for chemistry. Darwin took volume 1 of *Principles of Geology* with him when he sailed in HMS *Beagle* in the memorable voyage that, as far as the world was concerned, was to end on the Galapagos islands. Volume 2 caught up with him when the ship reached Montevideo. The problem of evolution was clearly posed in the book; his journey to South America presented the eager naturalist with a challenging variety of living forms and their modifications.

Two key principles were to guide Darwin in the elaboration of his idea of the development of species by natural selection. The first was derived from T. R. Malthus' theory of population growth. Malthus, a mathematician, had argued that, as human population tended to increase in geometrical proportion while food resources seemed to increase only in arithmetical proportion, there must be a relentless pressure on food resources so that the bulk of the population must always be faced by starvation. What was true of humans must also be true of all animals. The second principle was borrowed from the well-established practices of animal breeders, in particular those of Robert Bakewell. According to Darwin, Nature, in the form of the environment, can act, given a sufficiently long time, as an animal breeder. Characteristics which favour survival in that environment will tend gradually to predominate in animals that are always under population pressure. As the variety of environments is virtually limitless, so the variety of species is enormous. There were, of course, problems with Darwin's synthesis. One was the development of the eye. In advanced forms its utility is obvious; in the initial stages of development the marginal advantage of a slightly light-sensitive membrane is far from clear. Another problem was that of the mechanism of transmission of advantageous modifications from parents to offspring. This was answered when Gregor Mendel founded the science of genetics.

Darwin's comprehensive and immensely influential synthesis drew, as we noted, on technological as well as strictly scientific sources. The mining industry and canal engineering contributed to the foundations of geology while the achievements of animal breeders offered a hint as to how Nature might operate. Certainly the publication of *Origin of Species* and the debate that followed marked one of the three pinnacles of public interest in science; the other two occurred following the acceptance of Newton's *Principia* and the publication of popular accounts of Einstein's theory of relativity in the 1920s and 1930s. A possible fourth occasion was the direct consequence of the dropping of the two atomic bombs in 1945. Whether Darwin's theory has been as useful, scientifically and practically, as the two main innovations in physical science of the same period is open to discussion.

The two key ideas introduced into physical science in the nineteenth century were those of field theory and of energy. Both were radical departures from Newtonian orthodoxy but this does not seem to have been realized at the time. Although Faraday's idea of the field pre-dated the energy doctrine, the latter was developed more rapidly. The man who must have credit for putting energy at the centre of physics was James Prescott Joule (1818–89), the second son of a brewer who was wealthy enough to enable his sons and daughter to live independent, comfortable lives without the disagreeable necessity of earning a living. For all that, Joule was a dedicated scientist who devoted his spare time to research. (Joule's siblings showed no interest in the brewery business. Joule worked at it until the father died, after which he sold the business.)

Joule[1] was one of those who were caught up in the 'electrical euphoria' – the enthusiasm for electric power – that had followed the publication of Jacobi's paper in 1835. Joule differed from the majority of those who participated in the electrical euphoria in the depth of his scientific insight, the soundness of his scientific method (he had been well taught by John Dalton) and his highly developed experimental skills. In the course of his systematic search for the best possible electric motor he studied the law of attraction between electromagnets, examined the suitability of different types of iron for making electromagnets and discovered the law that an electric current generates heat in proportion to the square of the current and the electric resistance (i^2r). Faraday had shown that the quantity of electricity generated is proportional to the amount of metal – usually zinc – used up in the battery. Joule demonstrated that the amount of heat generated by the current from a battery is the same as if an equal weight of battery metal had been burned in an atmosphere of oxygen; this was an experiment that demanded quite exceptional skill and, at the same time, a clear understanding.

Joule was disappointed to find, when he had built his best possible motor, that its maximum duty – Watt's measure: the amount of work that it could do for the 'combustion', or oxidation, of a given weight of fuel (zinc) – was only one fifth that of the best Cornish steam engine (at Fowey Consols) burning coal. And the zinc burned by the battery was many times more expensive than the coal burned in the furnace of the steam engine. Joule remarked that he almost gave up hope of the electric motor.

The back-EMF, caused by electromagnetic induction (Faraday) as the armature spun round, opposed the driving current so that the faster the motor revolved, the smaller the driving current. Joule found that as the speed was increased so the power increased up to the point when the driving current was reduced to one half its maximum value (which was when the motor was held motionless). Thereafter the power decreased. However, he found that the duty went on increasing steadily and it was clear that it would be greatest when the driving current was almost zero. That would be when the minimum of battery metal was being consumed. And the duty was, for engineers, the accepted measure of economic efficiency. The Cornish steam engines were well known to engineers for their very high duties.

Implicit in the sequence of Joule's discoveries was a fundamental question. The heat generated by the electric current from a battery was exactly accounted for by the chemical action – the combustion – in the battery. 'Electricity', he wrote, 'is a grand agent for carrying, arranging and converting chemical heat.' But where could the heat come from if the electric current was generated by a magneto? There is no chemical action, no combustion, in the action of a magneto.

According to the accepted theory heat is always conserved; it cannot be converted into anything else nor can it be created. It must either be tangible heat or it must be latent or combined (as, for example, in a fuel). Although von Rumford and Davy had insisted that heat could not be material and must be of the nature of motion, their arguments and experiments were inconclusive and were rejected by the scientific community. On the basis of the accepted theory, therefore, any heat due to a current must be the result of combustion (release of combined heat) in a battery or of a cooling of the armature of a magneto. Through a series of skilful and ingenious experiments Joule demonstrated that the armature was not cooled; on the contrary it was heated according to his heating law. There being no cooling anywhere and no chemical action, Joule concluded that the heat that appeared must be the mechanical effort, the *vis viva* used to drive the magneto in another form. Not, it should be noted, just 'motion' but *vis viva*, expressed in the engineer's measure of 'work'. He found the exchange rate between mechanical *vis viva* and heat by measuring the work done to drive his magneto and the amount of heat that the resulting current generated. Joule did not stop at this point. He realized that the exchange rate between heat energy and mechanical energy (to use modern expressions) must be constant, no matter what processes, no matter what materials, are involved. Consequently over the following years, 1843–47, he carried out experiments to show that the work done, or the *vis viva* expended, to compress air, to overcome friction between iron discs and to overcome the fluid friction of different liquids always produced the same quantity of heat. The exchange rate between the two forms of energy is fixed and invariable. The heat of any body must, he concluded, be due to the mechanical energy of the constituent atoms as they vibrate or, in the case of a gas, as they shoot about like tiny projectiles.

Newton, in the Preface to the first edition of the *Principia*, had expressed the hope that his system of mechanics would prove capable of solving the problems of physical science. Thereafter the possibility of finding a universal mechanics formed one of the publicly acknowledged goals of science, or 'philosophy'. However, the system of mechanics that was used to effect this correlation was not, strictly, that of Newton; it was that of Leibniz and, more particularly, that of the engineers.

There were a number of other claimants to have discovered the dynamical theory of heat and therefore to have established the energy doctrine. Of them J. R. Mayer has by far the strongest case. All of them gained their insights through one kind of observation and generalized from that standpoint without supporting evidence; still less did they show how the doctrine applied in other fields, such as electricity or electrochemistry. Interestingly, virtually all of these claimants were outside the academic establishment, the world of official science, and most had an engineering background. It was Joule, who falls into the same category, and Joule alone, who realized that the validity of the principle had to be clearly demonstrated over as wide a range as possible. And this he did. That is why the international unit of energy is known as the joule.

Joule had shown, as no one else had done, the full range of the energy principle. It was not enough to have had the right idea. He showed why the idea was right and how it applied in very varied circumstances. Although the wisdom of hindsight tends to conceal the fact, it was perfectly possible to have had the right idea and to have failed to understand how to apply it. Joule did not fail to understand its full application and, with his work, ideas and speculations about the electrical fluid and kindred dead ends disappeared from scientific discourse.

Joule's ideas and those of Carnot were reconciled, simply and effectively, by R. J. E. Clausius (1822–88), who retained the Carnot cycle as the ultimate limit of efficiency but added that some of the heat taken in actually disappears to reappear in exact measure in the form of the work done. Since heat is not conserved but is a form of energy (unlike water, which is always conserved in the working of a hydraulic engine), it follows that Carnot's argument that no engine can be more efficient than a reversible one must be restated. A more efficient engine can be imagined: one that absorbed heat energy from the cold body, or condenser, and converted it into useful work. Indeed, the oceans of the world are great reservoirs of low-grade, or low-temperature, heat. And an engine which could harness that energy without requiring a colder body would not violate the new axiom of the conservation of energy and could be more efficient than a reversible Carnot engine. Against this, Clausius pointed out that the net effect of such an engine, driving a Carnot engine in reverse, would be for heat energy to flow from a cold body to a hot body without any compensating change of any sort. And this is certainly against all human experience. Heat cannot flow from cold to hot without an equivalent compensation, somewhere, at some time. The Carnot engine therefore remains the criterion of ultimate efficiency.

Thermodynamics, as the new science and technology was called by William Thomson (1849), introduced an entirely new element into science. The relationship between heat energy and energy in other forms is not symmetrical. There is no limit to the conversion of mechanical energy into heat energy, but the reverse process is governed by the efficiency of the cycle. Heat can flow from hot to cold and the maximum of work can be generated by a Carnot engine; or it can flow by straightforward conduction and no work be done. The second case can be described as an engine of zero efficiency.

Energy is concerned in all the actions of Nature and technics. In every individual action – physical, chemical or biological – some energy is transformed into heat and that heat, of its nature, flows to the body or area of lowest temperature where it is unavailable to do work or for transformation into other forms of energy *unless a body of still lower temperature can be found.* After much thought Clausius found a way of expressing this natural tendency of energy to run to waste; he called his measure the entropy. Multiply the entropy change by the lowest available temperature and you get the amount of energy that has been lost in the operation. So the lower the lowest available temperature, the less the wastage. If you could find a body at the absolute zero of temperature (0°K, or – 273°C) there would be no wastage at all.

Science and the Steam Engine

The establishment of the principles of thermodynamics in the middle decades of the nineteenth century occurred at a time when the steam engine was being more widely and rapidly applied than at any previous time. Railroad building in all continents and most countries was at or near its peak; industry, in all economically advanced countries, was rapidly becoming dependent on the steam engine in its many varied forms; the introduction of a successful surface condenser (1855) and the development of the two-cylinder, compound marine engine were due initially to John Elder, a disciple of W. J. M. Rankine (1820–72), himself a pioneer of thermodynamics[2]. This made the steam ship much more economic and able to compete with sailing ships on profitable trade routes. Such was now the variety of engines designed for such a wide variety of purposes that, as Gustave Adolphe Hirn (1815–90) pointed out, a comprehensive theory of the steam engine had become impossible. Any steam engine was, he observed, a collection of necessary compromises: so much depended on the purpose for which the engine was wanted. To a considerable, if undetermined, extent the improvement of engines year by year was a consequence of the steady input of new and improved materials – packings for pistons, improved lubricants, superior-quality iron and steel – new and better machine tools and growing confidence on the part of engineers as they pushed their designs further forward in the light of their own and their colleagues' experience towards bigger, more powerful and faster engines. In short, advance was evolutionary and along much the same lines as had been the case with the Newcomen engine in the eighteenth century and the Cornish pumping engine of the early decades of the nineteenth century. Advance, while rapid, was achieved largely empirically.

The measure of 'duty' had served engineers well enough from the time of Watt to the period when Lean's monthly reports were most influential. As early as 1839, however, a Mr Parkes protested to the Institution of Civil Engineers in London that it was imprecise and could be misleading.

A good engine, badly operated, or supplied with steam from an inefficient boiler, would show a misleadingly poor duty. He therefore suggested that a different measure be used. By the end of the century the common measure used was the weight of feed water supplied (presumed equal to the steam generated), compared with the work done. Parkes' criticism, incidentally, did not affect Joule's conclusion about his electric motor, for the duty of the steam engine he compared it with had been measured under exacting circumstances and his measure of the duty of his electric motor was evidently precise. In fact, Joule's work was to provide the means to satisfy Parkes' demand.

Some time elapsed before thermodynamics could be systematically applied to the design and development of heat engines. The advances due to Joule, Clausius, Thomson, Regnault, Rankine and others in the two mid-century decades could not be rapidly assimilated and transformed into practice by working engineers. Largely due to Rankine, Carnot's basic law, that efficiency was proportional to the temperature range over which the engine worked, was soon accepted by Scottish engineers[3], most particularly by those engaged in the design and manufacture of marine engines, in which field Scottish engineers led the world. Joule's dynamical theory of heat indicated that the attempts made, early in the century, to economize in heat by using a vapour with a low latent heat were wholly misguided. The high latent heat of steam was an asset: it meant that the steam conveyed more heat, which is to say, energy from the furnace to the engine where it was converted into the mechanical form. (Appropriately enough, Joule was elected one of the first Honorary Members of the Institution of Engineers and Shipbuilders in Scotland.) Finally, thermodynamics explained a puzzling feature of the behaviour of saturated steam in the cylinder and predicted a remarkable property of such steam that was verified by experiment (see Note on Saturated Steam at end of chapter).

A particular problem at that time was the wetness of exhaust steam as it left the engine. In the days before thermodynamics this was ascribed to 'priming', or water bubbling over with the steam, just as water comes out of the spout of a boiling kettle. Thermodynamics gave another and sounder reason. When dry, saturated steam expands in a cylinder, doing work by pushing against a piston, some of the steam condenses as its heat energy – 'latent' and 'sensible' – is converted into mechanical energy and the steam cools down (the same process explains the formation of clouds as warm, moist air expands and cools as it rises up). Another – and wasteful – cause of wetness in the cylinder was the cooling of the areas round the exhaust ports where the hot cylinder was connected to the cold condenser. This caused some of the hot steam entering the cylinder to condense only to re-evaporate when the pressure fell on expansion and the exhaust ports opened. The heat associated with this re-evaporation was wasted as the steam rushed into the condenser.

There were four possible solutions to this problem and they were much discussed. The first and oldest was to revive Watt's steam jacket, which by using a separate supply of steam could certainly keep the cylinder hot; but it could be, and was, argued that any spare heat would be better used to generate more and hotter steam to drive the piston. Another and more modern solution was to heat up the steam by passing it through metal pipes exposed to hot furnace gases; that is, to superheat it, in which condition it has the properties of a gas. It is not known who first proposed the use of superheated steam. According to Rankine, writing in 1861, the engines of the American steam ship *Arctic* had, 'many years ago', been worked with superheated steam 'to good effect'. Superheated steam had two attractions for engineers: reduced cylinder condensation and, by Carnot's law, more efficient use of the steam by virtue of the greater temperature range over which the engine worked. In 1859 Hirn demonstrated that, other things being equal, superheated steam yielded more work than saturated steam. Again, it could be argued that the extra heat might have been better used to generate more and hotter steam; in addition, superheating raised serious mechanical and metallurgical problems. Pipes exposed to furnace gases were subject to severe corrosion and the animal and vegetable lubricating oils, then commonly used, dissociated at high temperatures.

It was fortunate that, at this time, the mineral oil industry began to develop. Limited quantities of mineral oil had been in use for some time; it seeped out of some rock formations or could be obtained by the distillation of shale oil. From it lubricating oil could be derived as well as a convenient, lighter component, called kerosine or paraffin. The latter found a ready market as a fuel for lamps and stoves. Hirn later claimed to have built the world's first refinery in 1856 and, at about the same time, a refinery for shale oil was opened in Scotland. In 1859 Colonel E. L. Drake discovered oil underground in Pennsylvania and with this discovery the great oil industry began its rapid growth. Lubricating oil from the new and abundant mineral oils allowed engines to operate at very high temperatures, whether by use of superheated steam or by other means. They were also to make possible the development of new and revolutionary heat engines.

For the present, however, we must return to the problem of cylinder condensation in steam engines. The third solution to the problem was to use more than one cylinder. It was pointed out that compound and triple-expansion engines (the latter began to appear in the early 1870s) were less susceptible to losses due to cylinder wetness as the pressure and temperature falls in each cylinder were smaller than in an equivalent single-cylinder engine; furthermore, only the last cylinder was connected to the cold condenser. The problems, then, were at what degree of expansion was compounding desirable and when should triple-expansion replace double-expansion or compounding? The fourth solution was simply to redesign the valve system.

These were difficult questions, beyond the capacity of the great majority of engine builders to tackle. As was soon pointed out – explicitly – what was required was a new approach: an engineering science that recognized a balance between theory and experiment, that treated the steam engine as an object for systematic study and at the same time took account of the, often conflicting, requirements that taken together must determine the best design for any given purpose. The centres in which the new approach developed were in Scotland, in particular the school founded by Rankine but also indebted to Joule and William Thomson; the American groups associated with B. F. Isherwood, of the US Navy, and Professor Thurston; the German school started by Clausius and Zeuner; and the Belgian school, centred on Liège and Ghent. Perhaps most important of all was the French school started by Hirn, of Mulhouse, and his two associates Hallauer and Leloutre. It was Hirn who carried out the classic – and very difficult – experiments to show that, in the working of a steam engine, heat actually disappears, to be converted into the *equivalent amount* of mechanical energy. By the end of the century the English engineer D. K. Clark admitted that the centre for steam engine research and expertise was Mulhouse.

The Invention of the Dynamo

A quite different practical application of Joule's ideas was the invention of the dynamo or, to give it its full name, the dynamo-electric machine to distinguish it from the magneto-electric engine.

One of Joule's important insights that has been overlooked for too long was his early recognition that the magneto converts mechanical energy into electrical energy[4]; it does not, as was believed at the time, convert magnetism into electricity. The latter belief necessarily implied an upper limit to the capacity of a magneto to generate current. In 1845, Wheatstone and Cooke patented the substitution of an electromagnet for a permanent magnet in a magneto for use in telegraphy. They specified that the current for the electromagnet should come from a battery (they therefore called it a 'voltaic magnet'). The result was very satisfactory, for '. . . a far greater effect is produced than [the] battery could produce if used in electric telegraphs without the intervention of such [a] machine'. They did not question why this was so. Joule could have told them: the mechanical energy used to drive the generator was being converted into electrical energy. The principle of the dynamo – as the etymology of the word indicates – is that mechanical energy is directly converted into electrical energy without any limitation imposed by a magnet. The steam engines of the day could supply immense quantities of mechanical energy cheaply and efficiently. In principle all, in practice most, of such energy should be convertible into electrical energy. Joule's ideas pointed to the way in which this could be done.

In July 1866 Henry Wilde, a wealthy Manchester inventor and fellow-member with Joule of the Manchester Literary and Philosophical Society, described his modified magneto. In his machine the permanent magnet of a magneto was replaced by an electromagnet energized by a magneto driven from the same shaft. This was much more than the replacement of Wheatstone's and Cooke's battery by a magneto; it was, as Wilde made quite clear, a recognition that unlimited mechanical energy could, in principle, be transformed into electrical energy. He pointed out that his machine could be used to energize the fixed electromagnet of a still bigger machine, and so on. No time was lost in developing the idea. Six months later, in January 1867, Werner von Siemens and Charles Wheatstone, almost simultaneously and certainly independently, announced the invention of the self-excited dynamo. In their machines the energizing magneto is dispensed with and the tiny residual magnetism of the fixed electromagnet generates sufficient current, as the machine is started up, to boost the effect of the fixed electromagnet. The process is cumulative until most of the mechanical energy applied to the machine is converted into the electrical form. All forms of dynamos and alternators work on Joule's principle of converting mechanical energy into the electrical form. So, too, do magnetos but in their case the strength of the permanent magnet is a limiting factor, in addition to such common elements as the thickness and conductivity of the wiring used.

That Wheatstone had, in 1845, got so close to the invention of the dynamo and yet, able as he was, had failed to take the final all-important step is confirmation of the key role of the energy doctrine in the completion of that step. Wheatstone, in effect, made amends for his earlier failure by contributing to the invention of the dynamo in 1867.

The Energy Question

The general doctrine of energy was soon understood to have economic and social implications. In his book *The Coal Question* (1865), W. S. Jevons (1835–82), a Manchester economist, explored some of the more disturbing implications as far as Britain was concerned. He had no difficulty in showing that Britain's prosperity depended on the steam engine, and everyone knew that the utility of the steam engine depended on a supply of cheap coal. The recently established Geological Survey made possible a very rough estimate of the economically recoverable coal reserves of the country. The reports of the – also recently established – Mining Records Office gave the tonnage of coal mined per annum. These showed that the rate of extraction was increasing at 3 per cent every year. No advanced mathematics was needed for Jevons to prove that by 1965 Britain would be mining more coal every year than lay under the ground, which, of course, was nonsense. Long before then, as the best seams were worked out, the price of coal would rise and go on rising until the competitive edge of British manufacturing industry was completely eroded and industrial supremacy went to countries, like America, with abundant supplies of cheap coal.

What could be done? Jevons had no difficulty in showing that none of the sources of dilute energy – wind, tides, solar power – could possibly compensate for high-cost coal and he argued that it would be uneconomic to import coal from the United States; far better to use the coal where it lay, in the States, for the Americans are as ingenious, as hard working and as businesslike as the British. With that relish that Victorian economists reserved for their most gloomy forecasts, Jevons concluded that Britain's only choice lay between a short life and a merry one or a protracted decline into mediocrity and poverty.

This was the first authoritative prediction of an energy crisis and it is easy to conclude that Jevons was remarkably prescient in his conclusion. Certainly his attitude, and that of his many contemporaries who were alarmed by his prediction, showed much more responsibility than was evident during the extravagant years that preceded (and followed?) the oil crisis of 1973. But how, we may wonder, would Jevons account for the more recent prosperity of Japan, South Korea, Taiwan, Singapore and Hong Kong, all countries with little or no indigenous coal or oil – or, indeed, Holland, the country with negative energy resources? Perhaps a lack of cheap fuel is an incentive to invention and therefore to economic advance.

The practical development of the concepts of energy may have been delayed in the case of steam engines by a combination of technical difficulties and sociological factors, in the case of the dynamo by technical and economic factors. In one field, however, none of these difficulties was significant. This was in telegraphy. The new breed of telegraph engineers, initially without acquired prejudices, were aware of the need for some internationally agreed standard of electrical resistance, a standard that would be as fundamental and as accurately reproducible as possible. At the 1861 meeting of the British Association (BA) the telegraph engineers Sir Charles Bright and Mr Latimer Clark proposed that such a unit should be called the 'ohmad', in honour of Georg Simon Ohm, the discoverer of the law of electrical resistance. The BA immediately instituted a committee comprising, among others, William Thomson, Charles Wheatstone and Fleeming Jenkin to find the best way of establishing such an international unit. With little delay they concluded that agreed units should be established for electric current and electromotive force as well as for resistance, and that all the units should be expressed in terms of the 'French' (i.e. metric) measures of mass and length. The sufficient precedent had been the 'absolute units' proposed by Wilhelm Weber ten years earlier for electrical units based on − or reduced to − the principles of mechanics. Weber had shown, and the BA Committee accepted, that electrical measurements should be expressed in mechanical units. Thus a unit electrical charge was one that exerted a force of one mechanical unit on an identical charge situated a centimetre away. A unit magnetic pole was defined in the same way and a unit electric current was one that, flowing along a wire one centimetre long and bent into an arc of a circle of one centimetre radius, exerted a unit force on a unit magnetic pole at the centre of the circle. By this means all other electrical measurements − electromotive force, resistance, capacitance, etc. could be standardized[5].

The resulting system of units should, the Committee stated, bear a definite relation to the unit of work, 'the great connecting link between all physical measurements'. This last recommendation brought Joule into the business. His law of electrical heating, involving both resistance and current, formed the link between electricity and energy in its general form. In addition, since the heat generated by a current can be expressed in mechanical measure, the law offered an independent check on the value of a resistance as determined by Weber's method.

The progressive refinement of absolute measures of electrical units was carried on by the BA Committee for year after year up to the outbreak of the First World War. The practical importance of the amp, the volt and the ohm for science need hardly be stressed. For our purposes it is enough to point out that the exact relationship between these units and energy, asserted and established by the Committee, expanded to include Joule and James Clerk Maxwell, *made possible the buying and selling of electrical energy for whatever purpose the customer desired.* In other words, *the electrical supply industry of every nation could only come into being after the quantified energy relationship had been established.* And for this the world is indebted to Thomson, Joule, Maxwell, Wheatstone, Fleeming Jenkin and the other early members of the Committee.

The middle years of the nineteenth century were characterized by scientific synthesis in the physical sciences as well as in biology. Joule had shown how heat could be expressed in terms of – could be reduced to – mechanics, the mechanics of energy. Electricity now underwent a kindred transformation. The process began with the development of Michael Faraday's idea of the electric and magnetic field and its relationship to the idea of the ether. Abundant experimental evidence had made the acceptance of the undulatory theory of light and of radiant heat incontrovertible by the 1850s. If light, travelling from a bright source to the eye of an observer, is supposed to consist of very rapid vibrations, or high-frequency waves, it is only common sense to ask what it is that is waving or vibrating? In the case of breakers on a beach it is sea water that moves up and down; sound waves are rapid vibrations in air, for we know that sound cannot travel through a vacuum where there is no air. What is it that vibrates to give us light? The answer that nineteenth-century science gave with increasing conviction was that light vibrations took place in the elastic ether, a medium that pervaded all matter and space and whose only known function was to carry light and radiant heat waves. The ether was supposed to be very rigid and, at the same time, very elastic in order to transmit the ultra-rapid vibrations of light and radiant heat. In origin the notion of an ether was of respectable age; Descartes, we saw, had used the idea in the seventeenth century, but it can be traced back to classical antiquity.

However by mid-century evidence had been found that there must be a close relationship between light and electromagnetism. Faraday had shown that a powerful electromagnet could produce a marked effect on a ray of light: a strong magnetic field deflected the 'plane of polarization' of a suitably polarized ray. This, Faraday realized, showed that light was, in some way, electromagnetic in nature. From realizing this to being able to show how and in what way light was electromagnetic was an enormous step that not even Faraday could take.

A further clue was no less tantalizing and even more unexpected. Weber's rationalization of electrical units led to two distinct systems. A unit current, exerting unit force on a unit magnetic pole, is one that carries unit electric charge in unit time. A unit charge is also defined as one that exerts unit force on an identical charge. As it happens the first, 'electromagnetic', unit charge turned out to be much greater than the second, 'electrostatic', unit charge. Or, in other words, there are many more electrostatic units than electromagnetic units in the same charge. In 1856–7 Weber and R. H. A. Kohlrausch carefully measured the same charge, first in one set of units and then in the other. They found that if they divided the size of the charge in electrostatic units by its size in electromagnetic units they got a number very close to the velocity of light, as recently measured by Hippolyte Fizeau. Moreover, the dimensions of this number were those of a velocity.

The man who put together the pieces of this natural jigsaw was James Clerk Maxwell (1831–79), a member of a land-owning family in Dumfriesshire, in southern Scotland. Educated at Edinburgh and Cambridge Universities, his was one of the most fertile minds in nineteenth-century science[6]. His work for the British Association Committee testifies to his interest in technological problems and in all probability influenced the elaboration of his seminal field theory. Maxwell began by supposing the magnetic field round a wire carrying a current to be essentially dynamic, for the lines of force vanish the moment the current is turned off. He suggested that the moment a current, consisting of individual charges, starts to flow along a wire it causes a series of vortices, rather like smoke rings, to spin round in the surrounding ether (figure 13.1). The axes of these vortices or ether rings are the lines of force and the directions in which they rotate – clockwise or anticlockwise – determine the direction of the lines of force. The ether rings exert a centrifugal force as they spin and so press outward on each other; at the same time they tend to contract so there is always a tension along their axes; along, that is, the lines of force. (The Scottish engineer W. J. M. Rankine had, with great ingenuity. already explored the possibilities of atoms associated with vortices in his attempts to establish a mechanical theory of heat.)

Once the ether rings nearest to the wire have been set spinning, their motion must be transmitted to the next series, and so on. This must be done in such a way that all the ether rings rotate in the same direction. To ensure this, Maxwell supposed that each ether ring was separated from its neighbours by small particles that acted like idle wheels: that is, they transmitted motion from ring to ring in such a way that all rotated in the same direction.

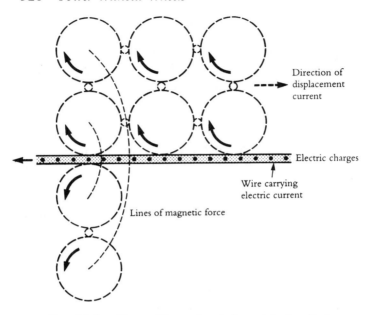

FIG 13.1 *Maxwell's mechanical model of radiation*

However, the ether has to be elastic. The idle wheels, therefore, must have a limited degree of freedom of motion. When a current starts to flow and the rings nearest the wire start to turn, the particles round their edges start to move orbitally, like epicyclic gears. Those furthest from the wire tend to move in the opposite direction to the current but the restraining forces of the elastic ether soon stop them and they begin to rotate, transmitting spin to the next outermost ring.

This simple model suggested how electromagnetic induction, discovered by Faraday, could be explained mechanically. The outermost particles gave rise to the transient secondary current, flowing in the opposite direction to the primary current. And when the primary current and the innermost rings stopped, the continued motion of the outer rings caused a momentary current to flow, this time in the same direction as the original, primary current. Maxwell explains:

It appears therefore that the phenomena of induced currents are part of the process of communicating the rotatory motion of the vortices from one part of the field to another.

In this way Maxwell constructed a model of the electromagnetic field that could account for the observed phenomena of induction. It resembled a busy Victorian workshop with its pulleys, gears and line-shafting, all industriously turning. It was also very British and, in its apparent crudity of conception, it shocked philosophically inclined Continental scientists. But Maxwell was no Victorian philistine; he was developing his profound ideas on the nature of electromagnetic fields:

The conception of a particle having its motion connected with that of a vortex by perfect rolling contact may appear somewhat awkward. I do not bring it forward as a mode of connection existing in nature or even as that which I would willingly assent to as an electrical hypothesis. It is however a mode of connection which is mechanically conceivable and easily investigated . . . I venture to say that anyone who understands the provisional and temporary character of this hypothesis will find himself helped rather than hindered by it in his search after the true interpretation of this phenomena.

Maxwell, in short, was following much the same procedure as that outlined by Rankine in his paper (much admired by Pierre Duhem) on hypothetical and abstractive theories. He went on to consider the analogous case of the application of electric tension, or force, to the ether or to any other non-conductor. The small particles, or idle wheels, will be displaced from their rest positions until restrained by the limited elasticity of the medium. When they are in the process of displacement, when the electric tension is being applied, the drift of particles constitutes a brief, transient electric current. This short-lived 'displacement' current (as Maxwell called it) will generate a transient magnetic field. In its turn this transient magnetic field will, as Faraday had shown, generate a transient electric tension, or electromotive force. And so the process will continue, outwards from the original source of the disturbance, the electric field generating a magnetic field and vice versa.

In his paper of 1865, 'A Dynamical Theory of the Electromagnetic Field', Maxwell dispensed with the ether rings and idle wheels but retained the idea of the displacement current. He recapitulated the simple equations that expressed the way in which an electric current, which could be solely a displacement current, could generate a magnetic field and the way in which a changing magnetic field could generate an electric tension. When these simultaneous equations were simplified by leaving out the primary generating current, he found that the resulting electromagnetic disturbance was propagated with a velocity that was the same as the ratio of the two sets of units and therefore equal to the velocity of light. Light must therefore be an electromagnetic disturbance, with the lines of electric and magnetic force at right angles to each other and to the direction in which the disturbance is travelling. Maxwell offered no explanation as to the electrical or magnetic processes that must take place in the different sources of light, or radiant heat.

Maxwell's views can be summed up very briefly: only in a steady state can a magnetic field exist without causing an electric field and vice versa. When one is changing, it automatically brings the other into being for as long as the change continues. Always these mutually generating fields must be at right angles to each other (as his original model indicated) and they must both travel with the same velocity, which is equal to that of light.

Few, at that time, were competent to understand the abstruse theory of a diffident Scottish academic and fewer still could have appreciated its fundamental importance. Maxwell was not a public figure in the way that Faraday was. He was no brilliant lecturer, capable of charming a fashionable audience that might include royalty – or fascinated children. For the great majority the most exciting event in the world of electrical science and technology at that time was undoubtedly the successful completion of the great Atlantic cable in 1866. This owed much to the enterprise of the American businessman, Cyrus Field, and to the scientific skill of William Thomson, who was knighted for his share in the work. And in this fine enterprise the *Great Eastern* found, at last, a useful and successful role as a cable layer. In the years that followed cables were laid across all the great seas and oceans so that the time required for world-wide communications shortened from weeks or months to seconds. With the advance of science and technology, with the immense improvement in communications, reasonable people could hope that international misunderstandings would be a thing of the past and great wars no longer possible. The Atlantic cable had come into use just as a terrible war ended.

The American Civil War, 1861–65, between the northern and the southern States, was fought out at appalling cost in dead and wounded. It was the first total war and its material consequence was the strengthening of the industrial base of the northern States. The railroad was, for the first time, an important factor in war; cavalry, almost for the last time, also had their part. Techno-logically, there were few war-inspired innovations. The most famous weapon of the time, the Gatling gun, was invented at the very beginning of the War but was not adopted by the US army until after the War was over. The best-known cannons used in the war were the outwardly identical guns designed, independently, by Admiral John A. Dahlgren[7] for the US Navy and the 'Columbiad', designed for the US Army by Major T. J. Rodman. As Professor Johnson has remarked, it is surprising that these guns were developed without any evident co-operation between the two defence departments concerned.

One of the most famous episodes of the War was the drawn fight between the USS *Monitor* and the Confederate armoured vessel *Virginia* (originally the USS *Merrimack*). The *Monitor* – which was to give its name to a class of highly specialized warships – had been designed by John Ericsson and built specifically to challenge the *Virginia*. It incorporated two muzzle-loading Dahlgren guns of eleven-inch bore carried in a revolving turret. Such was the thickness of armour and the relatively low muzzle velocities of the guns on both sides that neither vessel was able to penetrate the armour of its opponent. As both vessels had little freeboard and could operate only in calm water, the immediate lessons of the encounter were limited[8]. In the longer term the fight revealed the early problems of developing steam-propelled iron or steel warships. Advances in the iron and steel industry made bigger guns and thick armour possible; steam meant greater speed and the use of auxiliary machinery; with these came the introduction of the explosive shell and the invention of the torpedo, driven by compressed air (1864); all these raised difficult problems for naval tacticians and naval architects. One indication of the resulting uncertainties was the belief that these changes revived the possibility of ramming as a tactical expedient and, up to about 1910, ships were designed with this in mind. By 1914 most of these problems and uncertainties had been resolved and the big-gun warship had reached a state of near-perfection. This was just at the time when the submarine and, more particularly, aircraft were about to make it obsolete.

In spite of the terrible casualties – recorded in the

photographs of Matthew Brady as well as in numbers – on both sides in the Civil War, there were few innovations in military medicine. Perhaps Miss Nightingale had exhausted all possibilities, given contemporary technological and scientific knowledge, in that field. The significant advances in medicine in the period of the American Civil War were made in Europe and they sprang substantially from the work of Louis Pasteur (1822–95). However, one innovation, significant for the future, during these disastrous years was the Act of Congress, passed in 1863 at the height of the War, establishing the National Academy of Sciences. The duty of the Academy, which had limited membership, was to assist the US government.

Science, Technology and Medicine

Trained as a chemist, Pasteur's initial researches were in organic chemistry, particularly in the optical action of organic crystals and in isomerism. His observation of the different properties of the same substance formed from natural products and by chemical synthesis led him to study the action of alcoholic fermentation, which he believed was due to the action of yeast as a living plant. Cagniard de la Tour and Theodor Schwann had held this belief but their opinion had been discredited by the influential Liebig, who believed that the action was purely chemical. Pasteur's experiments reinstated the theory (1857), which he was able to generalize. He showed that putrefaction, long believed to be of the same basic nature as fermentation, was also due to the action of living organisms – 'germs', or microbes. And in this way he established the germ theory of disease and the science of microbiology.

It was one thing to have established a new and comprehensive theory of disease; it was another thing to apply it effectively in practical medicine. The credit for this belongs to the English surgeon, Joseph (later Lord) Lister

(1827–1912), at the time Professor of Surgery at Glasgow University. A humane man – not all surgeons were humane in those days – he was appalled by the suffering and high rate of mortality in the surgical wards of Glasgow Royal Infirmary. He was studying suppuration, so closely associated with unsuccessful surgery, when the Professor of Chemistry, Thomas Anderson, called his attention to Pasteur's publications of 1861 and 1864. In the latter year Lister heard of the apparently successful use of 'carbolic acid' in the treatment of sewage at Carlisle.

Carbolic acid, or phenol, was a derivative of benzene. It had been discovered and prepared by a group of Manchester chemists, W. Crace Calvert, Alexander McDougall and Angus Smith, as a disinfectant and as a means of suppressing the stench (and miasma!) associated with sewage. Lister at once adopted it as a disinfectant to kill germs: wounds were to be carefully swabbed with carbolic acid and instruments sterilized. The results of its use in cases of compound fracture – a common enough occurrence in an industrial city like Glasgow – were most promising, for the mortality rate dropped sharply. Over the years from 1865 onwards Lister invented and refined techniques for antiseptic surgery, including the famous spray method. In this way modern surgery was born and the ancient craft of medicine, at last, began to acquire a scientific foundation. The common points here are the substitution of a scientific for an empirical base – Semmelweis had no science to back up his intuition – and the importance, which we have stressed in other cases, of development as well as discovery.

Technologically and scientifically, then, the 1860s were immensely fruitful; one of the most creative decades in human history. The American Civil War, however, contemporary with such great advances, cast the shadow of total war that was to disfigure the next century.

A NOTE ON THE SPECIFIC HEAT OF SATURATED STEAM

Steam rising from boiling water is said to be saturated. This does not mean that it is wet; saturated steam is normally quite dry and wholly transparent. It becomes wet as it begins to condense into tiny droplets of water that make it appear cloudy. The specific heat capacity of saturated steam is negative; this may seem a contradiction, certainly a paradox. However, the important word is 'saturated'. In order to expand, saturated steam must do work, either against a piston or similar mechanical device or against atmospheric pressure. The necessary energy can only be got from the heat, latent and 'sensible', of the steam, some of which will necessarily be condensed. The temperature of the wet steam will have fallen. But in order to restore the steam, plus water, to the state of all being saturated steam, heat will have to be *added* to evaporate the moisture. Conversely, if a volume of saturated steam is to be raised in temperature it must be compressed to the pressure it would have if generated from a boiler. But compression will raise the temperature above that which it would have if it came directly from a boiler. It will, in other words, be superheated and heat must be abstracted from it so that it will have the appropriate temperature and pressure for the state of saturation.

14

A Second Industrial Revolution

Werner von Siemens[1], who, with his brothers Carl and Wilhelm (Sir William), contributed so much to the electrical supply industry in Germany, Britain and Russia, was concerned with the cable laying for the Britain to India telegraph line. He later recalled that: 'In my dealings with Englishmen and Frenchmen during the cable layings [in the 1860s] I had often the painful occasion to be convinced in what low esteem the Germans were held as a nation by other peoples.' The successful wars of national consolidation waged by Prussia against Denmark, Austria and, finally, France changed the international status of Germany from that of a loose collection of minor states, inhabited by professors, poets and musicians, to that of a militarily powerful and efficient empire. Change, or rather the acceleration of change, was also taking place in technology and science. The social institutions for what is now called 'high tech' had been assembled in Germany and were working with unrivalled efficiency. Von Humboldt's universities, greatly expanded, were supplemented by state technical colleges, or polytechnics, the eight largest of which were, in 1899, formally elevated to the status of degree-awarding universities of technology. More perceptive observers than Siemens' Englishmen and Frenchmen had long appreciated the changes taking place in Germany: the German Ph.D. degree system was adopted by American universities well before the end of the century.

In 1862 Justin Morrill's Act was signed by President Lincoln. It empowered the States of the Union to bestow large grants of land on state colleges to be established primarily for teaching agriculture and mechanics. A subsequent Act of 1890 gave the land-grant colleges an assured annual income. These colleges subsequently became the great state universities and therefore the backbone of American higher scientific and technological education. In 1865 the Massachusetts Institute of Technology (MIT) – a private foundation – was established, followed in 1876 by the Johns Hopkins University, Baltimore, where the emphasis was on research and advanced scholarship. These institutions constituted the foundation for subsequent American industrial leadership. In Britain development was not so rapid. In 1868 Matthew Arnold's eloquent *Higher Schools and Universities in Germany* compared British higher institutions, including technical colleges, unfavourably with their German counterparts. The book inspired a long campaign on behalf of university reform in Britain.

Change was not restricted to the western nations. Japan had chosen to isolate itself from the rest of the world for many centuries until, in 1853–54, Commodore Matthew Perry, a US Naval Officer in command of a small squadron, was able peaceably to persuade Japan to join the comity of nations. With astonishing speed the Japanese modernized their country. By 1890 they had established the largest school of electrical engineering in the world, having recruited British teachers to staff it.

The rise of the electrical engineering profession and of the academic study of the subject followed the success of the electricity supply industry from 1881 onwards. The contributions of the academic world to the older mechanical and civil engineering professions owed a great deal, at least as far as the English-speaking world was concerned, to the textbooks written by Rankine. In a comparatively short working life Rankine, apart from his engineering and scientific work, wrote four major textbooks beginning with his *Manual of Applied Mechanics*

(1858) which went through 21 editions up to 1921. His other books were on the steam-engine, civil engineering and machinery and millwork[2]. As far as technical education was concerned the evident success of American industry and of American colleges and universities was a considerable influence on Germany and on those countries that followed German practice. The rise of new, or reformed, academic disciplines was not, however, limited to technology. The ancient discipline of physics was recast, it might almost be said founded, on the basis of the new doctrines of energy and field theory, soon to be supplemented by sub-atomic physics. This was a far cry from the old natural philosophy, still more from the ancient Aristotelian physics.

German predominance in the manufacture of synthetic dyestuffs[3] began to be apparent in the early 1870s. Between that time and the end of the century German dyestuffs firms consolidated and extended their leadership. A. von Baeyer succeeded in synthesizing the colouring element – indole – of indigo in 1880. However, to turn this into a commercially feasible manufacturing process proved very difficult; seventeen years of research and development work and the expenditure of a vast amount of money were required before Badische Anilin und Soda Fabrik could market synthetic indigo in 1897. One consequence was that indigo growing virtually ended in India and other tropical countries, and from that time onwards German synthetic dyestuffs dominated the world's markets. Another consequence of this extended German triumph was that British and French troops marched to war in 1914 wearing khaki and blue and red uniforms dyed with German dyestuffs. The latter made good targets for German riflemen.

Another indirect consequence of the rise of the German organic chemicals industry was the discovery by Paul Ehrlich (1854–1915) that aniline dyes used as staining agents for sections under the microscope had differential biological effects. This led him to study the therapeutic effects of such dyestuffs and derived compounds. His best-known discovery was that of a compound he named salvarsan that was to be the first effective remedy against syphilis. He was therefore the originator of what he called chemotherapy. Another, if less imposing, discovery made by the German fine chemicals industry was that of the analgesic known as aspirin.

Photography was a French invention, but for a short time, in the 1850s and 1860s, Manchester took over the leadership. It was there that the first flashlight photographs were taken and John Benjamin Dancer, the scientific instrument maker, invented microphotography[4]. In 1876 there was an exhibition of scientific instruments in London. Ernst Abbe, of Jena University, visited the exhibition, which brought home to him the need for improved optical glass, a need that no one seemed able to satisfy. In 1881 Abbe and Schott, in collaboration with Carl Zeiss, started research at Jena. After a great deal of work they succeeded, achieving a great advance in precision microscopy. Before 1880 Germany had imported optical glass; thereafter she became increasingly a net exporter. Before 1914 she had dominated the European scientific instrument trade. The name of Zeiss became world famous for cameras; that of Leitz for microscopes.

German industrial progress was accompanied and, no doubt, supported by the expansion of universities and technical colleges as well as by the foundation of specialist institutions. In 1895 the Reichsanstalt was established, at Helmholtz's suggestion; the first director was Professor Kohlrausch. The Reichsanstalt was the inspiration for the British National Physical Laboratory (1902) and the US Bureau of Standards ; unusually, France was slow to establish a similar institution. Other German state and industrial research organizations were the Centralstelle for explosive research, the Koniglich Mechanisch-Technische Versuchs Anhalt at Gross Lichterfeld, the Chemisch

Technische Versuchs Anhalt, the laboratories of the Brauerbund and, of course, the laboratories of the dyestuffs firms that were to combine, towards the end of the First World War, to form the Interessen Gemeinschaft Farbenindustrie, or 'IG Farben'.

A writer in the American journal *Science* remarked, in the course of an article on the Reichsanstalt:

> Germany is rapidly moving towards industrial supremacy in Europe. If England is losing her supremacy in manufactures and in commerce, as many claim, it is because of English conservatism and the failure to utilize to the fullest extent the lessons taught by science.

Plainly, a German 'economic miracle' is not a new phenomenon.

New Sources of Power

Charles Babbage, writing in 1851[5], had identified a large market that the steam engine could not satisfy. A small motive power was required, ranging from fractional to about two horsepower, to meet the needs of the multitude of small workshops. It should '. . . commence as well as cease its action at a moment's notice, require no expense of time for its management and be of moderate price both in original cost and in daily expense. A small steam engine does not fulfil these conditions.' Having described the requirements Babbage offered a solution. This was, in effect, the column-of-water engine. A steam engine, working steadily 24 hours a day, seven days a week, was to pump water up to a reservoir from which water under high pressure was to be distributed by underground mains to consumers whose water-pressure engines could work all sorts of machinery, including teagles (lifts or elevators), cranes, hoists, presses and machinery of all kinds. The water-pressure engines would take up little space, would be available instantly as required, need little maintenance or supervision and be clean in operation. Furthermore, the steam engine, working steadily at a uniform rate, would necessarily be economical and the reservoir would enable the system to meet, for short periods, peak demands far in excess of the power of the steam engine.

Independently of Babbage these ideas were developed by W. G. Armstrong (1810–1900), who substituted hydraulic 'accumulators' for a reservoir. The accumulators were huge vertical steel cylinders with pistons carrying heavy iron ingots. A steam engine pumped water into the cylinders, raising the pistons up to the top. The accumulators were also connected to high-pressure mains that ran under the streets to the various customers. At first sight this was less economical than the reservoir scheme but it had the great advantage that the hydraulic power stations could be located in the most convenient places; in fact, in town centres. Hydraulic power was simple, reliable and safe. The only drawback was that high-pressure mains were expensive and, because of fluid friction, economic transmission was limited to a distance of a few kilometres. Beyond that it was not economical. Nevertheless, hydraulic power had a long life, reaching a peak in about 1890–1910 and not finally disappearing until the 1970s. It had one important offspring. This was the hydraulic control system that Armstrong developed for the big guns on the new ironclad warships of the later nineteenth century. And this has now branched out into many industrial applications, notably in the control of earth-moving equipment.

When Babbage was writing, gas mains were fairly common and were soon to become universal under the streets of the cities and towns of Europe and America. This meant that an alternative source of mechanical energy as well as of lighting was immediately available for all shops, factories and public buildings. There had been, as we noted, many attempts to use other fluids than steam to develop mechanical energy from heat but none was successful until J. J. E. Lenoir (1822–1900), a Belgian who became a naturalized Frenchman, and who had had a variety of jobs, invented his gas engine[6] in 1860; it was exhibited at the International Exhibition of 1862. Lenoir's engine was a horizontal, double-acting single-cylinder machine made by an experienced manufacturer of steam engines. The motion of the piston drew a mixture of air and coal gas into the cylinder where it was ignited by an electric spark, the exhaust gases being expelled from the cylinder by the return stroke. There was no compression. The engine was thermodynamically inefficient as it had been designed without reference to available scientific knowledge. And, as was inevitable with a new machine, there were various practical problems: there were difficulties with overheating, with ensuring uniform ignition and with the shock that followed the explosion of the mixture of air and gas.

The problems of a coal gas, or internal combustion, engine are different from those of a steam, or external combustion, engine. The most obvious differences concern the problem of heat. In the former the waste heat, leading to excessively high temperatures, has to be dispersed from the cylinder (hence the 'radiator' in a car); in the latter, the relatively low-temperature heat has to be conserved to prevent wasteful condensation in the cylinder (hence the lagging in all forms of steam engine).

In spite of its imperfections, Lenoir's engine was a modest commercial success. Between three and four hundred were made and sold in France; about fifty were made and sold by the Reading Ironworks in England. In the United States Messrs Coryell, of New York, made and sold about the same number. In all something like five hundred Lenoir engines were made. And their relative success inspired other inventors, among them F. Million and P. C. Hugon. But whether they had much – or any

– influence on the next innovator in the history of the gas engine is unknown, for he remains a shadowy figure. Alphonse Beau de Rochas published, in 1862, a short pamphlet[7] in which the basic tenets of thermodynamics were applied to the problems of all heat engines. The pamphlet was free of mathematics and of the language of advanced scholarship. Nevertheless, he laid down the main points briefly and clearly. He claimed that the furnaces of even the best steam engines admitted too much air for efficient combustion and went on to suggest that preliminary gasification of the fuel would enable much more effective use. This led him to a discussion of the combined gas and steam engine, in which the waste heat of the first engine generated steam for the second[8]. This, in turn, led him to consider the principles of the gas engine.

Beau de Rochas referred to a paper by Victor Regnault (in *Comptes Rendus*), concerned with the efficiency of heat engines and giving an account of Carnot's principles. This was the only reference in the whole pamphlet and it can be assumed that Carnot's arguments guided the subsequent discussion. Heat engines without preliminary compression of the air will never develop much power; he does not, therefore, hold out much hope for Lenoir's engine. Such engines violate the basic conditions for the most efficient use of the expansive force of gases:

> These conditions are, in effect, four in number. 1st the greatest possible volume of cylinder together with the minimum surface area; 2nd the greatest possible speed of expansion; 3rd the greatest possible expansion; 4th the greatest possible pressure at the beginning of the expansion.

The only practicable way of approximating to these basic conditions is to carry out a four-stage operation:

> 1st. Air and gas are drawn into the cylinder by one complete stroke of the piston.

2nd. By the following, return, stroke the air is compressed.

3rd. Ignition at the dead-point and expansion in the following, powered, third stroke.

4th. Expulsion of the burned gases from the cylinder during the third and last stroke.

This is the specification of the universal four-stroke cycle.

Intuition can, as we have noted, guide an engineer to an efficient design although he may lack a theory to guide him. The career of Nicolaus August Otto (1832–91) exemplifies this; and it confirms that radical innovations are usually brought about by outsiders. Neither Lenoir nor Otto had had any connection with the motive power – which, in effect, meant steam engine – industry prior to launching their innovations. Otto was a commercial traveller, in the grocery business, and it was not until he was 29 that he became interested in the possibilities of the gas engine.

The main drawback to the Lenoir engine was that the shock of the explosion made it run erratically. To avoid this Otto reverted to an earlier design patented by two Italians, Eugenio Barsanti (1821–64) and Felice Mateucci (1808–87), a physicist and an engineer respectively, in 1857. Unfortunately they were unable to perfect their promising design for Mateucci fell ill at a critical time and, shortly afterwards, Barsanti died of typhoid. Their, and Otto's, engine consisted of a tall vertical cylinder with a 'free' piston carrying a long toothed rack in place of the conventional piston rod. A compressed charge of gas and air under the piston was ignited electrically; the resulting explosion drove the piston up with enough kinetic energy to carry it almost to the top of the cylinder, leaving a near-vacuum underneath. The ascending rack turned a pinion that rotated freely, like the 'free wheel' on a bicycle. When the piston started to descend, partly under its own weight but mainly because of atmospheric pressure, the free pinion engaged a friction clutch so that the rack caused two fly wheels to spin round. When the piston reached the bottom a fresh charge of air and gas was injected and the cycle repeated.

This outline description fits either the Barsanti and Mateucci proposal of 1857 or the working Otto engine. But when Otto applied for a patent in 1866 it was granted on the grounds of the (significant) mechanical improvements he had made; notably to the friction clutch and to the valve mechanism. He had made considerable personal sacrifices in the course of his work and was therefore fortunate to find an able collaborator in Eugen Langen (1835–95), an engineer who had been trained at the Karlsruhe Technische Hochschule. Langen was able to put some money into the firm they established in 1864: N. A. Otto and Company. At the Paris Exhibition of 1867 they displayed an improved version of the engine in which a small pilot flame was used to ignite the air and gas mixture. The engine was awarded a gold medal; deservedly for its duty was more than twice that of the Lenoir engine. It was noisy (not excessively so) and heavy – it weighed about one tonne – for its output of about two horsepower; but, as the cylinder was vertical with the fly wheels mounted on top, it took up little floor space (about one square metre), an important consideration in a small workshop. It was efficient, for practically all the energy expended to drive up the piston was recovered on the down, or working, stroke. Furthermore, the expansion was rapid and was as complete as possible, thereby meeting Beau de Rochas' requirements. Following the success in Paris came a flood of orders that the little enterprise was hard pressed to meet.

In 1869 a businessman with Manchester connections, L. A. Roosen-Runge, joined the partnership and the name was changed to Langen, Otto and Roosen. In the same year they moved to Deutz, near Cologne, and foreign licensees began to manufacture the engine. First were Crossleys, of

Manchester, followed quickly by Sarazin, of Paris, and the Schleicher brothers, of Philadelphia. In 1872 the firm changed its name for the third time, to Gasmotorenfabrik Deutz (GFD) and they were joined by two talented engineers, Gottlieb Daimler (1834–1900), as Production Manager, and Wilhelm Maybach (1846–1929), as Design Engineer. In all the firm and its licensees made about six thousand of these little engines, ranging between one and three horsepower. For the first time a heat engine that was an alternative to the classical steam engine had been successfully launched. For 150 years British engineers and British engines had dominated the world of heat (in practice, steam) engines. That monopoly was now broken. By coincidence, the monopoly had begun with one atmospheric engine and it ended with another.

The success of the atmospheric gas engine enabled Otto to go on with the development of a radically different type of engine. This was the 'silent Otto', an engine with a horizontal cylinder, conventional piston, connecting rod and fly wheel. Unlike the Lenoir engine it was single-acting and ran on a four-stroke cycle, although Otto always denied that he knew anything about Beau de Rochas' pamphlet (few copies of it were produced). It has been suggested that Beau de Rochas' paper was written after Otto announced his four-stroke engine; the purpose being to stop Otto's patent. Otto's invention was, he claimed, entirely independent. His quiet, smooth-running, efficient engine was an immense success. The various licensees took it up at once and great numbers were built; Crossleys alone built about forty thousand between 1877 and 1900. The 'silent Otto' and the various forms that derived from it were employed in a huge number of industries and they ranged in size from fractional horsepower – the sort of engine that could be used in a small retail shop – to giants that developed many thousands of horsepower. Babbage's demand had been fully satisfied. The future seemed to belong to the gas engine. And in a sense it certainly did.

Abundant mineral oil in the last third of the century meant efficient lubricants for gas engines, as well as for high-pressure steam engines. It also made possible the development of the oil engine. The gas engine, versatile as it was, was tied to gas mains. There was clearly a vast market for engines of the same type on farms, in villages and on boats. The lighter fractions of mineral oil, when vaporized by heat and a jet of air, constitute a highly flammable gas. This vapour can be used as fuel for an engine. Accordingly, in the last quarter of the nineteenth century a number of successful oil engines were invented and developed: the Brayton, the Priestman, the Hornsby and the Ackroyd Stuart were among the most successful. There remained the residue of mineral oil refining: this was the lightest fraction, the final product after lubricating oil and fuel oil had been extracted. It was the volatile and highly dangerous gasoline, or petrol.

Daimler, who seems to have been a quarrelsome man, left GFD after a dispute in 1882. He took Maybach with him and used the money he had made with Otto to develop an engine that ran on the lightest fraction, a high-speed gasoline, or petrol, engine. In 1883 he and Maybach built an engine with a simple carburettor to vaporize the gasoline and with incandescent ignition, the mixture of air and vapour being ignited by the insertion of a red-hot metal bar. Two years later they made the world's first motorcycle and, in the following year, 1886, their first automobile. In 1890 the Daimler Motor Company was established. Peugeot, in France, were soon building automobiles with Daimler engines. Karl Benz was another who took up the development and manufacture of the new engine. He began by making gas engines in 1883; these worked on the two-stroke principle invented by the Scottish engineer, Dugald Clerk. Benz used electric ignition and this soon became universal in liquid-fuelled automobile engines. More than the automobile followed from this invention: within twenty years of the first motorcycle, the new high-speed, lightweight gasoline engine was used to make the first flight in a heavier-than-air machine. It is paradoxical that while Daimler, Benz and Peugeot are household names, the name of Otto is familiar only to historians of the heat engine. It was much the same in the cases of Thomas Newcomen and James Watt.

The beam engine remained a common form of stationary steam engine until well past the middle of the century when compounding and higher pressures became more common[9]. Two main types of compound engine were developed: the horizontal and the vertical. In the former, the high- and low-pressure cylinders could be arranged either in tandem, i.e. in line with the pistons sharing the same piston rod, or side by side with the pistons linked to cranks on opposite sides of a fly wheel. The latter form was called the cross-compound; a hybrid was the tandem cross-compound with a pair of cylinders on each side of the fly wheel. In this advanced arrangement there would be one small, high-pressure cylinder, two medium size, intermediate-pressure cylinders and one large, low-pressure cylinder. Large horizontal engines were most commonly used in multi-storey textile mills. The drive continued to be by vertical and horizontal line-shafting until the last quarter of the century, when rope drive was introduced in the United States. Another innovation from America was the efficient valve mechanism invented by George H. Corliss (1817–88).

The vertical engine had its cylinders aligned in a row and inverted over the crankshaft. It was most commonly used in factories, foundries and rolling mills where economy of space was important. For obvious reasons this form of engine was particularly well adapted for marine use, whether in merchant ships or warships. The triple-expansion (three or four cylinders) marine engine of the early 1870s was immediately successful; engines of this type were in common use up to the time of the Second World War. Quadruple-expansion engines were built for very large ships; the doomed *Titanic* had a pair of such engines.

The rapid advance in the power, economy and reliability of marine steam engines following the introduction of compounding, high-pressure operation and superheating, together with the use of Bessemer or open-hearth steel for ship building, brought about a continued substantial reduction in the cost of sea transport. This, as much as anything, helped to falsify Jevons' prediction. It accounts for the cheapness of the imported corn fed to British horses, for by the 1880s most of the corn fed to horses in Britain was imported from North America. And it explains the sad decline of once proud British farming, a decline that lasted from the 1870s to 1939. The high-efficiency steam ship made it cheaper to import wheat from the mid-western states of the USA and the prairie provinces of Canada than to grow it in much of Britain. The introduction of the refrigerator ship in the 1870s enabled the importation of cheap refrigerated beef and mutton from Argentina and Australasia. So dependent on food imported by sea had Britain become by 1914 that German U-boats came close to starving the country into defeat in the First and Second World Wars.

The pacemaker in the development of the steam engine, and, indeed, in many other things as well, had been the railroad system. The railroad was the strategic industry of the nineteenth century. As mining and textiles had previously stimulated innovation, invention and growth in ancillary and related industries, so did the railroad in the nineteenth century. The feasibility of horizontal cylinders was fully established following experience with locomotives, in particular with *Planet*. The complex problems involved in developing a reliable and efficient fire-tube boiler were solved by Séguin and Stephenson. The demand for large numbers of locomotives after 1830 encouraged developments in machine tool technology. The marine engine was, for a time in the mid-century, a pacemaker for the development of the steam engine, but the influence of the railroad was more general. As we saw, the railroad was the biggest single factor in the creation of a telegraph network and had the side effect of establishing the general acceptance of standard (Greenwich mean) time in place of local time. The elaboration of signalling networks represented the first essays in control systems, including recognition of the fail-safe principle. The need to organize a multitude of freight wagons, through railway clearing houses, had implications for commercial enterprises outside the transport industry. Finally, the establishment of railroad networks had enormous, irreversible consequences for society. Things could never be the same again after the Stephensons. The railroad made possible great continent-wide nations such as Australia, Canada, the USA and Russia.

By the end of the century the steam locomotive had reached a point of near-perfection. Beyond this only piecemeal, evolutionary improvement could be foreseen. At the same time major improvements no longer came from British engineers. American engineers produced the bogie car, the Westinghouse vacuum brake, the Pullman car and the vestibule; Henri Giffard, from France, invented his efficient steam injector[10]; Mallet, from Belgium, designed the first articulated locomotive; Belpaire, also Belgian, designed an efficient boiler. With André Chapelon, France was later to have one of the most distinguished of all locomotive engineers. In Britain railways had been so long established that railway engineering had acquired the characteristics of an old craft, with its elders and masters and, perhaps, a distrust of radical innovation.

Such was the diversity of power sources; so widely adaptable and efficient were they, covering an enormous range of human requirements, that it would be quite reasonable to ask what possible scope could there be for the electric motor on which great hopes had been built earlier in the nineteenth century?

The Rise of the Electrical Supply Industry

Nevertheless, hopes of discovering a source of extremely cheap energy were endemic among inventors interested in electricity. A second, and brief, false dawn came in 1881 with the introduction of the accumulator. As early as 1803 the German chemist J. W. Ritter had made the first secondary cell but he was poor and lacked the means to develop it. In any case its utility in the days before an efficient mechanical generator of electricity was severely limited. In 1859 Gaston Planté invented a rechargeable battery that consisted of two lead plates immersed in dilute sulphuric acid. It was greatly improved by Camille Fauré in 1881, who coated the surfaces of the lead plates with lead dioxide (PbO_2). The new accumulator aroused great enthusiasm in Britain; a second electrical euphoria. Here, it was claimed, was the answer to Jevons' *Coal Question*: the energy crisis need never occur! Every little stream, the wind blowing over every little hill, could, day in, day out, drive dynamos to charge up accumulators that would act as reservoirs of energy. From these reservoirs current could be drawn to provide power as required. Unfortunately, although the introduction of Fauré's accumulator came at a time when the market for electricity was expanding, the revolution foreseen by the enthusiasts never occurred. Small was not so beautiful.

In the third quarter of the nineteenth century a further practical application of electricity was added to telegraphy and electroplating. This was its use for lighting[11]. Davy's arc light remained a scientific curiosity until large magnetos became available. In 1844 Leon Foucault substituted longer-lasting carbon for charcoal; shortly afterwards, the arc light was used in the Place de la Concorde, Paris, and to illuminate the National Gallery in London. In 1858 the South Foreland lighthouse and, in 1862, the Dungeness lighthouse were equipped with arc lights, current being supplied by large magnetos with many permanent magnets and driven by steam engines. Much ingenuity was shown by many inventors in improving the arc light. At the same time the magneto was replaced by the dynamo, first in Siemens' original version and then, in 1871, by the improved Gramme machine. The arc lamp was now increasingly widely used, particularly for main streets and squares in big cities and for such commercially important places as harbours and railway marshalling yards. Before the end of the 1870s a flood-lit football match had been played under arc lights at Sheffield.

A source of very brilliant but harsh light, the electric arc had its practical limitations. It was a clumsy, complicated affair. A pair of electric relays was required to keep the tips of the carbon rods at the correct distance for the arc; the carbon rods had to be replaced at frequent intervals; when working the arc lamp emitted a hissing noise. In short it was not a practicable proposition for use in private houses or offices. The market was, therefore, limited.

The carbon filament incandescent bulb was invented independently and simultaneously by Thomas Alva Edison[12] (1847–1931) and Joseph Swan (1828–1914) over the years 1878–80. It was familiar knowledge that the greater the current flowing through a high-resistance wire, the more incandescent the heated wire became. There were many attempts to find the best wire for the purpose but none was found that was reasonably cheap and that would resist oxidation when heated to incandescence. The obvious solution was to put the wire in a glass bulb evacuated of air. Here the problem was that of getting a sufficiently good vacuum. It was one thing to exhaust a vessel of air to such an extent that a flame or a small animal in it would go out or die; it was much more difficult to produce a 'hard' vacuum in which there was so little residual oxygen that the wire would not be destroyed by oxidation after a relatively short time. Fortunately the highly efficient Sprengel pump was invented in 1865 and improved to such an extent that, by 1875, a sufficiently hard vacuum could be obtained.

The carbon filament lamp gave a soft light and was inefficient by later standards in that a relatively high proportion of the energy was dissipated in the form of heat. It had the merit that the filament had an extremely long life and was resistant to shock; accordingly for many years carbon filament lamps were used in warships for they could stand the shocks of heavy gunfire. With the carbon filament lamp the domestic market was open to the new electricity supply industry. Here was a source of light that was compact, safe, free from odours and free from the dangers of fire or explosion. What were needed next were much more efficient power stations and distribution systems to compete with the long-established and highly efficient gas supplies.

The properties and limitations of the arc lamp had decreed that lighting systems were sold as complete units – lamps, steam engine, belting, dynamo, switches, wires and insulators – to individual users, who might be municipal authorities, railway companies, factory owners, etc. The sale was decided on the simple ground, was electricity cheaper than the existing gas, or other illuminant; or, if it was more expensive, did the improved illumination justify the extra cost? With the extension of the market to private individuals, made possible by the incandescent lamp, a number of fresh problems arose. Safety was the most urgent problem; high voltages (much in excess of 100) were quite unacceptable. Simplicity of operation and, of course, a rational system of charging were essential. Fortunately the British Association Committee on Electrical Units had, over the previous twenty years, cleared the way by establishing internationally accepted units of voltage, current, resistance and, through the work of Joule, energy. Electrical energy could, as we saw, be sold in rational and legally defined units measured by the familiar meters to be found in every home.

The world's first power station, selling electricity to the public, was opened by Edison at Pearl Street, in New York in 1882. It transmitted direct current at 110 volts, for Edison shared the view of many others, including William Thomson (Lord Kelvin), that direct current was preferable to alternating current. The preference was soundly based. Direct current was immediately applicable to electrochemical processes and raised no problems when used to drive electric motors; moreover it could be transformed quite easily to high voltages for transmission, thereby minimizing waste of energy by joule heating of the transmission lines. Large accumulators could be charged up in parallel and then switched to series for transmission so that the voltage was increased as the current (remember the i^2r law) was reduced. The system was not as simple and convenient as an alternating current transformer but it had the compensatory advantage – in days when generator breakdowns were quite frequent – of a built-in reservoir of energy, always available to meet sudden peaks in demand or breakdowns in supply; the accumulators, in other words, acted rather like gas holders as well as transformers.

As late as 1883 Osborne Reynolds[13], a friend, colleague and the first biographer of Joule, failed completely to foresee that electricity was potentially the most efficient of all transmitters of energy. In contrast Friedrich Engels[14], who was no engineer and who, as far as is known, never met Joule or Wilde, predicted that electricity at high voltages could efficiently transmit energy over great distances and that this would revolutionize industry. As early as 1876 Sir William Siemens had suggested that the energy of Niagara Falls could be harnessed to generate electricity that could be economically transmitted to the cities of New York, Toronto, Philadelphia and Boston.

The growth of the electrical supply industry was so rapid that by the last decade of the nineteenth century three large firms had emerged in the United States. They were the

Edison General Electric Company, the Thomson–Houston Company and the George Westinghouse Company. Edison, the best-known and perhaps the most prolific of all American inventors, had a remarkable capacity for creating legends about himself that would have delighted the heart of Samuel Smiles. Elihu Thomson was a dedicated electrical inventor while Westinghouse was an able and versatile engineer. In 1894 the Edison Company and Thomson–Houston amalgamated to form the General Electric Company, leaving Westinghouse to constitute the rival firm. In Europe the leading companies were Siemens–Halske, the enterprise founded by another versatile inventor S. Schukert, and the Austro-Hungarian firm of Ganz (Budapest). The Siemens brothers also established autonomous firms in Britain and in Russia. With the emergence of these large firms, some of them 'multinational', electricity became an advanced technology, characterized by the replacement of the empirical inventor, or 'electrician' as he was called in the days of Sturgeon, Henry and Wheatstone, by the electrical engineer.

The advantages of high-voltage transmission had been stressed by Thomson in his Presidential Address to the Mathematics and Physics section of the 1881 British Association meeting. In 1883, at Grenoble, electricity at 3000 volts was successfully transmitted a distance of 14 km; three years later, at Paris, the voltage was raised to 6000 volts and the distance to 56 km. Three years on again, in 1889, the Deptford Power Station, on the Thames estuary, went on-stream. This, in several respects, was to be the prototype for all subsequent power stations. Location was dictated not by proximity to the market but by convenience of coal supply, brought by ship from the north of England; alternating current was generated at 86 Hz and transmitted at 10 000 volts by means of a specially designed cable to substations in the heart of London, some 12 to 13 km away. The Deptford Power

Station was the creation of a remarkable young man, Sebastian Ziani de Ferranti, who was 25 years old when it first went on-stream.

Thereafter the upward trend in transmission voltages continued and even accelerated. As part of the Frankfurt Exhibition in 1891 the voltage was raised to 25 000 volts, the distance to 171 km and 114 horsepower was transmitted at 75 per cent efficiency. By 1897 in California and in India transmission voltages of well over 30 000 volts were being used and by the end of the century this had been practically doubled to 60 000 volts, again in California.

A most opportune invention, as far as the growing electrical supply industry was concerned, was that of the electrolytic process of extracting aluminium by electrolysis from the ore, bauxite (1886). The process was invented, simultaneously and independently, by the American, Charles Martin Hall (1863–1914), and the Frenchman, P. L. T. Heroult (1863–1914); they later joined forces to perfect it as the Hall–Heroult process.

Aluminium, previously a semi-precious metal, became available in great quantities and this had two effects on the electrical supply industry. The extractive process depended on very heavy electric currents; it was, therefore, an excellent customer for the supply industry. On the other hand, aluminium – strong, light, resistant to corrosion and with excellent electrical properties – proved ideal for the construction of transmission lines.

The last major advance was due to the Yugoslav–American engineer, Nicola Tesla, who showed that the AC motor was feasible and who worked out the system of polyphase distribution. Henry Wilde had long ago found that two alternators, coupled in parallel, tended to pull each other into phase, although the theoretical explanation had to await the mathematical skills of John Hopkinson. Comprehensive distribution systems for electric power, linking distant towns, cities, even countries, could now be created.

As the invention and improvement of individual components made a public electricity supply increasingly competitive, particularly as regards electric lighting, so great efforts were made to design high-speed steam engines to drive generators. They were to be made redundant, however, by the appearance of the steam turbine in 1884. In the two major forms that this engine took – impulse and reaction – and in its general purpose, that of converting the energy of steam (heat energy) into useful mechanical effect, the steam turbine closely resembled the hydraulic version. Steam, unlike water, expands when passing from high- to low-pressure; accordingly, the diameter of the steam turbine increases progressively from the high to the low pressure end, just as the diameters of the cylinders of an ordinary compound steam engine increase from high to low. The turbine was soon to take over the job of driving electricity generators and, at much the same time, its suitability for propelling high-speed ships – torpedo boats and ocean liners – was appreciated. The electricity supply industry now became the leading strategic industry.

The possibility of using electric motors to power locomotives on railroads had aroused little interest after the limitations of the electric battery as a source of power were fully understood. The invention of the power station, however, renewed speculation as to the possibilities of electric locomotives. In 1879 Siemens and Halske exhibited a narrow-gauge, passenger-carrying electric railway at the Berlin Exhibition of that year. Three hundred metres of track were laid out in an oval, current being drawn from one of the running rails and a third rail. In 1880 they built a short, permanent railroad or tramway at Lichterfelde. But the effective beginning of the electric railroad was at Richmond, Virginia, in 1881.

Frank J. Sprague (1857–1934) resigned his commission in the US Navy to become an assistant to Edison, who was then experimenting with electric traction at Menlo Park.

Sprague visited London, in 1882, as secretary to the jury assessing dynamos and gas engines at the British Electrical Exhibition of that year. In London he had the opportunity to sample the pleasures of travel on the steam-hauled London underground railway. This may well have turned his thoughts to the advantages of electric railways. With great energy Sprague set to work; by 1885 he had perfected the motor, mount, gearing and truck for street-cars. And in 1888, in Richmond, Virginia, a Sprague system of electric street-cars was inaugurated. This marked the real beginning of railway electrification. In further confirmation that electrotechnology was the strategic technology of the period, we may postulate that Sprague's electric street-cars stand in the same relation to subsequent rail electrification as the Stephensons' Liverpool and Manchester Railway did to steam railroads.

The advantages of smooth electric traction were clear to many railway engineers, but the costs of building power stations and of installing the necessary distribution system of masts, cables and overhead wiring were excessive. One way of cutting the costs was to put the power station on the electric locomotive. J. J. Heilmann, of Alsace, pioneered the steam–electric locomotive with modest promise – no more – of eventual success. From 1893 onwards experimental steam–electric locomotives were built and tested in France, Germany, Britain and the USA. None was successful; the gain from smooth running and reduced damage to the track was more than exceeded by the much greater cost of a complicated locomotive whose overall thermal efficiency was, inevitably, reduced. Only if a heat engine could be made that was much more compact and thermodynamically more efficient than a conventional steam engine could such rail 'electrification on the cheap' hope to succeed.

Rudolf Christian Karl Diesel (1858–1913) was born of German parents in Paris and spent his early years there. His interest in technology was aroused by the Paris museums where he saw machines such as Cugnot's steam carriage. Subsequently he went to Germany for his education and attended the Munich Polytechnic where he heard von Linde's lectures on thermodynamics. These lectures, and in particular von Linde's account of the Carnot cycle, inspired him to try to design a heat engine[15] working on a cycle that would approximate to that of a Carnot engine (apparently he did not know of the Stirling engine). Diesel has a unique place in the history of heat engines: his is the only name firmly and irretrievably attached to a particular form of oil or gas engine. Historians and specialist engineers might talk of the Newcomen engine, the Watt engine, the Otto engine, the Wankel engine, the Carnot cycle; everyone has heard of the diesel engine.

There is a certain irony about this. Diesel's patent (1892) and his explanatory book *Theory and Construction of the Rational Heat Motor* (1894) relate not to one but to a family of engines which, working on the same principles, could use different fuels: gas, powdered coal or other solid fuel and oil. In the diesel engine the first stroke of the piston draws air into the cylinder and the second stroke, the return, compresses the air to such an extent that the temperature rises, 'adiabatically', far above the flash point of the fuel. It may be that the 'fire piston' he had seen at the museum in Paris was the catalyst for this insight. When the piston reaches the end of its travel and the pressure and temperature of the air are at their highest, a port opens and the injection of fuel under pressure begins. The fuel immediately ignites. As the piston is now moving downwards the air would normally be cooled but the burning fuel prevents this. Instead, the temperature of the air does not change and the heat energy of the burning fuel is converted into useful work driving the piston down the cylinder. This accords with Carnot's axiom that there must be no useless flow of heat (energy) from the hot body (the burning fuel) to a cold body (the air). The essential difference between Diesel's engine and all other internal combustion engines is that, in his machine, the burning fuel does *not* heat up the air – compression has already heated it up to the temperature of the burning fuel.

Following the isothermal expansion of the air, the fuel supply is turned off and the expansion continues, 'adiabatically', so that the temperature and pressure now fall as the heat energy of the air is converted into useful work. When the temperature and pressure have fallen – ideally – to those of the atmosphere, an exhaust port opens and the piston expels the air and burned fuel, after which it returns, drawing in the next charge of air. Although it involves a degree of repetition it is worth quoting Diesel's own account of his cycle:

> 1. Production of the highest temperature of the cycle (the temperature of combustion) not by and during combustion but before and independently of it, entirely by compression of ordinary air.
>
> 2. Gradual introduction of the fuel into the . . . highly compressed and highly heated air during the return stroke. The fuel is added in such a way that there is no increase in temperature of the air . . . After ignition combustion should be controlled . . . to maintain the right proportions between pressure, volume and temperature.
>
> 3. Correct choice of the proper weight of air in proportion to the thermal value of the fuel . . .

In this way Diesel proposed to extract the maximum possible energy from burning fuel. His first engine, a single-cylinder oil engine, was built by Maschinenfabrik Augsburg-Nurnberg (MAN) in 1893. The use of pulverized coal and other fuels had proved impracticable. And experience soon showed that modifications to the cycle were necessary. For example, it was found that during the fuel-burning stage it was better to work at constant pressure rather than constant temperature, and it was not possible to continue expansion right down to atmospheric pressure and temperature – an enormously long cylinder would have been required. Nevertheless the Diesel engine was to prove far more economical than its rivals. It had other merits, too. Electric ignition was dispensed with and it was far less susceptible to fire than the high-octane gasoline or petrol engine. Its disadvantages were that it was heavy and noisy as well as being more expensive to manufacture.

Diesel foresaw, in his book, that his engine would replace steam locomotives on railroads as well as be used for street-cars and other road vehicles; he supposed that it would be applied to the propulsion of all sizes of ships and boats and that it would meet requirements for large and small stationary power sources on land.

These prophecies have proved, in retrospect, remarkably accurate. The problem for the historian, however, is to identify the particular market in which the diesel engine would surpass all possible rivals – the basic requirement of inventions. His engine came at a time when the electric motor was, at last, beginning to show great promise for the future, when the gas engine was proving itself economical, convenient and reliable for a multitude of purposes, when oil and gasoline engines with flame or electric ignition were available and when the marine steam engine, in its two main forms, was well established. The steam turbine, with steam now generated in oil-fired boilers, was a near-ideal power source for large, fast ships, while the simple, sturdy and cheap triple-expansion engine was well suited for smaller ships down to fishing boats and tugs. Diesel died tragically when he was lost overboard from the Hoek van Holland to Harwich ferry steamer before his prophecies came true. There is a parallel here to Sadi Carnot's sad career.

The Efficiency of Heat Engines

As the century ended, British and American engineers ended a long debate about the ideal steam engine. There were too many different forms of steam engine serving too many different purposes to hope for one type of engine that would, whether big or small, fast or slow, be the best for all requirements. Nevertheless, the best practical cycle, a standard for all engines, could at least be prescribed. The Carnot cycle was too abstract for this purpose. At the end of the century the engineers agreed that the practical ideal was to be an engine working on what Professor Thurston called the Rankine cycle[16]. It was originally called the Clausius cycle but historical research showed that Rankine had described it a few months before Clausius. A few years earlier the entropy–temperature diagram had been introduced; this showed how efficiently (or not) heat energy was being utilized by the engine. Willard Gibbs had made use of the entropy–temperature diagram in his classic papers of 1873 while the Belgian engineer Th. Belpaire had given a brief account of it in the previous year. In 1898 Captain Riall Sankey RE, who had played a leading part in the efforts to define the ideal practical engine, produced the first energy flow diagram, now called, appropriately enough, the Sankey diagram.

In the same period the problem of cylinder condensation was effectively resolved. The Willans high-speed engine achieved this by exhausting the steam through the middle of the piston. The Stumpf 'uniflow' double-acting engine (1908) had entry ports at each end of the cylinder with exhaust ports in the middle. In this way the exhaust ports were only exposed, or opened, when the piston, half as long as the cylinder, reached the end of its travel in the middle of the cylinder. In the uniflow engine therefore (as in the Willans engine), the steam always flowed in the same direction with no wasteful reverses of direction and with greatly reduced cylinder condensation. However, both engines were already obsolescent, for the turbine combined both advantages as well as giving the smooth, direct rotation that James Watt had sought more than a hundred years before.

If the invention and development of the steam engine in the eighteenth century was almost entirely due to British engineers, it would be reasonable to assert that in the nineteenth century continued progress was due to French, Belgian, German and American as well as British engineers[17]. The leading names are those of Sadi Carnot (this is confirmed by every textbook on heat engines), Victor Regnault (as Kelvin recognized) and Hirn with his associates in Mulhouse (as admitted by D. K. Clark in 1898). Finally, to be remembered are W. J. M. Rankine, Sir Charles Parsons and Riall Sankey together with G. Corliss, B. F. Isherwood, R. H. Thurston and the designer of the exemplary American locomotive 50 000. In the twentieth century the evolutionary improvement of the steam turbine has been a multinational enterprise and in the steam turbine the machine can be fairly said to have reached a form beyond which no major improvement can be foreseen.

The first invention and subsequent improvement of the steam engine was in response to the needs of the mining industry. It was successful because it met the requirements outlined at the beginning of chapter 6. Its application to mills and factories, from the end of the eighteenth century onwards, meant the incorporation of additional inventions to achieve rotative motion. The invention of the steam locomotive and the rapid spread of railroads during the first half of the nineteenth century resulted in a compact, high-pressure engine with a thermally efficient boiler. The development of long-distance steam navigation in the middle of the nineteenth century set new requirements that the engine triumphantly met: maximum fuel economy, compounding up to quadruple expansion and surface condensation. By the end of the century the desideratum of the new electrical supply industry was for a very high-speed engine that was also smooth running. One response was the Willans engine; another, more successful, was the steam turbine. Thus the steam engine proved remarkably adaptable, responding successfully to every challenge by evolving different, specialized forms, with cross-fertilization between the forms.

The understanding and efficient exploitation of energy were the main achievements of the nineteenth century. Early in the century people were impressed by the fact that a lump of coal which could easily be held in the hand could, if burned in the furnace of a good Cornish steam engine, do work equivalent to raising a man of average weight (say, 170 pounds or 80 kg) from sea level to the summit of Mont Blanc. During the course of the century there were, as we saw, four occasions when it was widely believed that abundant, cheap energy would soon be freely available. It is hardly necessary to add that on every occasion expectations were disappointed. Furthermore, to put it colloquially, if ungrammatically, experience has been that energy doesn't come cheap. Every source of energy exploited by humanity so far has been highly concentrated: wood, coal, oil, natural gas, uranium, waterfalls (Niagara is the exemplar) all represent concentrated energy. And even before the heat engine and the hydraulic turbine humanity always exploited concentrated *animal* power: oxen, the strongest horses, elephants. Only in exceptional circumstances, in Arctic or Antarctic regions for example, are dogs used as sources of motive or locomotive power. No one, to my knowledge, has ever attempted to harness the strength of cats. The long-established emphasis on concentrated power has some bearing on the contemporary debate about 'renewable' sources of energy[18]. Wind, tide, waves, solar heat may come free of charge but they represent 'dilute' energy, equivalent to many thin coal seams. Large-scale exploitation of dilute energy sources would require extensive installations: would these be environmentally acceptable or economically viable?

A NOTE ON THE 'FIRE PISTON'

The 'fire piston' was invented in about 1804 by Joseph Mollet. It consisted of a short brass tube, closed at one end and fitted with a piston. The piston rod ended in a flat disc. A piece of tinder was placed in the cylinder, which was held in one hand while the flat of the other hand was brought down sharply on the disc. The sudden compression raised the temperature of the air so much that the tinder ignited (see note 4, chapter 10). 'Fire pistons' were manufactured commercially in France but never proved popular. The explanation of the operation of the 'fire piston' was given by Joule: the mechanical energy used to compress the air is converted into heat; conversely, in the case of the 'Heronic engine', the energy of the compressed air is converted into mechanical energy to raise the water, the air being cooled as it loses energy.

It has been claimed that the 'fire piston' was invented independently in China but Professor Fox discounts this claim. It is possible that the 'fire piston' was the source of a story, widespread earlier in this century, that someone had invented a perpetual match, but the international controllers of the match industry had suppressed the invention (and the inventor).

The Century of Wars

The expansion of major industrial cities and towns that had caused such appalling social problems early in the nineteenth century had been followed by an accelerating application of technology by reformed and new municipal authorities[1]. Manchester, the shock city of the Industrial Revolution, had built a large and efficient water supply system, drawing water from the Pennine hills and the 'Lake District', some 100 km to the north. Liverpool drew its water from reservoirs in north Wales. But it was Birmingham, largely under the influence of the Chamberlain family, that set the pattern for what was called 'municipal socialism'. Local businessmen took a pride in their city having efficient ambulances, fire service, police, transportation, schools, gas works, water and electricity supplies, etc. A record achieved by Glasgow City was that at one stage it ran a tramway or street-car service on which it was possible to travel '14 miles for 1 [old] penny' (or about 22 km for half a new penny or just under a cent). It was said to have been the cheapest public transport ever achieved. And Glasgow trams made a profit.

  Although the twentieth century cannot rival the seventeenth century in that there have been more years of peace in Europe since 1901 than there were between 1601 and 1700, few would deny that the wars that took place during this century have exceeded all others in ferocity, barbarism and destruction. Science and technology were fully utilized in these wars but played no part in beginning them or for their overall conduct. The causes were mainly the imperialism that brought about the

South African war (1899–1902) between Britain and the Afrikaner republics, political miscalculations and malice that lay at the root of the First World War and the debased political philosophies that led to the Second World War as well as to the atrocities of the Holocaust and the earlier 'liquidation of the kulaks'. In one way science did contribute, although indirectly, to racial persecution. Neo-Darwinism tended to emphasize different 'racial qualities'. Lecturing in 1901, on 'Darwinism and Statecraft', the English biometrician Karl Pearson scornfully rejected democracy, maintaining that the struggle for existence ensures progress and the survival of the fittest race[2]. Without war humanity stagnates. From the hottest furnace comes the finest metal. 'Kaffirs, negroes, Red Indians' are so low they must be destroyed. A good stock must not live alongside a bad one. In this way were misapplied and perverted the teachings of Bakewell, the thought of Darwin and the experiments of Mendel. To their credit Pearson's fellow scientists rejected his arguments with contempt. Whether he had in him the makings of an Eichmann or of a de Stogumber is an intriguing if unanswerable question. What are quite clear are the consequences of very similar beliefs held by others who had the power to put them into effect.

After the progress of science and technology during the nineteenth century it is understandable that there were attempts to forecast the course of technology in the twentieth century. H. G. Wells, for example, had some interesting speculations about the future of transportation[3]. He was sceptical about aeronautics as a possible means of transport. If he was thinking about lighter-than-air craft, as he probably was (1901), he was justified; two years were to pass before the Wright brothers made the first airplane flight. He made some shrewd comments on a subject that is now particularly fashionable. Horse traffic, he observed, with its cruelty and filth, with animals that exhaust and pollute the air, must give place to motor carriages in a few years. Wells painted a graphic picture of the hazards to be faced by a well-dressed young lady, wearing a long skirt of the period, as she crossed a Piccadilly made filthy by horse dung (to add to the problem, horse dung was a fertile breeding ground for flies). In his 'Anticipations of the reaction of mechanical and scientific progress upon human life and thought' (1902), Wells imagined that 'special motor tracks and individual motor cars will, to a large extent, replace railways . . .'. He foresaw that the rapid development of transport would reverse an historic trend. For centuries people had been flocking from the countryside into the towns; now, improved transportation would effectively diffuse great cities out into the countryside. His generally optimistic outlook was clouded by one vision. At a time of growing and abrasive nationalism, the prospects of a war with modern weapons in which English generals, with all the virtues and defects of their class, would act as 'the polished drovers to the shambles' of great armies of trusting young men, was a vision that, he said, 'haunts my mind'. This, while remarkably accurate, should be contrasted with his earlier expressed belief that knowledge of the future was becoming possible, thanks to inductive sociology. He admitted that natural catastrophe, pollution of the atmosphere, drugs or collective madness may end life on earth, but he still believed in human destiny: '. . . there stirs something within us now that can never die'. In his *A Modern Utopia*, published in 1905, he outlined his plan of a modernized Platonic republic in which the functions of Plato's Guardians were taken over by a voluntary nobility called the Samurai; a tribute to the dramatic rise of modern Japan.

A Revolution in Transport

For many people the first automobiles were regarded as 'horse-less carriages', vehicles to convey affluent folk from their homes to and from the nearest railway station, or for leisurely jaunts about town. Before it could become universal, a number of changes and improvements were necessary. British roads had, since the triumph of the railroad, fallen into disrepair; American roads were undeveloped; only perhaps in France were roads satisfactory. Roads would have to be renovated and repaired before motor cars could use them. Here, the path was smoothed for the automobile by a remarkable social and technological event of the last third of the nineteenth century: the vogue of the bicycle. This, not surprisingly, seems to have begun in France in the late 1860s. By 1870 the 'ordinary' velocipede with its huge, pedal-driven, front wheel and small trailing wheel had appeared; it was popular with athletic and daring gentlemen, although of less appeal to others. In 1885 J. K. Starley, of Coventry, produced the safety bicycle. With pedals, chains, gearing and most of the features of the modern bicycle, it could be comfortably propelled at a reasonable speed. It allowed, indeed encouraged, women as well as men to take to the road and the bicycle boom was soon well under way. The social and political consequences of the bicycle, particularly as regards the progressive emancipation of women, have never been fully assessed. Technologically the consequences were also important.

As a result of the bicycle boom efforts were made by many people, some of them influential, to improve the roads, to provide reasonable wayside facilities for cyclists and, of course, to bring down the price of bicycles. In 1888 J. B. Dunlop, an Irish veterinary surgeon, invented the pneumatic tyre which, although not intended for the purpose, added greatly to the effectiveness of bicycles, and a variety of techniques were developed for its production. The suspension wheel, with its echoes of Cayley and of Hewes, had already been introduced and light tubular steel construction standardized. Methods of braking and head and tail lamps were perfected. A key point was made by Hiram Maxim; Professor John Rae quotes him[4]:

> It has been the habit to give the gasoline engine all the credit for bringing in the automobile – in my opinion this is the wrong explanation. We have had the steam engine for over a century. We could have built steam vehicles in 1880, or indeed in 1870. But we did not. We waited until 1895.
>
> The reason why we did not build road vehicles before this, in my opinion, was because the bicycle had not yet come in numbers and had not directed men's minds to the possibilities of long distance travel over the ordinary highway. We thought the railroad was good enough. The bicycle created a new demand which went beyond the ability of the railroad to supply. Then it came about that the bicycle could not satisfy the demand it had created. A mechanically propelled vehicle was wanted instead of a foot propelled one, and we know now that the automobile was the answer.

Following the popularity of the bicycle, roads were renovated, signposts erected, repair facilities created (easily extended to serve the early motorists) and road maps printed. The bicycle having paved the way technologically, administratively and psychologically, the automobile could come into its own, granted, as Professor Rae pointed out, the right social and economic conditions in Europe and America. Three possible motors were available: the steam engine, a strong candidate following the invention of the flash boiler by L. Serpollet in 1899; the electric motor, made feasible for limited runs by the introduction of accumulators and the availability of mains electricity for recharging; and finally the gasoline engine of Daimler, Maybach and Benz. In spite of its relative complexity the speed, power and light weight of the last ensured its ultimate success.

The first gasoline automobiles were wooden horse carriages minus the shafts, the motor being discreetly hidden under the bodywork and with a steering tiller in place of reins. However, the new industry of bicycle making had more affinity with the construction of automobiles than had the old craft of carriage building. Light metal tubes, brakes, suspension wheels, pneumatic tyres, gear wheels, ball bearings and the other components of bicycles together with the associated specialized machine tools could be adapted for the manufacture of automobiles. This meant that in England, Coventry and Birmingham, long associated with gun – that is, tube – making, became the centres of the new industry and for the manufacture of machine tools.

Although Germans, beginning with the disciples of Nicolaus Otto, had pioneered the automobile, the light, popular car was originally developed by Frenchmen. They substituted the steering wheel for the tiller; they put the motor boldly and unashamedly in front; and they invented motor sport. In short, they replaced the motor-driven horse carriage by the motor car. In the first decade of the twentieth century the leading names in the automobile world were De Dion, Panhard, Peugeot, Levassor, Mors; and the pioneering role played by the French is perpetuated by the now familiar names they coined: automobile, chassis, chauffeur, garage. Although Americans had, at first, shown interest in the steam car – the Stanley 'steamer' was fast and relatively successful – leadership in the manufacture of gasoline motor cars was soon to go to the United States. America had certain advantages; among them a high standard of living, a large population, an abundance of mineral oil and an innovative spirit equal to that of any European country.

Although many American automobile pioneers who gave their names to firms that are active today were in business during the first decade of the century, the most representative name is that of Henry Ford (1863–1947).

Born on a farm in Michigan, his mechanical talents led him to take a job in Detroit where he began experiments with automobiles. Ford, an intuitive engineer and an inspired businessman, built his first automobile in 1896 and, in 1903, founded the Ford Motor Company.

Ford appreciated the enormous size of the market awaiting exploitation and he understood that the task facing him was to design a car for the multitude and then to discover a way of making it as cheaply as possible. The first problem was solved by 1908 when the famous Model T appeared. This was a basic, 20 horsepower machine with two forward gears and one reverse; it was sturdily built, easy to drive and easy to maintain and service. Starting was by crank; electric starting, when introduced in 1912, was made an option on the 'Tin Lizzie', as the Model T was nicknamed.

The Model T was immensely successful, but Ford believed that if he could reduce the price still further an even greater market would be opened up. The answer was the moving assembly line (1913). Rather like Boulton and Watt, Ford had succeeded in attracting talent to his firm. In Professor Rae's view, the invention of the assembly line was probably due to a group that included Ford and some of his talented collaborators. Mass, or at least, large batch production with interchangeable components had been practised in the textile machinery industry early in the nineteenth century and had been further developed in the small-arms industry. In the case of large machines – iron framed power looms, mules, carding engines, etc. – workers would take the components to the machine which would be assembled in one place, just as builders take wood, sand, cement, bricks, plaster, etc., to the place where a house is being built. With Ford's assembly line, however, each worker stays in one place; the automobile under assembly moves along the line so that each worker can carry out one simplified job, all (interchangeable) components to be added, having been brought to the workers concerned. The whole operation required immensely detailed planning to ensure the exact synchronism of job performance, supply of components and movement along the assembly line. Skill in building, or rather assembling, automobiles, was finally banished to the tool room.

The Model T remained in production until 1927. Its success had been bought at the, no doubt reasonable, cost of inflexibility. Over the years the Ford Motor Company had manufactured just one automobile; a sturdy, reliable vehicle that the vast majority of Americans could afford. With the last Model T the line closed down for a year while planning went ahead for the next model. Only the largest of enterprises could afford to do this.

The great automobile industry, rather like the textile industry, stimulated development in ancillary or related industries, notably in the oil industry, in highway and bridge construction and in the business management of large engineering concerns. In addition, its associated technology has been strategic: one that has had a profound impact on, for example, military technology. The automobile has certainly radically changed the way of life in all advanced and most developing countries. It is often criticized for causing pollution and for the loss or life and injuries due to traffic accidents. As regards pollution, we noted that unacceptable pollution was ascribed to intensive horse traffic; indeed, pollution is indissolubly linked to humankind. And, as far as deaths and injuries on the roads are concerned, before the motor car is condemned it should be remembered that, every year, many lives are saved and much suffering alleviated or prevented by the use of motor cars. On the other hand, in pre-automobile days many deaths and injuries were caused by horse traffic and much suffering endured by draught animals. These facts, as well as the pleasure automobiles give to many people, must be taken into account in any assessment of the value of motor transport.

The British, who had pioneered the railroad, played a relatively minor part in the development of the automobile. Whereas the Stephensons had foreseen the universal railroad, no engineer or businessman called attention to the possibilities of the universal motor car. It was, perhaps, indicative that the best known, if the least 'popular', British car of those (and subsequent) years was the Rolls Royce. Had the innovative spirit in Britain been seduced by the strident claims, the honourable (if well rewarded) burdens and the golden promises of Empire? It was, after all, the age of Cecil Rhodes, Dr Jameson and Barney Barnato as well as of the highly competitive Indian Civil Service examination.

The Origins of Electronic Communications

The automobile is one of the most distinctive features of the twentieth century; so, too, are radio and television but their origin is quite different. It began with Maxwell's theory which had remained speculative and unproven for more than twenty years. Men argued that his theory was not necessarily correct; other theories, consistent with Newton's doctrine of action at a distance, might be formulated. After all, the electromagnetic nature of light had been indicated by Faraday and by Weber and Kohlrausch's experiments. Maxwell's theory could be dismissed as the wisdom of hindsight based on dubious assumptions and odd ideas about the ether. Maxwell himself had given no thought to experimental testing of the theory; still less had he proposed any practical use for his electromagnetic waves. Some scientists familiar with the theory, such as G. F. Fitzgerald, Oliver Lodge and H. A. Lorentz, considered the possibility of generating and detecting Maxwell-type radiation but found the practical difficulties insurmountable.

In 1879 the Berlin Academy of Sciences offered a prize to the experimentalist who could demonstrate that a changing, or transient, electric field generates a transient magnetic field, and vice versa; this would be an acid test of Maxwell's theory. Heinrich Hertz (1857–94), a pupil of Hermann von Helmholtz, saw that the problem would be solved and Maxwell's theory confirmed if he could show that electromagnetic waves, generated by a changing or oscillating electric current, travelled through space with the same velocity as light[5]. According to the old, pre-Maxwell, theory based on Newton's principles, the communication of inductive effects must be instantaneous and the velocity accordingly infinite. The point was explained by Henri Poincaré, the distinguished French mathematician:

> . . . *according to the old theory the propagation of inductive effects should be instantaneous.* If indeed there be no displacement currents and nothing, electrically speaking, in the dielectric (in effect, the open space) that separates the inducing circuit from that in which the effects are induced, it must be admitted that the induced effect in the secondary circuit takes place at the same instant as the inducing cause in the primary; otherwise in the interval, if there were one, the effect is not yet produced in the secondary; and as there is nothing in the dielectric that separates the two circuits, there is nothing anywhere. Thus the instantaneous propagation of induction is a conclusion that the old theory cannot escape.

This suggested a decisive experiment, an *experimentum crucis*, on the outcome of which Maxwell's theory would stand or fall. The requirements seem formidable, but Hertz had certain known facts to help him. In 1850, Fizeau and, in 1875, Werner Siemens had shown that electricity travels along a wire at a velocity almost equal to that of light. And Kirchhoff had deduced, on the old theory, that electricity should travel along a wire at a velocity that increases as the resistance is reduced until, when the resistance is zero, it becomes equal to that of light. In the case of alternating currents the higher the frequency (or the more rapid the oscillations) the more closely the velocity of the electricity approaches that of light. The same inferences can be drawn from Maxwell's theory.

The problem was to find a way of generating high-frequency oscillations and to compare the velocity of their propagation along a wire with that of the electromagnetic waves that, according to Maxwell, should radiate into space with the velocity of light. As early as 1842 Joseph Henry had pointed out that the discharge of a Leyden jar (or condenser) can be oscillatory. Generally, the electric charge on one plate of a condenser will surge through the connecting circuit and 'overshoot', piling up on the other plate until it starts to flow back, overshooting once again and piling up on the first plate. This pendulum-like process continues, with diminishing amplitude as resistance causes the energy of the charge to be dissipated as heat and the condenser is discharged. In 1853 William Thomson gave a mathematical analysis of oscillatory discharges; he showed how the frequency of the oscillations depended on the capacity of the condenser and the inductance of the circuit. Finally, in 1858 Feddersen detected oscillatory currents using a spark gap, a rapidly revolving mirror and a camera. All that Hertz needed was an induction coil to provide very high voltages, two small brass plates to form a small condenser, two short lengths of wire joined to the plates with a small gap between them for the spark to jump over and a suitable detector to reveal the electromagnetic waves in space and the oscillations along a long wire. The induction coil was to be connected to the two short lengths of wire, so that when a spark took place an electric charge surged from one plate to the other with a frequency that depended on the capacity of the plates plus the pieces of wire. The oscillations were to be picked up – induced on – a third small brass plate joined to the long length of wire.

The capacity and the inductance were small so that the frequency was very high indeed.

In 1886 Hertz invented a simple detector for high-frequency – and therefore very short-wave – oscillations. It consisted of an adjustable loop of wire with a small gap in it for sparks (figure 15.1). If the detector was brought near an oscillating circuit, sparks would leap the small gap when the loop was held in the right position to obtain the maximum inductive linkage. The detector could be varied in size and so tuned to the frequency, or the wavelength, of the oscillator; in this way the maximum response was obtained by the 'resonator', as the detector was called.

With his simple apparatus consisting only of an oscillator, a long length of wire and a tuned resonator Hertz carried out a decisive series of experiments in 1888 and

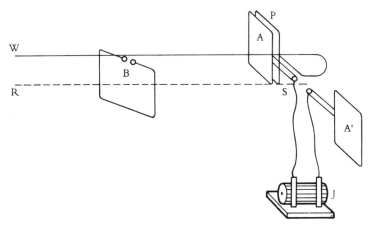

FIG 15.1   *Hertz's apparatus*

A—A', brass plates (capacitors); P, brass plate connected to wire W; R—S, centre line; B, resonator; J, induction coil. When the plane of B coincides with the vertical plane through R—S no current can be induced in it from electrical oscillations between A and A'. The sparks across the gap in B will be solely due to waves in the wire W. In this way the lengths of the standing waves in W can be measured. When B is at right angles to R—S waves in W will have no effect on it. At intermediate angles (as shown) the interference effects appear

1889. He measured the length of standing waves transmitted along a long open wire and showed how waves travelling along the wire interfered with vibrations in space from the same source. The wavelength as measured on the wire and the calculated frequency of the oscillator enabled him to calculate the velocity of the waves on the wire. From the interference pattern he was able to show that the velocity of the vibrations, or waves, in space was that of light. And he demonstrated that the waves could be reflected from the walls of a large room to give the effect of standing waves in space. The *experimentum crucis* had therefore decided in favour of Maxwell's theory and the reality of Maxwell-type electromagnetic waves in empty space had been established.

Hertz went on to show that these waves had the same properties as the vibrations that caused the sensation and phenomena of light. They could be reflected and refracted; they were 'normally polarized' and could be made to display the characteristic phenomenon of interference. Hertz had enormously extended the spectrum, from visible light through radiant heat to the new Maxwell-type waves whose length, from crest to crest, was measured in metres, rather than in tiny fractions of a millimetre. Significantly, too, Hertz showed that the field strength of electromagnetic waves diminished much less rapidly with distance than did the field strength of Faraday-type induction.

There was no rush to make practical use of 'Hertzian waves', as they were called. Why should there have been? The countries and continents of the world were covered by telegraph lines; telegraph cables ran under all the seas and oceans. And in 1873 the ingenious Mr Graham Bell had invented the telephone – surely the ultimate perfection of telegraphy? There had been a few desultory attempts to use Faraday-type induction for special communication purposes, for example from shore to light ship, but the aerials required were far too large and the range limited. Electrostatic communication had also been considered but it, too, was found to be unsatisfactory.

Hertz died, tragically young, on 1 January 1894. Five months later Oliver Lodge, in a lecture in London, outlined Hertz's achievements and discussed techniques for detecting and studying Hertzian waves. The 'coherer', as Lodge called it, had been invented by Eduard Branly. It was a tube of powdered metal the particles of which cohered when exposed to Hertzian waves. As the resistance fell sharply, a coherer in a circuit containing a battery and an electrical bell served as an effective detector of Hertzian waves. The vibration of the bell set the instrument for the next signal.

Although much work was being done on Hertzian waves, it was in physics and not technology. It was left to a wealthy young Italian, barely out of his teens, to convert these scientific developments into a revolutionary method of transmitting information. Guglielmo Marconi (1874–1937) had attended Augusto Righi's lectures at Bologna University on Maxwell's theory and on Hertz's experiments; and he had read Lodge's London lecture. He could, he said later, hardly credit that the great men of science had not already seen the practical possibilities of Hertzian waves; but, as Lodge later confessed, they had not. Marconi, in short, was the typical outsider who, having no prior connection with an art or technology, revolutionizes it.

Marconi's first experiments were carried out on his father's estate. He systematically improved the coherer and developed the aerial, or antenna, in place of the two lengths of wire, or dipole, that Hertz had used. He soon realized that for long-distance communication waves of much greater length than those used by Hertz were necessary. For Marconi, the wavelength from crest to crest was measured in hundreds, even thousands, of metres, not the four or five metres that Hertz had employed. He found, too, that he could greatly extend the range of reception by 'earthing' his aerials and by using high transmitting and receiving aerials. Convinced of the commercial possibilities of his inventions the young Marconi left for England. In June 1896 he lodged his first patent – the first radio patent – and founded the Marconi Company with an issued capital of £100 000. The issue was immediately subscribed and cable and telegraph shares fell, quite unjustifiably, on the Stock Exchange. Marconi was after a very different market.

Marconi was fortunate; his mother was a wealthy British heiress who could open many doors, commercial and official, in London. He never knew those long periods of failure, frustration and poverty that seem to have been the lot of many inventors and innovators; he was, on the contrary, successful from the start. Just like Boulton and Watt, he had immediately recognized the importance of finding the right market for his invention. In his case it was that enormous sector that telegraph or cable could not serve: shipping. He had come to Britain because that country was then the foremost maritime nation in the world. More than half the world's ships were registered in Britain. British ship-owners should therefore be the first to realize the advantages of 'wireless telegraphy'.

In the closing years of the century there was widespread interest, most of it due to Marconi, in the practical possibilities of Hertzian waves. In 1899 two engineers had suggested that Hertzian waves could be made to guide a torpedo (presumably travelling on the surface)[6]. By the turn of the century French engineers had succeeded in transmitting a message over a distance of six kilometres to a balloon at a height of 800 metres. Less dramatic was Charles Nordmann's attempt, on the Mont Blanc range, to detect the possible radiation of Hertzian waves from the sun. He did not succeed but concluded that it was possible that they were radiated from sunspots and solar flares. This was the first suggestion of what is now called radio astronomy. In the meantime Marconi was extending the range of wireless communication. He sent wireless messages to the USS *Philadelphia*, steaming across the Atlantic, and from Poldhu, in Cornwall, to an Italian cruiser visiting Cronstadt. In 1901 he claimed to have sent messages across the Atlantic, between Poldhu and Newfoundland. These stations, about 4000 km apart, were separated by the curvature of the earth, a 'mountain' of sea water about 400 km 'high'. The Hertzian waves had not behaved like light rays; they had somehow managed to bend round the earth. According to received theory they should not have done this, any more than light waves could bend over Mont Blanc to allow people in Chamonix to see what was going on in Courmayeur. Marconi's great advantage was that he knew enough but not too much. It was the successful transmission of Hertzian waves round the curvature of the earth that led O. Heaviside and A. E. Kennelly to postulate the existence of an electrically conducting layer in the upper atmosphere so that Hertzian waves were somehow channelled or guided round the earth. This layer, composed of charged particles or ions, was later to be called (by R. A. Watson Watt) the ionosphere.

In 1903 an international conference was held in Berlin to settle the allocation of priorities and rights; safety at sea was an urgent question. Another valuable application of Hertzian waves, it was suggested, would be the transmission of time signals to ships. By 1904 the Marconi system was being used to transmit regular news bulletins to the Cunarder, RMS *Campania*, on her voyages across the Atlantic.

Wireless telegraphy was used for military purposes during the Russo–Japanese war, 1904–05. The brilliant Japanese Admiral, Togo, devised a scheme to lure the surviving ships of the Russian Pacific fleet from the security of Port Arthur by sending a weak force of Japanese ships to manoeuvre off-shore, mines having been surreptitiously laid in the fairway. When the Russian ships emerged the weak force was to signal, by wireless telegraphy, to a much larger Japanese force lurking well out of sight. But the Russian flagship struck one of the mines and sank, whereupon the remaining Russian ships returned to harbour. Wireless telegraphy was not, therefore, tested on this occasion. It was, however, used for news gathering during the same war. The London *Times* hired a steamer, the *Haimun*, and fitted it with a transmitter using the somewhat different system devised by the American radio pioneer Lee de Forest (1873–1961). The idea was that the *Haimun* would shadow the Japanese fleet and send reports to a receiving station on the Chinese mainland from whence they would be cabled back to London. The Japanese soon realized that the Russians, too, could receive and interpret these reports. Naturally they took steps to restrict the *Haimun*'s movements.

The two events that helped to fix the utility of 'wireless' in the minds of the British public were its use in effecting the arrest, in 1911, of the celebrated murderer Dr Crippen and his mistress when they were trying to escape from the country (cf. the use of the telegraph in the arrest of the Slough murderer, see chapter 10), and its invaluable contribution to the saving of lives in the *Titanic* disaster of 1912. This event publicized the value of wireless telegraphy in the United States as well as in Britain.

There are two other candidates for the title of inventor of radio. They are the Welshman, D. E. Hughes, and the Russian, A. S. Popov. Hughes demonstrated, in 1879–80, a form of short-range wireless telegraphy that depended on Hertzian waves, but this was before Hertz's experiments. In any case he misunderstood the principle and failed to develop his invention. He was not, therefore, a pioneer of radio. Similarly Popov, who had read Lodge's lecture, invented a receiver for Hertzian waves but used it to detect radiation from lightning flashes. He was therefore the founder of an important branch of meteorology. But he does not seem to have invented a transmitter, nor to have appreciated the possible use of

Hertzian waves for the communication of information. The next steps in the progress of wireless telegraphy were the use in receivers of resonant oscillatory circuits, as analysed by William Thomson, and the invention of an efficient generator of continuous oscillations at a steady frequency to replace the spark. Much clearer reception was possible using a resonant circuit, tuned to the frequency of the transmitter, and the energy of the transmitter could be concentrated on the common frequency. The problem of generating continuous oscillations was not so easy. Two immediate solutions were the 'singing arc' of Duddell and a high-speed alternator, such as that devised by Alexanderson, which could generate an alternating current at 22 000 Hertz (or cycles), although this was suitable only for large shore stations[7].

The Importance of the Electron

The establishment of John Dalton's atomic theory led some people to speculate that if solids, liquids and gases were atomic in structure then perhaps electricity was, too. This speculation was strengthened by the discovery of Faraday's laws of electrolysis (1835) which related quantity of electricity to chemical equivalents; in effect to atomic weights. It was not until 1874, however, that the Irishman G. J. Stoney (1826–1911) identified the 'atom' of electricity that he called the 'electrine' and made a plausible estimate of its charge[8]. Before long he changed the name to 'electron'. Paralleling Stoney's researches others were investigating what were called cathode rays. A glass tube, exhausted of air, is fitted with a negative terminal and a positive terminal – the 'cathode' and 'anode' respectively. If there is a high voltage difference between them, rays seem to travel between these terminals, causing the glass walls to fluoresce. There was much debate about these rays; were they waves, like Hertz waves, or were they a stream of particles? In 1897 J. J.

Thomson (1856–1940), showed conclusively that cathode rays consisted of streams of particles. He measured the charge to mass ratio (*e/m*) of the particles and showed that it was the same no matter what metal the cathode was made of. He found, too, that identical particles were emitted from various hot wires and, once again, the *e/m* ratio was always the same. Thomson went on to prove that these particles were constituent parts of the atom – which was seen to have a structure – and to be the same as Stoney's electrons[9].

In the course of these researches an important scientific instrument was developed by K. F. Braun in 1897, the cathode ray tube. A narrow stream of electrons is accelerated down the tube by the positively charged anode, further down the tube. They overshoot the anode and impinge on a screen coated with zinc sulphide to cause a point of whitish light; which point çan be made to reproduce the patterns of varying voltages that an engineer might wish to study. Although the practical application of cathode ray tubes lay in the relatively distant future, the application of the thermionic emission of electrons was more immediate. Edison had discovered (1880), in the course of his researches into electric light bulbs, that a small current flowed across the empty space between the heated filament of the bulb and a wire probe inside it. J. A. Fleming, in 1904 in England, used the discovery to invent a simple rectifying device, which was later called a 'diode'. A cathode inside an exhausted glass bulb was heated by an electric current so that it emitted electrons. These were attracted to a small cylindrical anode surrounding the cathode and kept at a positive voltage. Fleming wrote to Marconi explaining how the device could be used in wireless telegraphy. His diode would pass an electric current in one direction only, since electrons, being negative in charge, can travel from a negative cathode to a positive anode, but not in the reverse direction.

The diode worked on a different principle from the

coherer; it rectified the rapidly oscillating currents induced in antennae by Hertzian waves, thus turning them into direct currents. Of course sound waves at the same frequency as Hertzian waves would be far above the human audible range. But if the incoming Hertzian waves and the resulting rectified currents could be modulated, like the currents flowing through a telephone system, then audible sounds would be produced in an earpiece.

Fleming's diode was, however, almost immediately superseded by the 'cat's whisker' crystal detector (1906). It had been found that a crystal could act as a rectifier, and as crystals were very cheap and reliable they soon replaced diodes. The second stage in the development of wireless telegraphy and, a little later, telephony was therefore that of the 'crystal set', still remembered by the elderly and the very elderly.

The thermionic tube or valve was not to be forgotten. Lee de Forest made a major modification to the diode. He found that by adding a spiral of thin wire between the cathode and the anode – a so-called 'grid' – the versatility of the device could be greatly increased. The current flowing through this device, called a triode, could be controlled by varying the voltage on the grid, for this had the effect of varying the number of electrons that could flow past the grid. The triode, in fact, acted just like a valve used to control the flow of water or gas.

The triode (or valve) could, de Forest knew, be used to amplify weak oscillations; it could therefore greatly increase the distance over which wireless telegraphy could be effectively received. And its ability to amplify signals opened up another possibility. An alternating current in a tuned, or resonant, circuit could go on oscillating for ever without being fed from outside if the inherent resistance of the components could be reduced to zero. (In much the same way a pendulum would go on swinging for ever if the friction and air resistance could be completely eliminated – the escapement compensates for these retarding factors). De Forest realized that by combining a triode and a resonant circuit the amplifying power of the triode could fully compensate for the joule heating loss of electrical energy due to the inherent resistance in the resonant circuit. The alternating current in the resonant circuit would therefore continue indefinitely. Here was an efficient method of generating continuous wave transmissions. The method was soon to replace generation by spark, arc or alternator.

The continuous high-frequency oscillations generated by a triode and its associated tuned circuit could be amplified by another triode and suitably modulated to correspond with sound waves impinging on a microphone. Transmitted as Hertzian waves, the modulated signals could be received, amplified, rectified by a diode and the resulting varying electric current transformed into sound by a suitable loud-speaker on the same principle as the telephone receiver. This was the system of radio telephony developed by Lee de Forest by 1910. In the 1920s the 'radio set' replaced the crystal set and the triode was followed by a series of multi-electrode valves or tubes: tetrodes, pentodes, hexodes and a wide variety developed, later on, for special purposes such as television and radar. The realization that this could be used as a means of entertainment was arrived at only gradually; after all, no one had thought of the telephone as a means of entertainment. Hertzian waves were first thought of as a means of supplementing the telegraph system, particularly for providing a service to ships at sea, and then as a means of supplementing telephone services. Even prophets such as H. G. Wells did not foresee 'wireless' as a great medium of education and entertainment.

The First Airplanes

Although Sir George Cayley (1773–1857) had laid down the necessary specification for the design of a successful airplane at the beginning of the nineteenth century, for the remainder of that richly creative era the 'conquest of flight' was a happy hunting ground for conmen, lunatics, enthusiastic amateurs and the occasional devotee: that is, one who would try to view the problems objectively and resolve them systematically or scientifically[10]. The main difficulties were threefold. First, there was the lack of a suitable lightweight power source; the limitations of the steam engine and the gas engine were obvious. Secondly, there were the problems inherent in flight itself. Finally, in view of the common opinion of engineers and scientific men that heavier-than-air flight was impossible, there was no economic or military incentive for 'research and development'. For these reasons few devotees were attracted to aeronautics in the way that Joule and others had been attracted by the possibilities of the efficient electric motor.

The invention and development of the gasoline or petrol engine removed the first and most serious obstacle to the airplane. This made the prospect more attractive to devotees although the second obstacle still remained. There had, for many years, been no serious difficulty about building a simple flying machine that could become airborne, rise to about a metre, fly for a hundred metres or so and then land. Sir Hiram Maxim had built such a machine in 1894; it was powered by a steam engine with a high- pressure boiler without its furnace. This, however, was not an airplane. The necessary conditions to be satisfied by an airplane were that it should take off under its own power and land at the same height; that it should be capable of ascending and descending and of turning to left or right, all under the control of the pilot. These were the problems of flight.

There can be no question that the true airplane, a machine that not only met the necessary conditions but which was to show itself capable of virtually unlimited development, was invented by the brothers Orville and

Wilbur Wright (1871–1948 and 1867–1912 respectively). The sons of a clergyman of Dayton, Ohio, the Wright brothers were able mechanics who founded and built up a prosperous bicycle manufacturing business. Their interest in aviation was aroused by the widely publicized achievements of Otto Lilienthal (1848–96), who had made many successful flights on the hang gliders he had built and had thereby learned a great deal about the problems of flight. Lilienthal was on the point of applying power to gliders when he crashed and was killed. Besides Lilienthal, Professor Samuel P. Langley and Octave Chanute, two other devotees, influenced the Wright brothers.

Just as Joule, in his efforts to build the most efficient electric motor, had begun by trying to perfect the basic components and to master the principles, so the Wright brothers, well aware that you cannot run before you can walk, began by building a small biplane kite in 1899. This was to test Lilienthal's idea of 'wing warping' as it was called. If a pilot wishes to turn, say, to the right then he should twist, or warp, the wings in such a way that the left-hand leading edges are twisted up while the right-hand leading edges are twisted down, so that the left-hand wings rise and the right-hand wings dip down; the same effect is achieved today by means of ailerons. 'Wing warping' would also be used to correct an unintended roll to one side or the other. In short, this technique met a fundamental requirement in controlling an airplane or glider. In 1900 the brothers made the first of three gliders and set about making short, low-level flights with it over the sandy sea-shore at Kitty Hawk and Kill Devil in North Carolina. Gradually they taught themselves the elements of design and the principles of control in flight. They were able to correct the errors made and published by other and earlier would-be aviators and they taught themselves how to be the first pilots in the world. Their approach was, in a few words, systematic and scientific; they even built their own wind tunnel for testing the behaviour of wing sections under flight conditions. In 1902 they applied for a patent, later granted, for their system of control by the combined use of warping and rudder.

On 17 December 1903, a Wright biplane with two propellers, driven by a twelve horse power gasoline engine they had designed and made themselves, took off at Kill Devil and flew for some 120 feet, taking 12 seconds for the flight. Orville was at the controls. This, modest as it was, was the first true powered flight. More flights were made that day, the distance covered being increased to well over 800 feet and the time of flight to nearly one minute. In subsequent years the distances covered, the heights reached and the times of flight were steadily increased. In 1905 Orville and Wilbur Wright took their first passengers on flights and, in the same year, took their new plane to France for public demonstrations.

Everyone who has made and flown a model airplane, indeed anyone who has ever made and flown a paper dart, has some idea about aircraft stability, which was seen as the great problem in early aviation. The Wright planes were all inherently unstable. This was their deliberate choice: the more stable the plane the less sensitively it responds to the pilot. The Wrights wished to fly their planes; they did not want their planes to fly them. And, in those early days when so little was known about flying, this was a wise choice. In Europe the emphasis had been on stability: on planes that had tail units with flat horizontal surfaces for longitudinal stability. Accordingly, in the decade preceding the First World War, much effort was put into achieving stability without sacrificing sensitive control.

The Wright brothers have been seen by some people as unlettered mechanics who were lucky enough to be the first to make and fly an airplane. The above brief summary should dispel any such interpretation. They were essentially modest, unassertive men whose photographs, showing them clad in the stiff and uncomfortable clothes of the early part of the century, are unfamiliar. They belong in that shadowy area of history where are to be found other diffident men who achieved great things. In popular interest and esteem they were to be overtaken by the heroic and glamorous long-distance and trans-oceanic fliers of the 1920s and 1930s. In the long run their reputations are the more secure.

It is a surprising fact that after the Wright brothers the development of aviation languished in the United States in the decade before the outbreak of the First World War. The leadership was taken, unambiguously, by France. As with the automobile, a testimony to French leadership continues in such technical terms as aileron, fuselage, empennage, monocoque, nacelle. Following the Wright brothers, the next pioneers of aviation were Frenchmen or adoptive Frenchmen, among them Henri Farman, Ferdinand Ferber, Louis Blériot, Charles and Gabriel Voisin and Alberto Santos-Dumont. German efforts were initially in a direction that was eventually to prove a dead end: the rigid, dirigible airship. Nevertheless, before 1914 Germany had instituted scheduled commercial flights by airship between major cities and had carried over 42 000 passengers without a single fatality. This must be credited to the abilities of Graf von Zeppelin and his assistants.

No doubt many reasons can be found to account for European leadership in a technology where Americans had been the pioneers and which, on the face of it, should have appealed so strongly to the pioneering spirit of America. The single most important reason seems to have been – as later experience so abundantly proved – that state support was and is essential for a developing aircraft industry. In Europe, where powerful rival nations confronted each other across land frontiers and where the pressures making for war were steadily increasing, there was every incentive for state support even though no one could predict what role aircraft would have in future wars. In 1909, Britain, lagging behind her Continental rivals and with all her imperial eggs in the one basket of the Royal Navy, instituted a Board of Aeronautics. The initiator was Lord Haldane, one of a family that included two distinguished biologists and who himself was a man of academic attainment. The United States was even more tardy; the National Advisory Committee on Aeronautics (NACA) was not established until 1915.

The Impact of War

It is debatable whether the First World War stimulated aircraft development. It certainly did for specifically military items. Ways of mounting guns, even cannon, on aircraft and machine guns synchronized to fire between the blades of the propeller are two examples of war-time innovation. According to a well-informed, if highly polemical, aviation journalist C. G. Grey, the War added little, if anything, to the improvement of aircraft themselves. The multi-engined aircraft, the monocoque or tubular fuselage and the retractable undercarriage were all invented before the War. On the other hand, the requirements of mass production led to inflexibility, and the inevitable absence of co-operative competition plus maximum war-time secrecy imposed their own restrictions on progress. Assuming that all nations had continued to develop military aviation – there was little civil aviation for many years – each nation would, in the absence of war, have made its own distinctive contributions to eventual mutual benefit. Be this as it may, the combatant nations – Austria-Hungary, Serbia, Russia, Germany, Belgium, France, Great Britain and her Commonwealth nations – entered the War that was to demonstrate the immense changes that science and technology had brought about across the entire range of human activities, including the most evil.

The growth of aviation made public what had been quietly increasing for some decades: the leading role of the state in technology and, although less conspicuous, in science. This was associated largely with the defence industries during a sustained arms race that affected all the major western nations. It became very public in 1906 when a revolutionary British battleship, HMS *Dreadnought*, was launched. The backbone of the world's navies had been armoured battleships ('ironclads'), the largest displacing about 15 000 tons. Typical was Admiral Togo's flagship, the *Mikasa*, (15 200 tons), at the wholly conclusive battle of the Sea of Japan (27 May 1905). The *Mikasa* had four twelve-inch guns in two turrets, two guns firing forward, two firing aft, fourteen six-inch and numerous light and machine guns. She was relatively fast, her maximum speed being 18.5 knots. *Dreadnought*, a year later, had ten twelve-inch guns, of which six could fire forward, six aft and eight broadside. This formidable fighting machine made the navies of the world obsolete and ensured a naval armaments race as all major powers sought to equip themselves with dreadnought-type battleships.

The arguments of naval critics, for and against the dreadnought type, are hardly relevant to a discussion of the technology they represented[11]. Such a discussion must include assessment of a range of the most recent innovations: the introduction of turbine propulsion, hydraulic control of guns, new explosives, electric light and power and wireless telegraphy. The gyrocompass was another important novelty just before 1914. The axis of a rapidly spinning wheel or globe always points in the same direction, no matter how the spinning body is moved about. This is the salient feature of the child's spinning top; and also of the spinning earth whose axis, neglecting precession, always points towards the Pole Star (in the northern hemisphere) as the earth orbits the sun. J. B. L. Foucault (1819–68) had been the first to study and explain the properties of the gyroscope. By the beginning of the present century the advance of precision engineering and the development of electric motors had made possible the invention of the gyrocompass, a rival for, and improvement on, the old magnetic compass. Here the pioneer was Elmer Sperry (1860–1931), another versatile and prolific American inventor. By means of repeater motors (the electrical analogues of MacFarlane Gray's steam steering engine) the main gyrocompass could have 'slave' compasses at convenient places in the ship. It was also particularly suitable for use in submarines, another and menacing radical innovation of unknown potential in 1914. Later on the gyrocompass was to prove invaluable in aircraft and in space vehicles.

A particularly acute problem in the age of the dreadnought was slow to be recognized, so rapid had been the advances on so many different fronts. The theory of naval warfare was still, as it had been in 1800, that two lines of battleships would hammer at each other until one was overcome. This theory tended to obscure a serious problem in gunnery. In battle two dreadnought-type ships, turbine-powered and capable of well over 20 knots, could be moving in opposite directions at a relative speed of 50 mph (80 km/h), and, following improvements in guns and propellent explosives, firing at each other at a range of 10 miles (16 km) or more. Very clearly the gunnery techniques that were satisfactory in 1800 would not be satisfactory under these new conditions. Taking account of time of flight, a shell that would have hit a stationary target fired from a stationary ship, would fall as much as half a mile, or a kilometre, astern of a fast-moving dreadnought, if aimed directly at the target dreadnought and not ahead of it. How far ahead was one question? And how to allow for a change of course was another? What was needed was a computer that could take account of these variables so that the guns could be aligned to hit a fast-moving target.

The answer was found in the development of the tidal predictor invented by James Thomson, the younger brother of William Thomson (Lord Kelvin). This was a mechanical integrator, or analogue computer, which, in turn, was a development of the Morin *compteur* and so traced an ancestry back through the medieval clockmakers to the Anti-Kythera mechanism. The Argo clock, as the British predictor was called, was produced at the instigation of Arthur Pollen, a wealthy British businessman engaged in the engineering industry and with influential connections in governing circles. Once again an outsider had played a key part in a significant innovation. But the British predictor was at some disadvantage *vis à vis* its German rival, for the latter was fed with information from the superb range-finders that were the products of the advanced German optical industry. This particular German advantage lasted up to the Second World War.

At the outbreak of the First World War in July – August 1914 the navies of the combatant nations reflected contemporary technology to a far greater extent than did the armies or even the still experimental airplane units. From dreadnought battleships and battle cruisers down to destroyers and the new submarines the navies were thoroughly modern. It was the requirements of the submarine that gave the diesel engine a market in which it had clear advantages over its three rivals: the steam engine, the oil and the gasoline or petrol engine. The diesel engine was, unlike the steam engine, compact; and unlike the gasoline engine its fuel was not dangerously explosive; more, its fuel economy greatly increased the cruising range of submarines.

As the War proceeded, technological and scientific shortcomings became apparent in the combatant nations. In Britain, the disappearance of German dyestuffs led to an acute shortage of dyes for soldiers' and sailors' uniforms. In 1915 therefore the government made capital available for a new British dyestuffs firm at Huddersfield – 'British Dyes'. The government, through the Board of Trade, was not, it seems, willing to allow a chemist to sit on the board of the new company; for if one did the other board members, being ordinary businessmen, would be entirely in his hands. On this the journal *Nature* commented:

We are thus authoritatively informed from his seat in Parliament, by the Secretary of the very Board which is entrusted with the duty to look after the commercial and industrial interests of the country, that the first qualification of a director of a public company subsidised by the Government, is that he must know nothing of the business in which that company proposes to engage.

Besides dyestuffs, metallurgical, electrical, optical and other technologically advanced products, previously imported from Germany, were in short supply. Immense and successful efforts were made to catch up in leading areas of modern technology. In 1916 the government took an important step when the Department of Scientific and Industrial Research (DSIR) was established. Its function was to subsidize research in industry, to carry out scientific research on behalf of government and to encourage the formation of industrial research associations, subsidized jointly by government and industry. The War was a tragedy and a disaster for Britain, as for all the combatant nations, but at least it gave that country a chance – which on the whole it took – to modernize its industry and technology. An apparent – but dangerously delusive – bonus for Britain was that, after the War, German competition was, for a time, virtually eliminated. The United States, with a progressive economy, a world lead in automobile manufacture, highly progressive electrical and chemical industries and a society that welcomed and encouraged innovation (witness the early film industry), had much less reason for concern about technology and industry that had war-time Britain. Nevertheless, in 1916, President Wilson authorized the National Academy of

Sciences – with limited membership – to establish the National Research Council, a body with functions broadly similar to those of DSIR.

When the War ended in 1918 the governments of the leading powers were committed to, and were tending increasingly to dominate, technology – directly and indirectly. It was realized that technological leadership and industrial research were essential for victory in war and for prosperity in peace.

Paradigm Cases

The beginning of the twentieth century had seen two remarkable advances in physics that, taken together, amounted to a scientific revolution. In 1901 the German physicist Max Planck (1858–1947) enunciated his quantum theory. This was designed to reconcile the observed distribution of energy in the spectrum of heat radiated from a hot 'black' body (that is to say a body that will absorb all the radiation, of whatever wavelength, that falls upon it) with the accepted theory according to which the energy should increase without limit as the wavelength becomes shorter. In fact, experiments showed that the radiant energy diverged widely from that predicted by accepted theory, that it reached a peak at a wavelength of about 2/10 000 of a centimetre and that thereafter it declined at still shorter wavelengths. Planck found a formula that fitted the experimental results. His formula was based on his quantum theory, according to which radiant energy can be emitted or absorbed only in discrete units, which are determined by the product of the particular frequency and a constant h, known as the quantum of energy. Planck thought that his resolution of the problem was no more than a temporary expedient. On the contrary it became one of the major pillars of twentieth-century physics as it was found capable of explaining other important problems and as the extent of its predictive powers was increasingly understood.

Better known to the general public was the theory of relativity, the special theory of which was published by Albert Einstein (1879–1955) in 1905 (there were in fact several theories of relativity but Einstein's was the most successful and soon became the best known). Ever since Hertz had triumphantly validated Maxwell's theory of the electromagnetic field, and incidentally laid the foundations for radio and much else, the basic conflict between Maxwell and field theory on the one hand and Newton's doctrines on the other had been apparent. One or the other had to be modified. In the event it was Newton's mechanics that had to be changed.

Newton's postulates of absolute space and absolute time were, in their day, entirely reasonable. The fixed stars in the heavens had been identified, recorded and grouped into constellations in antiquity. Their 'proper motion' and that of the sun were not to be detected until after Newton's day. It was therefore to be expected that wherever and however you went in the universe a yard length would always be a yard long. Similarly in the clockwork universe, as conceived by the philosophers (scientists) of the seventeenth century, you would expect an hour to be an hour, sixty minutes duration, wherever you were. And a mass of one pound would, no less universally, be a mass of one pound wherever it was.

Bit by bit the basic assumptions were weakened until the increasing confirmation of Maxwell's electromagnetic theory brought things to a crisis point. Newton's basic postulates of absolute space and time and of the axiom of the conservation of mass were rejected by Einstein in his special theory of relativity (1905). According to Einstein there is no privileged position from which these can be measured with the assurance that the results represent the invariant truth. Space, time and mass vary with the relative motion of the measurer, or the observer; there are no absolutes against which these dimensions can be measured. The one constant is the velocity of light, which is always the same, however and wherever it is measured. According to Einstein a measuring rod moving, *relative* to an observer, would be shortened the faster it moved;

although only at very high speeds indeed would the shortening be appreciable. In the same way a mass would increase the faster it moved but, again, it would have to be travelling very fast for the increase to be noticeable. For speeds within human experience and detection Newton's laws were effectively valid (see the Note at the end of chapter 17).

Quantum theory and relativity found no practical applications for many years but they aroused great public interest and a veritable industry sprang up to meet the public demand for simple interpretations. In this respect popular versions of the theories were recruited, indiscriminately, to support a widespread revolt against past beliefs and codes of conduct; a revolt against what were felt to be the stuffy repressions and hypocritical respectability of the Victorian age. The best-known leaders of this revolt were certain literary figures of the first decades of the century, prominent among them being Lytton Strachey and Aldous Huxley.

The most remarkable and probably the most important industrial and technological development in the first third of the twentieth century was the rapid growth of the radio industry and its supporting electronic technology. By the 1930s practically every home in the industrially advanced countries had a radio set. To make this possible a great manufacturing industry was established that demanded new skills. To ensure the steady improvement of radio sets, extensive research and development was undertaken by a new class of technologists, recruited in part from the pool of physicists, who would otherwise have been teachers, and in part from electrical engineering. The 'cat's whisker' sets were replaced by vacuum tube, or valve, sets with intermediate amplification (superheterodyne) and power output to work loud-speakers. These sets were proudly, if somewhat misleadingly, advertised as 'superhets'. Incorporating the Edison phonograph, or gramophone, the radiogram offered a very complete form of home entertainment, to the detriment of the old-fashioned piano.

Although television had been predicted in the nineteenth century it was not until the 1920s that the first electromechanical systems appeared in the United States and Europe. One of the most interesting of the television pioneers was John Logie Baird, a somewhat eccentric inventor with a strong faith in his ideas. The Baird system was simple. A close spiral of holes, running from the edge to the centre, was drilled in a rigid disc. As the disc revolved rapidly, the sequence of holes scanned a brightly lit picture or scene, rather as a reader's eyes scan a page of print from top left-hand corner to bottom right-hand corner. The light passing through the holes fell on a photoelectric cell, the output of which varied with the intensity of the light falling on it. This varying intensity was transmitted by radio to a converse apparatus that was the receiver. The mechanical components of the Baird system meant that its development potential was severely limited. However, Baird rendered one lasting service: he most effectively publicized television. In this, his role was similar to that of Thomas Savery in the development of steam power.

The all-electric system that superseded the electromechanical system of television included, in the receiver, one of the most versatile of modern scientific instruments. This was the cathode ray oscilloscope, the practical form of Braun's cathode ray tube (chapter 15). In Braun's cathode ray tube the cathode is heated by a current, to give a copious supply of thermal electrons. Two pairs of parallel metal plates – one horizontal, the other vertical – between the cathode and the anode are arranged so that the stream of electrons passes between each pair. If now a uniformly increasing voltage is applied to the vertical pair, the point of light will be drawn at uniform speed across the screen from one side to the other to trace a straight line. A small increase in voltage on the horizontal pair will cause the next line to be traced just above the first one, and so on. In this way the screen can be covered by horizontal lines, just like lines of type on a printed page. This constituted a most efficient scanning system. By controlling the intensity of the stream of electrons, the point of light can be made faint or bright and by this means a picture can be produced on the screen. Precisely the same scanning can be performed by electromagnets in place of metal plates.

Experience had shown that the reception of 'short'-wave radio transmissions fell off very rapidly with distance from the transmitter. Accordingly short-wave bands were allocated, almost contemptuously, to an entirely new class of enthusiasts: radio amateurs, or 'hams' as they chose to call themselves. (Radio 'hams' were possibly the first truly amateur technologists. They formed legally and internationally recognized groups.) They were responsible for a remarkable discovery in the mid-1920s. The hams found that, although reception of short-wave transmissions deteriorated rapidly relatively close to a transmitter, at great distances, beyond the reach of good medium- and long-wave reception, short-wave reception was surprisingly strong; indeed, it was found to be ideal for trans-oceanic and trans-continental transmissions. The reason was seen to be almost obvious. High-frequency, or short-wave, signals transmitted along the ground are greatly attenuated, but transmitted upwards and reflected down from the ionosphere, whose existence had been predicted by Heaviside and Kennelly, they suffer far less attenuation. Realization that the ionosphere acted like a mirror, reflecting as well as guiding radio waves, was followed by experiments to find its height above the earth. In 1925 G. Breit and M. A. Tuve[1], of the Carnegie Institution in Washington DC, found a novel way to solve the problem. A short-wave (71.3 metres) transmitter was supplied with alternating current at 500 cycles per second (500 Hz) in such a way that it would transmit radio signals

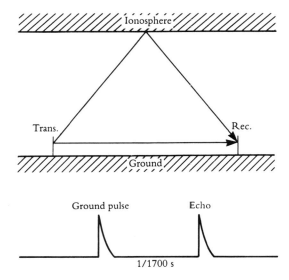

FIG 16.1   *Reflections from the ionosphere*

only during the positive half of each cycle. This meant that bursts, or pulses, of radio waves would be sent out 500 times a second with a period of quiescence following every pulse. A receiving station seven miles, or eleven kilometres, away picked up the pulses that came direct from the transmitter and those that were reflected back from the ionosphere (figure 16.1). The pulses, suitably amplified, were displayed on an oscilloscope on which the point of light crossed the screen 500 times a second. This meant that the signals due to the direct pulse and the reflected pulse always appeared at the same points on the scan and briefly displaced the line upwards. Their distance apart was proportional to the extra time taken by the reflected pulse and therefore to double the height of the ionosphere.

Breit and Tuve found that the time lag of the reflected pulse was about 1/1700 of a second. This meant that, knowing the velocity of radio signals, the ionosphere must be about 50 miles, or 80 km, above the earth. They found also that there was evidence for a second reflecting layer some 100 miles above the earth. Supporting evidence for their discovery came from the researches of E. V. Appleton, in England, who studied the variation of BBC broadcasts received at Cambridge over day and night. This he explained as due to interference between the direct radio waves and those reflected from the ionosphere. His calculations confirmed that the ionosphere was, in daytime, about 80 km above the earth. Later, Appleton used Breit's and Tuve's pulse technique to confirm that there was another reflecting layer above the ionosphere.

The development of pulse techniques in radio engineering and investigations of the ionosphere were of interest only to the new class of electronic engineers and to geophysicists. Even for electronic engineers the rapidly developing radio set with the prospects for television must have offered more challenges and opportunities. Towards the end of the 1930s, however, German engineers developed the Lorenz system for the guidance of airliners in bad weather. (Germany had built up a far bigger network of scheduled air services than any other European country and was, in this respect, second only to the United States.) In the Lorenz system a generator of radio waves is coupled to two transmitting aerials; the signals sent out by one aerial are modulated into short regular pulses while the signals from the other aerial are modulated into complementary and long regular pulses. The two aerials are highly directive so that the transmissions form narrow beams close together. The pilot of an airliner approaching the transmitter would hear, if he was on one side of the beams, a series of dots; if on the other he would hear a series of dashes. If he was exactly between the two narrow beams he would hear a continuous note as the long and short pulses came in together at the same strength (figure 16.2).

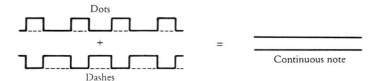

FIG 16.2 *Principles of aircraft navigation by radio beams*

The first general application of the pulse technique came with the invention of radar. This was an excellent example of simultaneous and independent invention, for radar was developed simultaneously in Britain, France, Germany, Holland, Italy and the United States. All these countries had highly developed electronics industries catering for domestic markets in radio and, to a very limited extent, television. In addition there were a number of different motives for developing radar[2]. It will be convenient to concentrate on the British development of radar; or RDF as it was first called, for in this case the motive was clear-cut and simple. (RDF, or radio direction finding, invented before 1914, was a system used by ships, world-wide, of navigating by taking bearings on shore radio stations. The expression would not arouse the curiosity of any enemy agent who came to hear of it; cf. the word 'tank' in the First World War.)

During the First World War air raids on London and a few British provincial cities by Zeppelin airships and then by Gotha bombers had inflicted minor damage and some casualties. They had had, however, a far more potent psychological effect. What were the enemy up to? Was it true that vast air fleets were ready to be launched after initial raids that were merely exercises in softening up? Rumour added its quota of exaggeration. The British government, while having no immediate answer to the bomber, took the step of setting up the Independent Air Force – later the RAF – with the specific objective of carrying the air war to German cities. The bomber remained a preoccupation of British defence experts after the War and a lurking terror to the British public; a terror that novelists and sensation-mongers exploited to the full. The bomber, a British politician remarked, will always get through. London was a huge target that could hardly be missed, only a few minutes' flying time from the coast. So great was the advance in the capabilities of aircraft after the War that it was believed by the public, and it seems by governments too, that within a short time after, or perhaps before, a declaration of war vast bomber fleets would reduce London and other major cities to rubble. Air raid deaths would be counted in millions, not the few hundreds of the First World War.

The possibility of using inductive heating to raise the temperature of bombers – or aircrew – to such a degree that they would be incapacitated was examined but found to be quite impracticable. No other kind of 'death ray' was possible. The only solution offered was a means of detecting and locating enemy aircraft at some considerable distance so that defending fighters ('interceptors') could be directed to shoot them down before they reached their target cities. This was to be done by transmitting short-wave radio pulses; the returning echoes from metal aircraft would be detected and displayed on cathode ray oscillscopes. From 1935 to 1939 the British Air Ministry developed and installed a chain of radar stations for the defence of the country. At the same time an elaborate system of fighter control was established so that radar information could be immediately translated into orders to be transmitted to defending fighters. Here, for the first time, was a complete defence system based on electronics. It made possible the successful outcome of the Battle of Britain.

British defence research had one other notable success during the Second World War. This was the invention of the cavity magnetron. The shorter the wavelength, or the higher the frequency, used in radar, the more accurate the set, the greater the discrimination over distance and the more difficult the transmission becomes to detect or to jam. The shortest wavelength possible, using more or less conventional triodes, was about 50 centimetres. There were available a number of special tubes or valves – klystrons and magnetrons, for example – that could generate oscillations of very short wavelength, but unfortunately their power output was feeble and any transmission would also be feeble and of very short range. Nevertheless, the British Admiralty set up a group under Professor M. L. E. Oliphant at Birmingham University to see if it would be possible to boost the power of any of these ultra-high-frequency valves. In early 1940 Dr J. T. (later Professor Sir John) Randall and Dr H. A. H. Boot discovered that it was possible greatly to increase the power output of the magnetron by drilling cylindrical cavities round the (thick-walled) anode (see Note at the end of the chapter). Randall[3] had been led to this idea by recollecting that Hertz had used a loop of wire as a resonating detector of electro-magnetic waves. The diameter of the loop was proportional to the wavelength. Reasoning from this Randall and his colleague drilled holes whose diameters were all proportional to about ten centimetres. They were rewarded with a valve that, incorporating, in effect, its own tuned, resonant circuit, generated immensely powerful pulses of ultra-short-wave oscillations. From this time onward radar sets working on a wavelength of 10 cm or less were built in great numbers. These sets, compact and powerful, played a major part – perhaps the major part – in the war against U-boats and in the mass bombing campaign. There can be little doubt – no doubt in the British case – that radar was invented and developed in response to the threat of war. The important invention of the cavity magnetron was made in the period known as the phoney war, before anything like the full scientific and technological resources of the country had been mobilized or before major challenges –

such as the U-boat threat – had emerged.

The main radar inventions must therefore be credited not to the War but to the earlier prospect of war. Refinements such as the bombing aids 'oboe', 'gee' and 'H₂S' were certainly made under the stress of and in response to the demands of war. The case of radar therefore supports our argument that, generally, the technological relationships of war are complex. Prospects of war are a stimulus to invention. Funding may be meagre but pressure is not usually intense; time is available to develop and refine the invention. As a rule inventions made under the immediate pressure of war tend to be *ad hoc*, temporary or corner-cutting devices such as Chappe's telegraph two hundred years ago or 'foxers' fifty years ago. An exception to this rule was the invention of the atomic bomb. As for the magnetron, it is now familiar to many people as the key component in the microwave cooker.

The magnetron was not the only major innovation in electronics over the period in which radar was developed. The 'cat's whisker' crystal rectifier had been superseded by the diode in the new radio sets. But there remained considerable academic interest in the electrical properties of crystals and research was carried out in Europe and America. During the War silicon and germanium crystals were developed as rectifiers for use in centimetric radar sets. After the War research continued, particularly at the Bell Telephone Laboratories, where there had been much interest in the field for a long time. In 1948 a programme directed by W. Shockley at the Bell Laboratories led to the invention of the transistor. Research had been concentrated on germanium and silicon as they were simpler to understand than most crystals. Eventually the analogue of the grid of the triode was discovered in a crystal. The crystal triode was developed, also at Bell Laboratories, by J. Bardeen and W. H. Brattain, who called it the transistor. This invention ended the rule of the thermionic tube or valve. Transistors were compact, simple, could be manufactured cheaply, and did not wear out. The result has been that radio and television sets together with a host of electronic devices are now built around transistors.

Civil Aviation

In the inter-war period the development of civil aviation in different countries was determined by particular political, economic and geographic factors. Private flying was limited to the wealthy, relatively few in number, who were also daring. The aircraft industries of all nations, except Germany, were kept in (much reduced) business by orders for military planes; these tended to be erratic, varying with the political climate. Civil air transport could exist only if subsidized. Perhaps the best way of doing this was by contract to carry mails. By 1919 British converted bombers were carrying air mail to troops stationed in Germany; the occasional passenger was also carried. In the United States an air mail service was established between Washington and New York. The same economic limitation applied to all nations. Germany, denied an air force by the Versailles Treaty, developed a large network of subsidized commercial air services; only in this way could she maintain an aircraft industry. German engineers were willing to design – and German governments to subsidize – very large aircraft. Examples of these were the post-war civilian Zeppelins and the huge Do X flying boats. The Do X was conspicuously unsuccessful. One took longer to cross the Atlantic than had Christopher Columbus and far longer than US Navy flying boats had taken in 1919. More successful were German planes with fuselages made of thin corrugated steel sheets on metal frames, and more successful still was Dr A. K. Rohrbach's fully stressed thin metal skin. Of necessity, France, Holland and Italy, nations with overseas colonies, established their own civil airlines.

There were high expectations of the airship. The Zeppelin airship had been developed in Germany before the War. Success in the bombing campaign, directed mainly against London, was temporary: it proved vulnerable to high-performance fighters. In peace time, however, the airship seemed to offer significant advantages over the airplane. Although slower it had an enormously greater range – 6000 to 7000 km – and far greater passenger-carrying capacity – a hundred or so passengers could be carried in comparative luxury. Transatlantic passenger flights by airship were quite feasible by 1930; for British Empire communications the airship offered major political and administrative advantages. It was the series of disasters, between 1930 and 1939, that overtook the British R-101, the American *Akron* and *Macon* and the German *Hindenburg* that led finally to the abandonment of a means of transport on which great hopes had been built[4].

The development of 'air liners', as they were grandiosely called, in the inter-war period initially followed the lines of the 1918 bomber: a biplane with two motors and a fuselage of doped fabric, or later plywood, stretched over a framework of thin metal tubes, the longerons. As three motors offered a much greater margin of safety, in case one engine failed, the low-wing, three-motor monoplane came to be favoured in the mid-1920s. Perhaps the best known in Europe was the Junkers Ju 52, with corrugated metal fuselage, of Deutsche Lufthansa and Deruluft; the military version became a reliable workhorse in war. A British state carrier, Imperial Airways, was founded mainly to strengthen the links of empire, although services to Paris and a few other European cities were operated. The carriage of air mail was a major duty of Imperial Airways; many, perhaps most, British families had relatives living or working in Australia, Canada, New Zealand, South Africa or India. Blue-painted mail or letter boxes carrying an oval sign announcing 'Air Mail' came to be a familiar feature of British streets (Britain never issued special 'air mail' stamps). The priority routes were, therefore, to Central and South Africa, India, Australia and, finally (not achieved), across the Atlantic. (The main hazard on the Australia route was, according to the Press, the four hundred miles of the 'shark infested' Timor sea.) Planes were mainly large and slow biplanes. Passengers were wealthy or otherwise privileged and high speed was not essential for Colonial governors; rather perhaps the contrary. It was intended that the airship, R-101, should make its first flight to India where, no doubt, its imposing size would have made a great impression. Imperial Airways planes bore classical names . . . Heracles, Scylla, Aurora, Cleo . . . and not, as one might have expected, the names of British flyers or engineers. The preference for classical names probably reflects the education and status of most of the passengers. It is hardly surprising that these planes were, with one notable exception, of no use to the war-time RAF. The exception was the 'Empire' flying boat, the military version of which, the 'Sunderland', was highly successful in the anti-U-boat campaign. Short distances and an extensive rail network meant that commercial airlines developed slowly in Britain. One route over which an airline could compete, on reasonable terms, was to the Channel Islands, popular with holiday makers but separated from England by more than fifty turbulent and unpopular miles of sea. Channel Islands Airways combined with other private airlines to form 'British Airways' just before the War.

As was the case with other countries, civil air transport in the United States[5] was aided indirectly by military orders and was maintained by air mail contracts. The great distances of continental America gave airlines a competitive advantage over rail and road transport that European airlines did not enjoy, while the wide variations in climate and topography were challenges to American engineering and scientific ingenuity. The most successful American airliner of the 1920s was a high-wing, tri-motor monoplane with a corrugated metal fuselage. It was made by the Ford Motor Company and was, accordingly, nicknamed the 'Tin Goose'. It was sturdy, easy to handle and reliable, but the small passenger cabin was extremely noisy. The Ford tri-motor showed its quality when, in 1929, Admiral Richard E. Byrd, USN, flew one to the south pole and back to base. Three years earlier he had flown a Fokker tri-motor monoplane over the north pole.

In the 1930s the accumulated experiences and extensive research and development of the previous decades came to fruition in a revolution in air transport that took place in the United States. American engineers had developed increasingly powerful series of engines with the cylinders arranged radially, like the spokes of a wheel. These radial engines enabled efficient air cooling without the need for complicated and heavy liquid cooling systems. Higher speeds, made possible by more powerful engines and smooth stressed skin metal fuselages, made retractable undercarriages economic. The high altitude of airports at important cities like Denver, Albuquerque and Salt Lake City (all between one and two kilometres above sea level) stimulated the invention and development of variable-pitch propellers. Finally, the introduction of wing flaps gave large, heavy and fast planes reasonably slow landing speeds.

The exemplars, or paradigms, of this revolution were three remarkable aircraft. The Boeing 247, the Lockheed Electra and, surpassing all others, the Douglas DC 2 and the bigger DC 3, the latter being, over its life span, the most widely used plane ever made (at their peak these planes carried some 95 per cent of American airline passengers; KLM, Royal Netherlands Airline, was the first European airline to use Douglas planes). All three were low-wing, metal monocoque monoplanes, built on the principle that goes back to the Britannia tubular bridge; they had twin radial engines and were equipped with variable-pitch propellers and retractable undercarriages. The radial engines and the propellers have been replaced by jet engines, but these details apart it would be fair to say that these planes and, in particular, the Douglas DC 3 (the 'C' stands for 'Commercial') stood in the same relation to subsequent planes, down to and including the jumbos, as *Planet* stood to all subsequent steam locomotives. Perhaps the best advertisement for the new Douglas planes was the performance of a DC 2 in the England to Australia air race of 1934. A special plane, the De Havilland 'Comet', of which three were built, was designed specifically to win this race. It was a low-wing, twin-engine plane carrying a pilot and co-pilot only. Inevitably, a Comet won. But a standard, airline, DC 2 entered by KLM came in a short time afterwards. The measure of this performance by the DC 2 is indicated by the fact that, as we now know, the Comet was, in effect, a prototype for the very successful De Havilland 'Mosquito' fighter-bomber of the Second World War.

The evolutionary continuity between planes of the DC 3 era and those of the modern period is confirmed by the fact that the first four-motor, pressurized passenger plane flew in America in airline service just before the United States entered the War. The remarkable improvement in the performance and reliability of commercial planes between the Wars and afterwards was due to advances in all aspects of aircraft construction and operation. Light and strong alloys, such as duralumin, were developed, the qualities of aviation fuel were improved, research greatly added to knowledge of the flow of air over wings and fuselages, progressively more powerful, efficient and reliable aero engines were built (about 1000 horsepower being reached just before the War), advanced radio communications and control systems were installed and, finally, there were many detailed inventions, among which the most important were wing flaps and variable-pitch propellers together with the widespread adoption of retractable undercarriages. It was symbolic, prophetic and tragic that when Mr Chamberlain went to Munich for his fateful meeting with Hitler he flew there and back in a Lockheed 'Electra' of British Airways. The airline and the plane foreshadowed the future; Mr Chamberlain's famous piece of paper – peace in our time – did not.

The carriage of mails had been the reason for the introduction of scheduled stage coaches in late-eighteenth-century England, thereby initiating a revolution in transport. Mail contracts had also played a key part in the introduction of – initially uneconomic – steam navigation across the Atlantic in the middle of the nineteenth century. And, now, in the first decades of the twentieth century, it gave a necessary boost to civil aviation in the United States and in the British Empire. One may surmise that, unless fax and electronic communications make mail wholly redundant, in the distant future the carriage of mail may be a factor in the establishment of regular interplanetary space flights.

Although the Russians had built four-engine bombers before the First World War and had operated them in the early stages of the conflict, and Germany, too, built and operated four engine-bombers later in the War, the bombers that were developed by western nations in the inter-war period were uniformly twin-engine. Monoplane bombers did not come into service until the 1930s and the prototype of the first four-motor bomber, the Boeing 'Flying Fortress' (B17), did not make its first flight until 1935. As far as the public were concerned the fighter plane was the more interesting. It was necessarily faster, accordingly more elegant and perhaps less threatening. The difference in name between the American pursuit fighter (P) and the British interceptor fighter indicated the different purpose and different design requirements of the two machines. While both were intended to destroy enemy aircraft, the latter was required to climb as fast as possible to intercept incoming bombers on their way to London and other targets easily accessible from the British coast. This problem was to have momentous consequences for the world's airlines, airforces and aircraft industries. The famous 'interceptor' fighter of the Second World War, the 'Spitfire', and its engine, the Rolls Royce 'Merlin', owed much to the accumulated skills of British aircraft firms[6]. It is a testimony to German engineering that the equally famous opponent of the Spitfire, the Messerschmitt 109, was so successful even though the German aircraft industry had been forbidden to build military planes after the Armistice and therefore lacked continuity in design and development. There was, however, an interesting difference between the two planes. The design of the Spitfire was much more a matter of intuition than that of the Me 109, whose design was far more systematic, scientific, optimal. This meant that the Spitfire allowed more scope for development. Yet when the Spitfire, the Me 109 and the Mustang (P51) (surely the apotheosis of the traditional single-seat fighter) were at their peaks of efficiency they, or rather their engines, were already obsolete.

The Jet Engine

Traditional German universities had, as we noted, concentrated on 'pure' science and had left technology to the technical universities. However, the divisions between 'pure' science and technology, never very clear, had become quite indistinct by the beginning of the twentieth century. This was particularly the case in chemistry and in the new science of electricity. Felix Klein (1849–1924), the brilliant mathematician, realized the artificiality of the division and persuaded the University of Göttingen to include technology within its faculty. Göttingen – Gauss's university and probably the leading university in the world for theoretical physics – therefore established a Department of Aeronautics. This Department did a great deal of research during the First World War but after the War aeronautical research contracted everywhere and the

Department looked for other fields to develop. Under
L. Prandtl one promising line was Anton Flettner's ideas
on the rotor ship. Three of these revolutionary vessels were
built and the results of their preliminary trials were
promising. Unfortunately the recession of the late 1920s
and the subsequent low price of oil brought development
to an end. Another promising innovation was that of the
jet engine, due to a young engineer graduate of the
Department and a student of Prandtl's, Hans P. von Ohain.
German engineers, it seems, were willing, and were
certainly given the means, to try out a wide variety of
experimental ideas. Von Ohain's jet engine powered an
experimental plane for a short flight in August 1939. For
several years thereafter the German authorities did not
show an urgent interest in the jet engine; possibly because
no immediate need for it was foreseen and Hitler did not
expect a long war, possibly because rockets and guided
missiles were accorded higher priority; probably because
of a combination of these two.

The course of development of the jet engine in Britain
began, as was so often the case, with a man who was
essentially an outsider playing the key role. Air
Commodore (Sir) Frank Whittle (born in Coventry in 1907)
inherited, according to his own account[7], his father's
mechanical aptitude and inventiveness. To these evident
abilities he added academic talent, determination and a
love of flying, at which he was to excel. He entered the
RAF as an apprentice and was, in due course, selected for
a commission. During his training at the RAF College,
Cranwell, he was required to write a number of essays,
or short theses; in his fourth term he chose, as a subject
for an essay, the 'Future Development in Aircraft Design'.

Young Whittle (he was 22 at the time) argued that high-
speed, long-distance flights would have to be at high
altitudes where air resistance was much less. Piston
engines would not, he believed, be suitable and he
therefore considered the possibility of the turbine, which

would perform far more satisfactorily at high altitudes and which promised other advantages as well. There was nothing new, or novel, about the idea of a gas turbine. At that time he thought of the turbine driving a propeller and he imagined the first use for his turbine engine would be to propel a small, high-speed mail plane. Whittle lodged a patent specification for his engine in 1930; eighteen months later the patent was published to the world. At last, after many attempts to persuade higher authority to take up his ideas, he was able to get a small company 'Power Jets Limited' established with the purpose of developing the jet engine. This was in March 1936. The 28 years old Flight Lieutenant Whittle was made Honorary Chief Engineer and Technical Consultant to the firm.

The jet engine is, in principle, a very simple machine; simpler even than the Watt or the Newcomen engine. It consists of a metal tube, open at both ends. At the leading end is a rotary air compressor that draws in and compresses air. This compressed air is then passed to combustion chambers where burning fuel maintains the air at a very high temperature and pressure in order to drive a turbine that is coupled to the compressor by a shaft (figure 16.3). After leaving the turbine the hot compressed air enters the tail pipe that ends in a nozzle. The function of this nozzle is to convert the high energy of the air (pressure and heat) into energy of motion, or kinetic

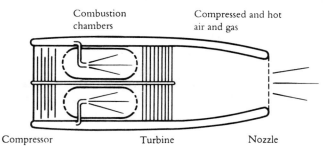

Combustion chambers   Compressed and hot air and gas

Compressor   Turbine   Nozzle

FIG 16.3   *The jet engine*

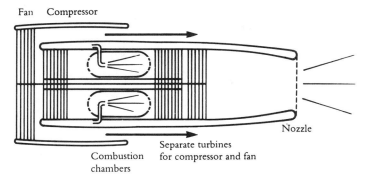

Fan   Compressor

Nozzle

Combustion   Separate turbines
chambers   for compressor and fan

FIG 16.4   *The by-pass or fan jet engine*

energy, or *vis viva* (cf. the hydraulic turbine, chapters 10 and 12). The reaction of this rushing air, escaping from the nozzle, drives the engine, and the plane to which it is attached, forward in the same way that a rocket is propelled.

The fan jet, or by-pass, engine is more efficient than the straight jet engine if the highest speed is not the desideratum (as is the case with commercial rather than military planes). In this form of engine only a proportion of the air sucked in is passed on to the combustion chambers, the remainder is blown out immediately, adding to the jet (figure 16.4). This means that the first section of the compressor must be of wider diameter than the rest, which is why the engines of wide-bodied planes have conspicuously large apertures in front.

As with Parent's water-wheel and de Pambour's steam locomotive, the thrust or force exerted by a jet engine, or engines, must, when the plane is flying level and at a uniform speed, exactly equal the air resistance the plane encounters (this includes the resistance due to the lift arising from the action of the wings in keeping the plane in the air). For a large commercial plane, flying at an economic, subsonic speed, the problem of design is to

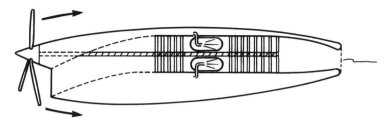

FIG 16.5   *The prop jet engine (the turbine converts most of the heat and pressure energy into the rotative form to drive the propeller)*

achieve the optimum thrust for the minimum fuel consumption. This can be achieved in the following way.

The thrust of an engine – the force it exerts – is proportional to the mass of air expelled per second and the increase in velocity of the air in passing through the engine. The amount of fuel consumed determines the power of the engine; this is proportional to mass of air expelled per second and the square of the velocity increase (i.e. to the energy expended). The fuel used is therefore proportional to the thrust times the velocity. Accordingly, if the mass of air can be increased for the same velocity, a greater thrust can be obtained for an economical rate of fuel consumption. The fan or by-pass jet is more economical than the pure jet, which gives the highest speeds, but not as economical as the turbo-jet, or turbine that drives a propeller (figure 16.5). In this last case little energy is left for the jet; most of it is used to drive the propeller so that as much air as possible 'by-passes' the combustion chamber and turbine.

The development of the jet engine necessitated materials that could stand unprecedented stresses and temperatures as well as new standards of machining. In this respect it followed a similar path to the development of all radically new engines. Indeed, the time lag, in years, between Watt's first conception (1755) and the first reasonably successful engine (1769–72) was of the same order of magnitude as Whittle's engine (1930–41).

One last point may be made concerning Whittle: he was not, at the time he made his invention and during the early period of its development, connected in any way with the aero engine manufacturing industry. Like Watt, he was an outsider to the engine building business. ·

New Materials from Chemistry

The chemical industry at the end of the nineteenth century comprised two main branches: heavy chemicals, chiefly concerned with the manufacture of sulphuric and other industrial acids, soda, inorganic fertilizers, etc., in great quantities; and the new organic chemicals industry that had sprung up round the synthetic dyestuffs industry and that had given rise to the manufacture of a number of related fine chemicals, including pharmaceuticals. In principle, at least, system (or science) replaced craft in the heavy chemicals industry when Dalton's atomic theory was accepted early in the nineteenth century. Towards the end of that century technology took a further step forward when chemical thermodynamics, originating from Gibbs' work, was used to optimize production. At the same time chemical engineering began to be important, and training for the new profession of chemical engineer was started up in the universities and technical colleges of the USA and (Tsarist) Russia. In both countries the needs of the oil industry were the main reason for the growth of the new profession.

A unique feature of the chemical industry was the large number of graduate chemists employed in it. This was true of all economically advanced countries, particularly of Germany. The German universities were well equipped with chemistry departments and the professoriate included many of the best-known names in chemistry. In most universities the chemistry departments were the biggest in terms of staff numbers and this remained true up to about 1940–45 when physics began to rival chemistry and in some cases to overtake it.

Perhaps the most familiar of the products of the chemical industry are what are commonly known as plastics. These substances, called polymers, are composed of extremely long molecules made up of repeated patterns, or chains, of relatively few atoms. There are many natural polymers, of which cellulose, a carbohydrate and a fundamental building block of plants, is one of the most common. Nitrocellulose, an explosive substance, and camphor were used, well over a hundred years ago, to make the first widely successful and familiar plastic, celluloid. This was followed by artificial silks, including the well-known rayon, and, finally, nylon and terylene or dacron. Nylon resulted from the work of the American chemist Wallace Hume Carothers (1896–1937). L. H. Baekeland (1863–1944), another American chemist, gave his name to bakelite plastics, which, among other things, are useful in electrical and electronic engineering. The more recent polythene and perspex (or plexiglas) are translucent and transparent plastics that have many different uses, at home and at work. Perhaps the most recent, teflon, with its familiar applications in the kitchen (and, some say, in politics!) is a by-product of space technology.

Although plastics represent a relatively new technology with enormously wide and numerous applications, they can hardly be classed as a strategic technology. Plastics are, almost invariably, used as more convenient or cheaper substitutes for materials that have long been in use: wood, ceramics, glass and metals. And plastics technology has not brought about major changes in other technologies and industries. On the other hand, the synthetics dyestuffs technology can fairly be termed a strategic technology. As we saw, it led to a wide range of changes in the fine chemicals industry, including pharmaceuticals, and it brought about the establishment of the first permanent industrial research laboratories.

These were instituted in German dyestuffs firms about 120 years ago. The industrial research laboratory is now a common feature of large firms, whether in chemical or in other industries.

The best-known development in modern medical and chemical technology has been that of antibiotics. (Sir) Alexander Fleming (1881–1955) at St Mary's Hospital, London, had for long been seeking something that would destroy pathogenic microbes without damaging the body. In 1928 he had been studying a colony of *Staphylococcus* when some mould, *Penicillium notatum*, accidentally landed on it and destroyed it. Unfortunately, although Fleming recognized the significance of this, there were great difficulties in developing the discovery and nothing further was done until 1938 when (Sir) Howard Florey began research at Oxford on antibacterial substances, among them being *Penicillium notatum*. Florey and (Sir) Ernst Chain succeeded in isolating penicillin and, after successful trials on animals in 1941, had enough at least to try on a human patient who was gravely ill, suffering from septicaemia. Penicillin brought about a dramatic improvement in his condition but unfortunately there was not enough to continue the treatment and the patient died. Britain, at that time, was under sustained air attack, there were other urgent demands on all branches of chemical industry and there were relatively few industrial biochemists. The large-scale manufacture of penicillin was therefore undertaken in the United States; by D-Day, in 1944, there was enough available to treat all casualties of the invasion of Europe. Here, again, the supply of highly qualified scientist/ technologists had been a critical factor in the development of an invention.

Computers

Although he held the Lucasian Chair of Mathematics – Newton's Chair – at Cambridge University for eleven years, Charles Babbage was, as his interests and life works show, an engineer of genius[8]. He was, however, an engineer of a most unusual, even unique, kind: an engineer of mental, or perhaps intellectual, processes. His book, *On the Economy of Machines and Manufactures*, was without precedent in subject matter and perception. If anyone was born out of his time it was Babbage; his ideas about the computer were only to find their rightful place by the middle years of the twentieth century. Like those of his near-contemporary, Cayley, to come to relative perfection these ideas required components missing at the time; in Cayley's case, an efficient lightweight internal combustion engine, in Babbage's case, electronics. During the first four decades of the nineteenth century Babbage planned two highly original mathematical machines: the difference engine (1823), which was a calculating machine intended for the reproduction of mathematical tables, and the analytical engine (1834), which was a computer in the modern sense. The analytical (in this context the word means mathematical) engine was, necessarily, entirely mechanical. It was to consist of complex gear trains arranged to give a memory facility and provision was to be made for what is now called conditional branching. The analytical engine was therefore to be unlike the difference engine in that it was a *universal* machine that could perform a wide range of different tasks and carry out the most complex calculations. Although the gear trains of the engine were based on decimal arithmetic, that is to base 10, the engine was to be programmed and instructions fed into it in binary code by means of Jacquard-type punched cards and the answers, too, given on punched cards.

Babbage had concluded that an analytical engine, or universal computer, could deal with algebra as well as arithmetic, that it could show foresight and, as we noted, that it had a function analogous to memory. Babbage's machines had columns of revolving wheels, each marked 0 to 9 round the circumference. If the lowest wheel is showing 7 and we turn it through 5 more digits we pass 0; on passing 0 a lever attached to the wheel turns the next wheel above from 0 to 1. The two wheels will show 1 and 2, or 12. This is no more than the simple arithmetic process of addition: on adding 5 to 7 we write down 2 and remember to 'carry' 1 (or 10) to make 12. In an analogous way the machine remembers to 'carry' the 1 to the upper wheel. For much more complex calculations and those involving many more digits a much more capacious memory will be necessary. It follows that for a universal computer as large a memory as possible will be necessary.

Babbage's machine, in its original, mechanical form, was never built, in spite of (relatively) generous government assistance. The reasons for this have been a matter of debate and need not detain us here. It is, however, interesting and relevant to note that Babbage's research associate was Ada Augusta, Countess of Lovelace and daughter of the poet Lord Byron. She was an excellent mathematician and wrote the best account of the machine and what it could do. In spite of this and her social distinction, Babbage's machines and his revolutionary conception lay dormant for the following hundred years and more. What need was there, people would have asked, for such a machine as the analytical engine? It would, however, be a mistake to assume that knowledge of it was lost. Babbage's ideas were well known in Turin, where the able Italian military engineer L. F. Menabrea was one of the first to understand them. Menabrea gave an account of them in the *Journal de Genève*, thus ensuring wide publicity in French as well as Italian circles. Babbage mentioned that he had held discussions about the analytical engine with F. W. Bessell and C. G. Jacobi, so that distinguished German mathematicians knew about it. His friend James MacCullagh, of Trinity College, Dublin, was familiar with his proposals. A model of the difference engine was displayed at the 1862 International Exhibition in London. Finally an affluent Swedish devotee,

G. Scheutz, assisted by his son and with some official support, made a difference engine, which he took to England. This machine won a medal at the 1855 Exhibition in Paris. In the following year it was bought for the Dudley Observatory in Albany, the state capital of New York. Discounting his publications, polemical and otherwise, Babbage's ideas were known, to some extent at least, in the international scientific world, and we might infer that this knowledge lingered on in the folk memory of mathematicians, astronomers and engineers as well as in entries in various encyclopedias.

In the intervening years the computer, or the computer-like machine, was represented by Morin's *compteur*, by the tidal predictor of James Thomson and then, after thirty years, by the gunnery predictor sponsored by Arthur Pollen. In the inter-war period came Vannevar Bush's differential analyser at MIT and similar machines were built at other universities. All these analogue computers were mechanical or electromechanical in operation, highly specialized, on the margins of technology and science. Each could perform only the one job for which it had been designed. None was a *universal* machine. If one was required to perform another task, it would have to be dismantled, the components changed or modified and then re-assembled.

Hermann Hollerith used punched cards in his machine designed to simplify work in connection with the US Census of 1890. Whether he knew of the Jacquard loom and its punched cards is not certain. The Hollerith machine – it was widely used and for a long time – was not the mechanical computer as conceived by Babbage, but it was the ancestor of a number of business machines used for routine purposes, and these machines spread knowledge of the principle to people concerned with calculating and computing devices.

Over the period that ended with the Second World War a variety of office and business machines had been developed to such an extent that the legion of clerks, on high stools, which had been an essential (and expensive) feature of nineteenth-century commerce and industry were largely made redundant. The most conspicuous and common of these desk-top machines was the typing machine, or typewriter. This was an American invention dating from 1873 (Remington) and made possible by skilled light engineering. Other inventions of the period included various adding machines, dictaphones, comptometers, cash registers and duplicating machines (Boulton and Watt had used a simple duplicating machine). Set beside the inventions of television, radar or the jet engine these simple, mostly mechanical, devices certainly lack the glamour of advanced science and high technology. Yet taken together they added up to radical change, effected over a few score years, in commercial and office practices that had remained virtually unchanged for centuries. As with all major innovations there was an accompanying, significant social change. The clerks and their beautiful copperplate writing vanished; young women were recruited to operate the typewriters and other office machines, rather as young women (and children) had operated the machinery in the classic textile mills of the earlier Industrial Revolution. Whether this furthered the independence and welfare of women must be left to feminist historians to consider. In one respect, though, progress had been virtually non-existent; by the 1930s the realization of Babbage's ideas, advanced as they were, seemed as far away as ever; but further discussion of this must be deferred to a later chapter.

A NOTE ON RADAR

Randall recollected that while on holiday at Aberystwyth in 1939 he had bought, at the University bookshop, a second-hand copy of the English translation of Hertz's original papers. He later added that he was glad that neither he nor Boot had read many of the extensive theories about the magnetron that had been published. Had they done so they would have been so confused that he doubted if they would have invented the cavity magnetron.

After the War Randall set up the Biophysics Unit at King's College, London, where Maurice Wilkins and his colleagues carried out their share of the research that led to the unravelling of the genetic code and for which Wilkins was one of the three recipients of the Nobel prize.

Foreshadowing the Future: Big Technology

Within the limitations of present knowledge we may speculate that, in centuries to come, the most important events of the present age will be considered to have been the first steps towards the conquest of space and interplanetary travel, achievements to rank in importance with the voyages of da Gama, Columbus, Magellan and the other navigators of the great age of discovery.

Voyages to the moon and beyond were barely conceivable before Galileo had demonstrated that the moon had mountains and seas and was therefore just like the earth; and, he inferred, the planets were just like the moon. The subsequent discovery that the moon was an arid body, without water and, indeed, lacked an atmosphere, made the prospect of a voyage to the moon impossible to consider, even had the means of transportation been available; except, that is, for entertaining story tellers like Cyrano de Bergerac (*Les Etats et empires de la Lune*, 1649) and Erich Raspe (*Baron Munchausen*, 1785). By the nineteenth century technology had advanced to a point at which the means for interplanetary travel were, at least in outline, conceivable. Although there were one or two suggestions of rocket propulsion, the method envisioned by the man who was the most famous of all imaginative science fiction writers, Jules Verne, in his *De la Terre à la lune* (1865), was a giant gun. How the occupants of the space capsule to be shot from the gun could survive instant acceleration to the

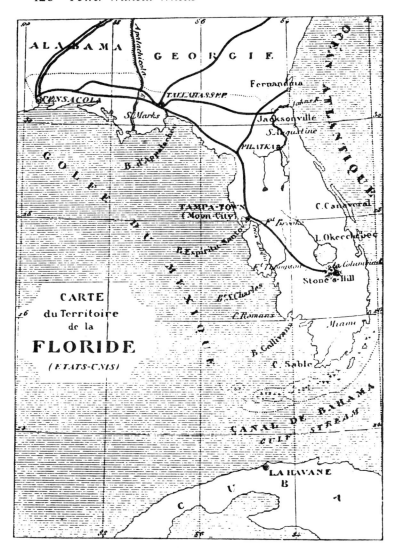

Carte de la Floride.

FIG 17.1 *Location of Jules Verne's space gun*

necessary escape velocity of 7 miles, or 11 km, per second was not clearly explained. On one point, however, Jules Verne could hardly be faulted: he placed his space gun just south of Lake Okeechobee, Florida, and quite near Cape Canaveral (figure 17.1).

In his book, *The War of the Worlds* (1898), H. G. Wells, the other great nineteenth-century writer of science fiction, described the invasion of the earth by technologically superior and utterly ruthless creatures from the planet Mars[1]. They, too, were carried by capsules – shells might be a better word – shot from a gun on Mars. In this case a little licence can be allowed the author: gravity on Mars is less than on earth and the Martians were, in all ways, inhuman. Wells does not seem to have changed his ideas about space travel; in his much later work, the film *The Shape of Things to Come* (1934), he imagined that the first moon shot would be by means of a space gun. Wells, however, differed from other writers of science fiction in that he used it as a vehicle for social criticism. Passages in *The War of the Worlds* foreshadowed the behaviour of ordinary people under occupation by a brutal enemy, an experience many were to share in the great wars that were to follow.

Although one or two writers had conjectured that rockets might provide a means of space travel and one or two had even talked about earth satellites, rockets remained firmly out of the mainstream of technological debate and effort. They were, no doubt, considered no more than toys; suitable for firework displays, the entertainment of children and nothing more. The Chinese had invented fireworks and were even reputed to have used rockets in battle. The Congreve rocket had been quite successful in the wars at the beginning of the nineteenth century, but after the peace of 1815 they fell into disuse, as did the visual telegraph networks. The rocket, after all, was not a precision weapon nor was it clear how it could ever be made into one.

Such interest as there was in rockets and the possibilities of space travel in the early decades of the present century was almost entirely limited to amateurs. Three figures stand out as particularly important in these early years. They are the Russian K. E. Tsiolkovsky (1857–1935), the American Professor Robert H. Goddard (1882–1945) and the German Hermann Oberth. Tsiolkovsky, a visionary, wrote papers at the turn of the century advocating rockets to carry men into space. Such rockets could act as earth satellites and could reach the moon, from which further exploration could be mounted. He summed up his ambition in memorable words: 'the earth is the cradle of mankind, but one cannot live in a cradle for ever.'

Goddard, an academic, was more modest in his ambitions but, arguably, his achievements were more valuable. From 1914 onwards he experimented with rockets, being the first to consider liquid fuels for them. He considered that rockets would be most useful – as indeed they proved to be – in high-altitude geophysical research. In 1920 he published the first theoretical study on the subject, *A Method of Reaching Extreme Altitudes*. Some years later he had succeeded in building and flying the first liquid-fuelled rocket. In addition he was able to apply gyroscopic steering to rockets; this was to be an important element in the development of the space rocket.

Hermann Oberth's interest in space travel was aroused by reading Jules Verne when he was a boy. After an exemplary scientific and technological education, he dedicated himself, after the First World War, to the development of the space rocket, which, he realized, technology had now brought within the realm of the possible. Unlike Goddard, he saw that the rocket could carry men as well as scientific instruments into space and, eventually, out to the planets. In 1923 he published his book, *The Rocket into Interplanetary Space*. This, too, was a theoretical study and it set out, strongly, the case for liquid as against solid fuels. Oberth's work and writings were influential and there were, in the 1920s and 1930s, some promising experiments with rockets, apart from those carried out by Goddard. Mails, as we might expect, were successfully carried short distances by rockets. And in several countries – particularly in Germany and America – interplanetary societies were founded. For full-scale development, however, more was needed than the efforts of dedicated amateurs, however talented. As with the telegraph,the interest and support of a major national institution or a very large corporation was required.

Under the terms of the Versailles Treaty, Germany was virtually disarmed. Her army, once the most powerful in Europe, if not in the world, was reduced to a Reichswehr of 100 000 men, all of whom had to be volunteers. She was not allowed to have tanks or heavy artillery and, as we saw, was denied an airforce. At the same time the German navy was severely cut back; submarines were banned and no major battleships or battle cruisers were allowed (hence the famous 'pocket battleships').

In the light of these restrictions and the fact that neither France nor the Soviet Union disarmed, it was hardly surprising that the commanders of the German army should seek some alternative weapons, not banned by the Treaty, to redress the balance. Influenced, no doubt, by Oberth's work, they decided to investigate the military potential of the rocket. In 1930 the German army initiated experimental work on liquid-propellant rockets at Kummersdorf, just south of Berlin. Two years later the commanding officer, Captain Walter Dornberger, was joined by the nineteen-year-old Wernher von Braun. Major-General, as he became, Dornberger later explained that their concern at that time was to get reliable data established by systematic experiments and to cut through the tangle of wild fantasy and unsubstantiated claims generated by the many enthusiasts. This, as far as we know, was the world's first research and development establishment devoted to rocket propulsion. America,

Britain and France showed no interest in rockets. Little is known about Russian activities at that time.

The rocket motor is basically simple and similar to the jet engine. In a combustion chamber liquid oxygen, or an oxygen compound, and a flammable liquid are vaporized and the mixture ignited. The heat and pressure energy of the resultant gas is converted into kinetic energy, or energy of motion, as it rushes out through a narrow nozzle. By Newton's third law of motion – action and reaction are equal and opposite – the action of the gas rushing out of the nozzle must be accompanied by an equal and opposite reaction on the motor, pushing it in the opposite direction. The same basic law accounts for the kick on the shoulder that the rifleman feels when he fires a bullet and for the (often) annoying behaviour of the garden hose when the water is turned on suddenly. The unique advantage of the rocket motor is that, being independent of an external air supply, it can operate in a vacuum; indeed, it is most efficient working in a vacuum.

When the first successful launch of the A4 (later V2) rocket took place on 3 October 1942, it marked the triumphant conclusion of years of detailed research and development work by Dornberger, von Braun and their colleagues[2]. At the heart of the project was the huge research and development establishment at Peenemunde, situated among pine forests at the north-western tip of the island of Usedom, just off the Baltic coast of Germany. The successful launch of the V2 rocket was a testimony to German engineering, all the more remarkable in that it was accomplished in spite of the arbitrary nature and casual brutality of the Nazi state. A long series of problems had been solved. The best aerodynamic form of the rocket had been determined, an elaborate and sensitive control system was developed, metals for the bodywork of the rocket were found that could stand the immense stresses and high temperatures of high-speed flight, reliable telemetry had been designed and the most efficient fuels found.

A particularly difficult problem was the control, or steering, of the rocket at the moment of lift-off. The rocket would be moving so slowly that the fins would be ineffective. In the same way a ship, whether driven by sail or propeller, must be moving at a minimum speed for the rudder to take effect: the vessel must have 'steerage way'. With insufficient steerage way a boat will just drift but a rocket would topple over. Von Braun solved the problem by fitting graphite steering fins in the driving jet from the rocket motor. Another key problem was that of injecting sufficient fuel and liquid oxygen into the combustion chamber to maintain the extremely high rate of combustion necessary. This was achieved by a high-power turbo pump, driven from another source of fuel. Indeed, the problems to be solved were so daunting that, even in the face of a steady stream of intelligence reports and photographs brought back by reconnaissance planes, highly qualified scientists and engineers on the Allied side denied that such a rocket was possible.

The V2 rocket was about 46 feet, or 15 metres, long and, without fuel and explosive head, weighed about 4½ tons. It could travel about 200 miles (320 km), rise to a height of about 60 miles (100 km) and reach a speed – quite unprecedented at the time – of about 3200 mph (4800 km/h). Between September 1944 and March 1945 over five hundred V2s were targeted on London, killing nearly three thousand people, and more than twice as many in Antwerp. These numbers are small compared with those of the casualties in the massive air raids of the War. And, indeed, when Albert Speer took into account the enormous technological and industrial effort put into developing and building V2s, he described the rocket programme as 'nothing but a mistaken investment'. However, the military–economic balance sheet does not concern us here. The impact of the V2 was psychological; it was a terror weapon. No warning and no defence were possible.

The V2 was small set beside the rockets that were developed after the War. Its importance, like that of Trevithick's simple steam locomotive, lay in the fact that it indicated what was possible. And for ten years, 1945–55, the western nations and Russia were interested in the new, space rocket solely as a weapon of war. It was the Russians who first saw that it could be used as a weapon of another kind[3]. On 4 October 1957 they launched the *Sputnik*, a small earth satellite, the size of a football. It orbited the world several times, bleeping out its message, seen on all cinema and television screens, that the Soviet Union was in the forefront of technology and science. Shortly after *Sputnik* the Soviet Union extended their experiments by sending a living creature, the dog Laika, into orbit round the earth. (They made no provision for bringing the animal safely down again; this made a much less favourable impression in the English- speaking world.) The final step in this series was taken on 12 April 1961, when Major Yuri Gagarin orbited the earth in a space capsule, *Vostok 1*. Gagarin was a passenger who took no part in controlling or navigating the capsule. But he did demonstrate that humans could survive the stresses of lift-off, weightlessness and descent to earth. In the meantime a series of space 'probes' was initiated by both America and Russia. Automated spacecraft were landed on the moon, while others were directed to Mars and Venus (in spite of its name, a singularly inhospitable planet). The Russian space authorities caused a sensation when they published the first photograph of the other side of the moon. And the Russian, Valentina Tereshkova, was the first woman to go 'into space' and orbit the earth. Russian achievements aroused the competitive spirit of America. After Gagarin's brave venture, President Kennedy called for a moon landing by 1970 and great resources were placed at NASA's disposal.

Sputniks and dogs might be disposable; human beings are not and, once put into orbit, must be brought back as safely as possible. The great problem here was that of re-entry into the earth's atmosphere. To keep an astronaut, or cosmonaut, in orbit at a height of about 150 miles (240 km) the rocket had to reach a velocity of about 18 000 mph (30 000 km/h). However, in 1848 Joule had proved that the immense kinetic energy of meteorites would be dissipated as intense heat on entering the earth's atmosphere, so that they would be completely incinerated. The rapid compression of the air in front of the meteorite and friction between the meteorite and the thin, bitterly cold air would produce enormous, furnace temperatures. The same intense heating would affect an orbiting capsule as it re-entered the atmosphere on its way back to the ground. Accordingly, much thought had to be given to the mode of re-entry and to providing the capsule with a 'heat shield'.

The basic principle of the rocket motor, and of the jet engine, was enunciated in the form of Newton's third law of motion. Newton, it is interesting to note, clearly foreshadowed – indeed he even illustrated – the earth satellite. The diagram in figure 17.2 is taken from the English-language edition of Newton's *Principia*[4].

Imagine an extremely high mountain, V, on a fictitious planet, C. A projectile thrown or fired from the top of V may reach the point D. If now the projectile is thrown or fired with progressively greater force it will follow the trajectories VE, VF, VG . . . until, if it is thrown or fired with sufficiently great force, its trajectory will take it right round the planet. If the planet has no atmosphere, or if V is so high that its summit is effectively above the atmosphere, the projectile will go on orbiting round the planet as a satellite (it is assumed that the peak of V may be lowered to allow the satellite to pass over it!). Of course, in Newton's day there was no means of lifting a projectile above the earth's atmosphere or of giving it sufficient velocity to follow a circular trajectory as a satellite. The rocket can do both of these.

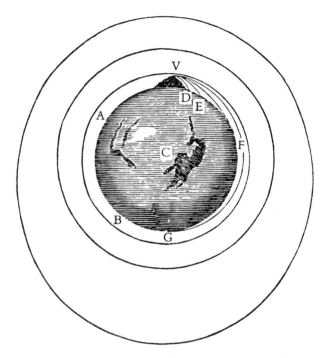

FIG 17.2   *Newton's illustration of a satellite*

Stimulated by Soviet propaganda success with *Sputnik* and the unfortunate Laika, America decided to enter the space race (as it was now called); a clear indication of this was the establishment of the National Aeronautics and Space Administration, NASA. Less than a month after Major Gagarin had orbited the earth, Cdr Alan B. Sheppard USN, made the first American journey into space. This was a 'sub-orbital' flight from Cape Canaveral, at the conclusion of which Cdr Sheppard was picked up when his capsule splashed down in the Atlantic. The project was publicized beforehand; it was televised and minute-by-minute accounts were broadcast. There was an agonizing delay as Cdr Sheppard waited in the capsule while faults were remedied and modifications were made. This openness was to characterize all subsequent American space ventures. Russian achievements were announced only after they had been accomplished. This made it difficult to know how efficient their programme was.

On 22 February 1962 Lt Cdr J. H. Glenn orbited the earth in a 'Mercury' capsule and so began a series of American orbital flights that were to lead to the first of the 'Apollo' missions. These culminated on 20 July 1969 with the moon landing of Neil Armstrong and 'Buzz' Aldrin and their safe return to earth when their capsule splashed down in the Pacific. The Apollo, three-man, series was interrupted by the dramatic adventure of Apollo 13 when a damaged spacecraft was, by a combination of cold courage and technical skill, enabled to bring its crew safely back to earth. The words of the mission commander – 'we have a problem' – stand as the understatement of the century. Apollo 17, however, marked the end of the series. A combination of the world-wide fuel crisis of 1973 and financial stringency led to the cancellation of such prodigiously expensive ventures and concentration on the less dramatic but financially and politically more rewarding 'Shuttle' projects. In 1975 there occurred a notable step forward in international co-operation with the Apollo–Soyuz link-up in space. In truth, the dedicated and extremely courageous men who led space exploration from America and from Russia had a great deal in common. There had already been tragic loss of life. Virgil Grissom and two of his colleagues had died in an 'oxygen fire' during an Apollo ground test[5], while, in 1971, the crew of three returning from Salyut 1 space station had died when their capsule had decompressed. Gagarin, the pioneer, had lost his life when a plane he was testing crashed.

The requirements and the limitations that the astronaut places on the design and operation of space vehicles has led some to claim that it would be much more efficient to replace the astronaut by computer systems. While it may well be that the astronauts of the present century will come to be regarded in the same light and as important as the Italian, Portuguese and Spanish navigators of the fifteenth and sixteenth centuries, there can be little doubt that the unmanned probes to Venus, Mars and the outer planets – Mariner, Voyager and the like – have relayed back more information of value to scientists. And there can be no question that the various automatic satellites orbiting the earth have been of great economic and social as well as military value. Earth satellites are used as aids in weather forecasting, in geological and mineralogical surveys and in terrestrial navigation.

The science fiction writer, Arthur C. Clarke, claims to have been the first to point out that earth satellites could bring about a revolution in communications and this has been the most familiar use to which satellites have been put. Telstar, the first communications satellite, was put into orbit in 1962. It was followed by syncom satellites, so positioned that they effectively remain above one spot on the earth's equator, and must therefore orbit round the Earth once every 24 hours. To do this they must be just over 22 000 miles, (35 000 km) from the Earth and travel with an orbital velocity of about 6000 mph (9000 km/h). Satellites of this kind are used to transmit telephone messages and television broadcasts between continents.

The V2 rocket may have been a misguided piece of military hardware – the Nazis would have been better advised to have invested their resources and skills in building jet fighters – but it pointed the way to space travel, as von Braun and his associates realized. That it also gave rise to a revolution in communications was surely something neither he nor his colleagues considered. This was another example of the unpredictability of technological change. The V2 is interesting from another point of view. Its development marked the first time that a state had devoted massive resources to one technological project. Nothing on the scale of Peenemunde had been seen before. Such an enterprise would hardly have been possible only a few decades earlier; there would not have been sufficient numbers of professional mathematicians, mechanical engineers, chemists, metallurgists or physicists. Organized science and technology made the V2 possible. And they made possible a roughly contemporary project that in some respects resembles the V2 project and in others forms an instructive contrast.

Atoms and Power

The revolution in physics associated with the names of Planck and Einstein had been preceded by W. K. Roentgen's discovery of X-rays in 1895 and Henri Becquerel's discovery of radioactivity from uranium in 1896. The value of X-rays for surgery was realized at once while radioactivity was used, following Marie and Pierre Curie's discoveries of very active radium and polonium (1898), in the treatment of cancer. These also had important consequences for our understanding of the structure of the atom. In 1911, a master experimentalist, the New Zealander Ernest Rutherford, working in Manchester, was able to announce that the atom consisted of a relatively tiny, concentrated nucleus, positively charged, surrounded by negative electrons, orbiting like distant planets round a tiny sun. Ironically, Rutherford announced his discovery to the same Manchester Society that had heard Dalton announce his atomic theory over a hundred years earlier. In 1913 the young Danish theoretician, Niels Bohr (1885–1962), working with Rutherford, accounted for the orbits of the electrons by using Planck's quantum theory. In this way nuclear physics, the most characteristic branch of twentieth-century physics, was born. Six years later, and just before he left Manchester, Rutherford announced that, using high-speed alpha particles from radioactive sources

(Rutherford had proved that the alpha particle was the nucleus of a helium atom), he had 'split the atom', changing a few nitrogen atoms into oxygen atoms. Soon there were speculations that the immense binding energies that held the nucleus together might possibly be exploited as a new source of power.

In the meantime Frederick Soddy had proposed that each chemical element is made up of large numbers of two or more atoms identical in all respects save that of weight. Chemically, they would be indistinguishable, but the existence of chemically identical atoms of different weights accounted for the fact that atomic weights are not, generally, simple multiples of the atomic weight of hydrogen, taken as 1. In 1919 F. W. Aston described his mass spectrograph, an instrument that made it possible to distinguish between and to measure the different atomic weights of the same element. In this way it was shown that chlorine, for example, whose atomic weight was 35.46, was actually a mixture of atoms whose weights were 35 and 37. These atoms, distinguished only by weight, are known as isotopes, and most elements in the periodic system have two or more isotopes (hydrogen, for example, has three).

Progress in nuclear physics was extremely rapid in the years between the World Wars. Big electrical machines were built to accelerate particles to speeds that would rival those of natural alpha particles and thereby split atoms to cause the artificial transmutation of elements. These machines worked on the same principle as the mass spectrograph. The atoms of an element were ionized; that is, they were given a positive electric charge by using alpha particles to knock off the negatively charged orbiting electrons. The ionized atoms, now being positively charged, could then be accelerated in the same way that electrons are accelerated in a cathode ray tube. The first of these machines was the electrostatic generator of J. D. Cockcroft and E. T. S. Walton (1932), who generated very high voltages by a procedure analogous to that used in the supply of high-voltage direct current (see chapter 14). Instead of batteries, electrostatic condensers were charged up in parallel and were then switched in series. In the Cockcroft and Walton apparatus protons – the positively charged nuclei of hydrogen atoms – were accelerated by the application of 700 000 to 800 000 volts to strike a lithium target; the lithium was disintegrated to yield alpha particles, or helium nuclei.

Better known was the cyclotron invented in 1930 by Ernest O. Lawrence (1901–58) of the University of California Radiation Laboratory. Instead of being accelerated in a straight line, as in the Cockcroft and Walton apparatus, the protons in the cyclotron were accelerated along a circular and therefore far longer path. This was done by using a powerful magnetic field together with rapidly changing electric fields. The magnetic field was so arranged that it bent the stream of protons into the circular path, while the electric field between two semicircular plates alternated at a high frequency in such a way that the protons were accelerated as they orbited. Next time round the protons would receive exactly the same acceleration and so on until they reached an enormous velocity. In much the same way the enthusiasts of the 'electrical euphoria' had hoped that an electric motor could be made to reach a near-infinite velocity. The alternate electric fields did not have to be enormous, as in the Cockcroft and Walton apparatus; it was enough that the stream of protons should receive the same acceleration for each orbit, irrespective of their velocity. Lawrence's cyclotron proved capable of remarkable and continued development. The later van de Graaff electrostatic accelerator worked on a different principle from the Cockcroft and Walton apparatus.

In 1932 James Chadwick discovered the neutron, of the same weight as the proton but without an electric charge. The absence of electric charge had made the neutron difficult to detect; it could pass right through atoms without knocking out electrons, or ionizing them, so that its passage would leave no track in a 'cloud chamber', the common detecting instrument in use at the time. The only thing that would stop or deflect a neutron was a direct collision with a nucleus. This meant that neutrons, being quite indifferent to the strong positive charge on the nucleus, could bring about atomic changes, or transformations, which would require powerful accelerators, such as the Cockcroft and Walton machine or the cyclotron, if positively charged alpha particles or protons were to be used. It was further realized that, with the exception of hydrogen, every nucleus consisted of protons and neutrons. In the same year H. C. Urey and two colleagues discovered the first isotope of hydrogen, which was called deuterium. Combined with oxygen deuterium forms heavy water. Later a third isotope, called tritium, was found. The third major discovery of 1932 was made by D. C. Anderson, of the California Institute of Technology; it was that of the positron, a particle with the same weight as the electron but carrying a positive charge. Three years later Yukawa predicted an intermediate particle, the elusive and transient meson; its tracks were detected some years later. The 1930s were, therefore, a most fruitful epoch in nuclear physics. The 'new men' who launched the new theories and made the startling discoveries were the professional scientists of America, Britain, Germany, France, Italy – and Japan and Russia. It was said to have been the age of innocence of physics.

The element uranium (U) had been identified and named over two hundred years ago; it was isolated over a hundred and fifty years ago, well before radium and the other radioactive elements. The heaviest of the known, naturally occurring, elements, its atomic weight had been found to be 238.07, but the application of the mass spectrograph showed that it was commonly made up of the isotopes ^{235}U (0.7 per cent), ^{238}U (99.3per cent) and a minute trace of ^{234}U. Enrico Fermi (1901–54), who had pioneered the use of neutrons to produce artificial radioactivity in ordinary elements, was the first to investigate the action of neutrons on the massive uranium nucleus.

One of Fermi's interests had been to see if neutrons acting on uranium could generate transuranic elements. Two German physical chemists, Otto Hahn and Friedrich Strassman, carried out research on these lines and found evidence not only of transuranic elements but also of other elements, one of which seemed to be barium, whose atomic weight was roughly one half that of uranium. Hahn and Strassman's paper of 6 January 1939, prompted the immediate suggestion by Lise Meitner and Otto Frisch, two refugee Austrian scientists working in Copenhagen, that the uranium nucleus, struck by a neutron, splits into two nuclei of approximately equal weights, one of which was an unstable barium nucleus; at the same time a considerable amount of energy is released.

This was intriguing. The slow transformation of a radioactive element by the emission of alpha and beta particles was well established but the splitting in two, the fission, of an atom was something entirely new. In mid-January 1939 Niels Bohr went to Princeton to discuss certain scientific problems with Albert Einstein. He had been told of Meitner's and Frisch's suggestion just before he left Copenhagen and he passed the information on to his former student, J. A. Wheeler, at Princeton. From that moment on the news spread rapidly among the members of the East Coast physics community. At a meeting in Washington DC at the end of January Bohr and Fermi discussed the problem of fission and Fermi put forward the idea that neutrons might be emitted during the process. The implications of this were momentous. Although by no means all the secondary neutrons emitted might cause fission – some might escape or be lost in other ways, some might be captured without causing fission – it was possible that enough would cause the fission of other uranium nuclei, causing them to emit still more neutrons so that a cascade, or chain reaction, would take place. Calculation showed that the energy released by a chain reaction would be enormous. The reasoning was straightforward and based on Einstein's special theory of relativity (1905). The binding energy, which holds an atomic nucleus together, is large in the case of a uranium atom. When a nucleus splits, that energy is associated with a change in the total masses of the constituent particles. Accordingly, the mass of a uranium nucleus is found to be greater than the combined masses of the particles into which it disintegrates. The difference, the 'mass defect', represents the immense kinetic energy released. ('Defect' is, of course, the wrong word. The expression should be 'mass deficit', but it is now too late to change.) According to Einstein this equals mc^2, where m is the mass that disappears and c is the velocity of light (see Note at the end of the chapter). The lost mass, m is very small per individual atom, but c^2 is extremely large; so here we are contemplating an enormous release of energy for the fission of a relatively small amount of uranium.

It became public knowledge, from the spring of 1939 onward, that enormous energy could, in theory, be derived from uranium for use as an explosive of unprecedented destructiveness or for controlled power. Realizing the implications of this American and British physicists with the support of Niels Bohr called for the voluntary censorship of publication. This was later put on a more formal basis although, throughout the Second World War, censorship remained basically voluntary. Nuclear physics was still part of the world of 'pure' science. As late as 1944, towards the end of the War, Max Born could briefly mention the possibilities of the fission process in his book *Atomic Physics*[6]:

> If it were possible to produce such a chain reaction
> the enormously condensed nuclear energy could be

utilized for practical purposes, such as driving power for engines or as explosive for super-bombs.

Alarmed by the possibility that Nazi Germany might be initiating research leading to the development of an atomic bomb, the scientific community, headed by Albert Einstein – a lifelong pacifist – raised the matter with President Roosevelt, who responded by appointing the small 'Advisory Committee on Uranium'. There was some cause for concern. Despite the persecutions of the Nazi state, Germany still retained an impressive corps of 'pure' scientists, including Werner Heisenberg, one of the world's leading nuclear physicists; there was some evidence that Germany was beginning research that could lead to the development of nuclear power; and, lastly, that Nazis controlled, directly and indirectly, extensive areas containing uranium ores.

In June 1940 the Advisory Committee became a subcommittee of the National Defense Research Committee and the investigation of key problems was contracted out to various universities and manufacturing corporations. These were formidable enough. It had been found that fission caused by slow ('thermal') neutrons occurred only in the uranium isotope ^{235}U, which amounted to 0.7 per cent of refined uranium. ^{235}U and ^{238}U are chemically indistinguishable. Their separation would be extremely difficult; analogies fail. In comparison, finding the proverbial needle in a haystack would be easy. An alternative that emerged was to use the transuranic and therefore artificial element, plutonium (Pu). Research revealed that the major isotope, ^{238}U, absorbs a neutron from the fission of ^{235}U to become ^{239}U which then quickly transforms itself into the fissionable element plutonium, ^{239}Pu. And plutonium, being a different (and very nasty) element, could, in theory, be chemically separated from uranium.

Britain had made a start in developing an atomic bomb

under the code name 'Tube Alloys'. Leaders such as Chadwick, Cockcroft and G. P. Thomson (son of J. J. Thomson) were convinced that an atomic bomb could be made and were well aware of the possibility that the Nazis might be working to the same end. However the collapse of France, the 'Blitz' and the demand for scientific manpower to meet other urgent commitments soon put a stop to work in that direction, and British nuclear resources, human and material, were transferred to the United States. Britain was in no position to pursue a project that was scientifically remote, technologically and industrially demanding and hugely expensive. All Britain's resources had to be directed towards defeating the U-boats and developing the strategic airforce; and this, in terms of physics, meant concentrating on centimetric radar and navigational aids. Whether Nazi Germany ever had a serious policy of developing an atomic bomb is still doubtful. For one thing, many of the best German scientists had become refugees in America or Britain. Remaining German nuclear scientists, unless they were dedicated Nazis, may well have had serious reservations about joining in such an enterprise, and even if they were willing to do so were soon informed that Hitler, if told about it, would demand one by next week. And Hitler was not a man to whom one could safely say that he was asking for the impossible. Whatever the actual situation in Germany, the potential threat could not be ignored.

A report from the National Academy of Sciences in November 1941 confirmed that uranium fission bombs could decide the outcome of the War. The mass of uranium required for explosive fission would be between 2 and 100 kg and, on conservative assumptions, the latter figure could have the same effect as 30 000 tons of TNT. Uranium is an unusual element; above a certain, critical size a mass of ^{235}U will disintegrate; the chain effect will result in the generation of great heat and radiation but no explosion. If, however, two subcritical masses can be rapidly brought together, then the much greater chain effect will result in an enormous explosion. Under pressure from the National Academy as well as from individual American, British and, not least, refugee scientists, the US government decided, in December 1941, to enlarge and reorganize the uranium programme. Dr Vannevar Bush decided that the Office of Scientific Research and Development, of which he was Director, should take over the project. Once the preliminary contracts for pilot plants had been settled, Dr Bush advised that the US army be brought in to supervise the construction of the main plants and to run the whole project. In August 1942 a new District of the Corps of Engineers, the Manhattan District, was established to carry out the necessary work in connection with the atomic bomb project. The project itself was designated DSM, 'Development of Substitute Materials'; a title hardly more informative than 'Tube Alloys'. On 17 September 1942 an able engineer officer, Brigadier Leslie R. Groves, was put in complete charge of work on DSM[7] or, as it was soon to be called, the Manhattan Project.

It had been decided that, such was the urgency, all the main possible routes to the atomic bomb should be followed. There were, in principle, several ways in which ^{235}U could be separated out from ^{238}U; chief among these were gaseous and thermal diffusion, the centrifuge and electromagnetic separation (basically a mass spectrograph on an immensely greater scale); alternatively fissionable ^{239}Pu could be chemically separated out after each ^{238}U atom, having absorbed one neutron, became ^{239}Pu. The overriding question on which all depended was whether a chain reaction was possible. It was Fermi who solved the problem and, at the same time, introduced a new method of deriving power: this was the atomic pile. On 2 December 1942 at the University of Chicago the world's first atomic pile was activated. With uranium slugs separated by graphite moderators, to slow the neutrons down, a chain reaction was initiated. On average slightly more than one neutron was produced per fission that was not either absorbed by ^{238}U, or by some impurity or lost to the pile. This meant that, for the first time, a chain reaction was in process. Atomic power and the atomic bomb were practical possibilities. The first pile was converting tiny quantities of ^{238}U into plutonium. By 12 December the pile was releasing enough energy in the form of heat to operate two domestic light bulbs (200 watts). To produce usable quantities of plutonium a pile releasing 500 to 1500 kilowatts would be required.

To 'scale up' to this extent without careful intermediate steps would be analogous to trying to build a supersonic jet airliner just after the first Wright biplane had flown. An intermediate pilot pile, to give 1000 kilowatts, was built at Clinton, a small town in Tennessee; it started up on 4 November 1943. One of the attractions of the Clinton site was the cheapness of electricity, thanks to the Tennessee Valley Authority. Another and much larger plant was built a few miles away at Oak Ridge; here diffusion and electromagnetic methods of separating ^{235}U from ^{238}U were employed. These different methods were often used in serial fashion, one after the other, to produce progressively purer ^{235}U. The electromagnetic separator was developed at Berkeley, University of California, and was called the 'calutron' after CALifornia University cycloTRON.

In the meantime Brigadier Groves had already decided that Clinton was not suitable for a large-scale plutonium plant. The site he chose was at Hanford, in Washington State, where the Columbia river could provide abundant cold water for cooling the piles. Construction began in June 1943 and the first pile was operating by September 1944. In the design, construction and operation of these and other plants, the DSM, or Manhattan, Project drew on the resources of manufacturing companies such as du Pont, Westinghouse, Standard Oil and many others, but it was all new territory; problems were met and solved as the whole project advanced.

Less than two years after the small Chicago pile went active, massive plants were at work producing quantities of ^{235}U and ^{239}Pu. The next problem was how to use the fissionable material to make a bomb. This necessitated the establishment of a large and specialized laboratory. Cheap power and abundant cold water were not essential; security was. The emphasis shifted from chemical, metallurgical and mechanical engineering to mathematics, physics and chemistry. Fundamental research had, of course, continued over the years 1940–43; the effects of slow, medium and fast neutrons on uranium isotopes had been extensively studied. The problem now, as supplies of near weapons-grade ^{235}U and plutonium became increasingly available, was the design, the triggering and the likely effects of detonating an actual bomb. Already, in 1942, the first steps had been taken when Dr J. Robert Oppenheimer initiated theoretical studies of an atomic bomb. At the end of that year the search for a suitable location for an experimental laboratory began. The place decided on was Los Alamos, in New Mexico, where before the War there had been a ranch school, run on the lines pioneered by Dr Kurt Hahn. Oppenheimer was familiar with the school and thought that, being remote, it would be suitable for a secure laboratory. Construction of new buildings began in December 1942 and at the beginning of the next year the University of California accepted a contract to run the laboratory, with Dr Oppenheimer in charge. He proceeded to recruit a remarkably talented team that included Fermi, Bohr, Chadwick and von Neumann. Here, apart from continuing research on uranium, were carried out investigations into the best shape of plutonium metal for a bomb, the most suitable 'detonator' to provide neutrons to initiate the chain reaction, the arming and firing of the bomb and many other things. One particular difficulty concerned the plutonium bomb. The uranium bomb could be exploded relatively easily by firing a subcritical mass of uranium out of a tube, or gun, to hit a second, subcritical mass. However, this technique could not be used in a plutonium bomb. The only solution was to implode a subcritical sphere of plutonium so that, compressed, it became critical. This meant that the plutonium sphere had to be surrounded by explosive charges that had to be detonated simultaneously.

By July 1945 these problems had been solved and there was enough uranium for one bomb and enough plutonium to make two bombs. There was some doubt, however, about the plutonium bomb, so it was decided to carry out a test, using up half the stock of plutonium metal. On 12 July 1945 a plutonium bomb was successfully exploded at Alamagordo, in the New Mexico desert and two hundred miles from Los Alamos. A month later a uranium bomb (called 'Little Boy') was exploded over Hiroshima and shortly after that the plutonium bomb ('Fat Man') was exploded over Nagasaki. These, it is reasonably claimed, brought the Second World War to an immediate end. The Japanese command was, of course, unaware that there were no more atomic bombs in the US arsenal.

The above brief sketch of the history of the atomic bomb illuminates a point I have made above. Only the wealthiest and most powerful nation in the world, engaged in a global war across the two major oceans, could find the many and complex resources to make the bomb; and, equally to the point, to run the risk that the project might prove a failure. It would seem certain that, had there been no war, the atomic bomb would have been developed much later than was actually the case. We may hazard the guess that the possibilities of cheap electric power through controlled nuclear fission, mentioned by Born and others, might have led to the eventual construction of nuclear power stations and they, in turn, to the making of bombs. In other words, in the absence of war the sequence might have been reversed, although the time needed for development would have been much longer. We recall, at this point, the enthusiasm that has often been aroused by the prospects of very cheap power. But these are mere speculations. Two things are quite certain. The atomic bomb, having been invented, cannot be uninvented. And any nation that has made and detonated an atomic bomb will always remain a 'nuclear' power; for, demonstrably, it possesses the necessary industrial resources together with the appropriate scientific and technological knowledge.

The history of the Manhattan Project and the development of nuclear power, civil and military, can be taken as confirmation of the argument, put forward by some scientists, that technology is science-dependent; that pure scientists makes the discoveries on which technologists base their inventions or innovations. Equally, however, the story of the liquid-fuelled rocket confirms the opposite thesis. The astronomical discoveries that have been made, thanks to satellites, shuttles and space probes, would not have been possible without rocket technology.

There are intriguing similarities between the Los Alamos and the Peenemunde establishments. Peenemunde was situated among pine forests with views across the blue waters of the Baltic sea; Los Alamos was also among pine forests with a view of the distant Sangre de Cristo Mountains. On both sides there was an enterprise headed by a major-general; on both sides, a corps of scientists and engineers led by two men of high scientific and technological achievement, von Braun and Robert Oppenheimer. And while there were well-known doubts on the part of American scientists about the morality of using the bomb without prior warning, there is evidence that von Braun and some of his associates also had their doubts. Their ambitions were directed towards space travel rather than the perfection of a terrible new weapon. These were, of course, fortuitous similarities and, undeniably, the scale of achievement that culminated at Los Alamos greatly exceeded that of Peenemunde. Yet, at a deeper level, there is an indicative similarity. America and, probably, Germany were the leading technological nations of the time. Los Alamos and Peenemunde demonstrated how the modern state can apply, on a massive scale, the organized science and technology that were developed in the nineteenth century to achieve remarkable ends. To that extent they represent something new in the history of technology.

It was known that at the opposite end of the table of atomic weights to that of uranium and plutonium, at the end occupied by hydrogen and its isotopes, the process of *fusion* would be accompanied by a mass 'defect' and a consequent release of energy. Hans Bethe, a distinguished member of the Los Alamos group, had published an accepted theory according to which the heat of the sun was due to the energy released by the fusion of hydrogen nuclei to form helium nuclei. (Helium was first detected on the sun by Sir Norman Lockyer before it was found on earth, hence the name.) As early as 1942 there was some discussion about the feasibility of a fusion bomb; the main obstacle was the immensely high temperature needed to initiate the fusion process. The advantage of the thermonuclear or 'hydrogen' bomb was that, unlike the fission bomb, there was no limit to the size. Criticality did not come into it, for fusible material could be added without limit, just as in the case of ordinary explosives. Such a bomb became feasible once the fission bomb had been built, for it could generate the necessary high temperature. As a result the first fusion bomb was detonated in the Pacific on 1 November 1952. Much more was needed than just putting a fission bomb and some fusible material together. To aid in some very complex theoretical work a remarkable 'electronic brain', or computer, was developed.

Britain picked up the reins of nuclear power again in 1946 when the Harwell research establishment, headed by Sir John Cockcroft, was founded. Other nuclear installations were at Aldermaston and in various defence establishments. A big diffusion plant was built at

Capenhurst, in Cheshire, and a big pile at Windscale, in Cumbria. The first British fission bomb was exploded in 1951 in the Monte Bello islands, off north-western Australia, and the first British thermonuclear bomb over Christmas Island in the Pacific in 1956. In the meantime Russia had exploded both fission and thermonuclear bombs. China and France followed some years later.

The energy released as heat in Fermi's little pile of 1942 and in the vastly greater piles at Oak Ridge, Hanford and elsewhere constituted waste to be got rid of by gas cooling or by the waters of the Columbia river. This waste heat could, of course, be used to raise steam to drive turbines and this made possible the first true submarine. Submarines in both World Wars had proved to be deadly and massively destructive weapons but they were not true submarines; rather they were submersible torpedo boats. Although there were sustained attempts to devise efficient closed-cycle engines, particularly by the designers of German U-boats, submarines remained dependent on batteries and electric motors for propulsion while submerged[8]. The 'schnorchel' breathing tube was a partial and very limited solution to the problem. At most submarines could travel about a hundred miles (160 km) submerged and that at a very reduced speed of about 5 or 6 knots. The true submarine was not built until 1957 when the American nuclear submarine *Nautilus* was launched. This vessel used the heat from a nuclear reactor (cf. Fermi's 'pile'), that did not need air, to generate the steam to drive her turbines. The *Nautilus* was faster under water than many surface warships and there was no limit to the distance she could travel submerged. Her most sensational – and historic – voyage was from America to Europe under the ice cap of the north pole in 1958.

Although the Germans are said to have fired (solid-fuel) rockets successfully from submerged submarines during the Second World War, the development of the Polaris submarine represented an enormous advance. These nuclear submarines carried long-range ballistic missiles armed with nuclear warheads. Thus the two revolutionary technologies of rocket propulsion and nuclear power came together to provide what is surely the ultimate weapon of deterrence. It remains, however, for future historians to assess and decide the part these formidable weapons, wielding sea and land power, played in substantially reducing the threat of nuclear war.

Turbines driven by steam raised by the heat of a nuclear reactor can also be used to generate electricity. The advantages of generating electricity in clean, labour-saving nuclear power stations seemed obvious. Yet the history of civil nuclear power since 1945 has been a chequered one. Initially there were widespread hopes of extremely cheap power followed by a veritable euphoria, enthusiastically promoted by press, radio and TV, when in 1956 the first nuclear power station, at Calder Hall[9], part of the Windscale complex in Cumbria, began to feed power into the public electricity supply. Britain was painfully dependent on increasingly expensive, locally mined coal. Jevons had not been forgotten. Here was salvation. It was not stressed, or even mentioned, that another purpose of Calder Hall was the production of military plutonium. In any case few, if any, realized that we had had our hopes raised in similar fashion several times before in the past 150 years. The failure to produce virtually free power combined with well publicized episodes such as the Windscale reactor fire that polluted an area of farmland in the north-west of England a year later, the Three Mile Island episode and the real disaster of Chernobyl have resulted in public disillusion with nuclear energy. To these factors should be added the atavistic fears about the taboo on the natural, recollections of the devastation caused at Hiroshima and Nagasaki and the impact of well-made films such as *On the Beach* and *The China Syndrome*. Nuclear power is now regarded as, at best, a dubious blessing; and the inability of engineers to give a cast-iron guarantee that there can be no future accidents or even disasters, coupled with the unsolved problem of the disposal of nuclear waste, go far to explain public concern.

The problems of nuclear waste and the risk of nuclear pollution would not apply if a practicable fusion process could be developed. This has not, up to the present, been achieved, although the United States, western Europe and Russia have invested a great deal of effort and money into research directed to that end. The principle of the Tokamak machine, developed from ideas put forward by the distinguished Russian physicist, Andrei Sakharov, involves the creation of an extremely high-temperature *plasma* of deuterium and tritium, the isotopes of hydrogen, in a closed, circular tube. The heated plasma is kept away from the walls of the tube and concentrated along the axis by a powerful magnetic field. Being a conductor of electricity, the plasma can be made to act as the secondary coil of a transformer so that a current can be induced in it that will raise the temperature of the plasma still further. If the temperature is high enough, fusion will result, with the release of enormous heat energy and the creation of helium as a by-product. At the Joint European Torus (JET) Laboratory at Culham, in Oxfordshire, the Tokamak machine has, for short intervals of time, reached and exceeded fusion temperature. It is therefore foreseeable that before too long it will be possible to generate sufficient heat energy by nuclear fusion to maintain the process and to leave a surplus for external use.

Deuterium and tritium can be easily obtained in quantity and many believe that, once the fusion process is achieved, safe and cheap power will be available. The first may well be true; on the second, past experience suggests that a degree of caution would be the wisest policy. It is most probable that, except in very special circumstances, dirt-cheap energy will never be a reality. The practical problems of developing fusion power will surely be formidable and the cost of the advanced machinery required will be high. It may be taken for granted that the first fusion power station will evoke the usual noisy protests.

And yet, taking the long view, nuclear power does offer hope for the future. If the fusion process can be economically harnessed, if the problem of waste can be solved (and why should it not be?) and if there are no further Chernobyls, the positive virtues of nuclear power are that there are plentiful sources of raw materials and that it is non-polluting. The reserves of fossil fuels are, we know, limited and are being used up at a rate that would have appalled Jevons[10]. It may be that sources of dilute energy can offer some alleviation, but their drawbacks have hardly been publicized and, once again, experience suggests that their benefits may have been exaggerated. Increasingly efficient use of fuel might be thought to be one answer to the energy problem but here we may invoke Jevons again. If, he pointed out, a manufacturing firm (or, for that matter, a family) makes an economy in its use of fuel (equals energy), then it will become more profitable (or have more money to spend), it will invest and expand its operations so that it will consume more fuel. This is a race that cannot be won. Quite simply, economic growth means greater energy consumption.

If and when the developing nations of Africa, Asia and South America, begin to climb the ladder of economic growth, as we all hope and intend that they will, reaching towards the levels of prosperity of America, Europe, Japan and the nations of south-east Asia, then the energy problem must become even more urgent. It would therefore be as well to try to tackle it now. And that will mean more intensive technological and scientific research, not least into nuclear energy.

A NOTE ON RELATIVITY

For many years after Newton's time the fastest things on earth, known to science, were cannon balls, shells and bullets. Their speeds were of the order of 1000 mph, or 2000 km/h and were negligible compared with that of light. It was not until the beginning of the twentieth century that atomic and nuclear physics provided direct experimental evidence of bodies possessing mass that moved with speeds that were not entirely negligible compared with that of light. Experiments on high-speed atomic particles confirmed Einstein's special theory.

According to the special theory (1905), the measured length, l, of a rod moving with velocity v relative to the observer will be

$$l = l_0(1 - v^2/c^2)$$

where l_0 is the length measured when there is no relative motion and c is the velocity of light. When v approaches the velocity of light l shrinks practically to nothing.

Conversely, the measured mass m of a body moving with velocity v relative to an observer will increase with the velocity

$$m = \frac{m_0}{(1 - v^2/c^2)^{1/2}}$$

this can be rewritten as

$$m = m_0(1 - v^2/c^2)^{-1/2}$$

which can be expanded by the binomial theorem to

$$m = m_0(1 + v^2/2c^2 + 3v^4/8c^4 + ...)$$

We can ignore the third and subsequent terms as they will be negligible at all but the very highest speeds. In which case, the last equation can be written as

$$m = m_0 + \tfrac{1}{2}m_0v^2/c^2 \text{ or}$$

$$m - m_0 = \tfrac{1}{2}m_0v^2/c^2 \text{ or}$$

$$Mc^2 = \tfrac{1}{2}m_0v^2$$

M being the increase in mass, $m - m_0$ and $\frac{1}{2}m_0v^2$ the kinetic energy. What this means is that energy – kinetic in this case – gained by a body results in an observed increase in mass; alternatively the increase in mass is equal to the energy gained divided by the square of the velocity of light. In the familiar, slow-moving, world around us these changes in mass are far too small to be observed. This is not the case in the atomic world. The mass of a uranium nucleus is greater than the total mass of the particles into which it splits on fission. The change in mass is exactly accounted for by the energy given up.

The complete theory was given by Einstein in 1905, long before the fission of uranium was discovered.

Technology and the Individual: Little Technology

The heat engines, whether steam or internal combustion, that drive the generators in power stations, large and small, are designed to work over the maximum possible temperature range, in accordance with the principle laid down by Sadi Carnot and consistent with practical design requirements. To expand the steam or gas right down to atmospheric temperature would require enormous cylinders so, as a compromise, some of the heat energy has to be sacrificed at the lower end of the temperature scale and the residual heat in the steam or gas is carried off by condensing or cooling water. The rejected heat could have added little useful power to the engine, its temperature would be too low, but it could still be useful for many other purposes. This was realized early in the nineteenth century when waste heat from Cornish steam engines was used to preheat the boiler feed water, thus improving the overall fuel economy of the engines. Later in the same century the hot air and fumes rising up factory chimneys were used to drive small-horsepower air engines, working on Stirling's principle. There were many other early instances of the utilization of waste heat, perhaps the most intriguing being the one mentioned by Lynn White: a very old practice was to use the hot air rising up the chimney to turn a spit so that the joint of meat mounted on it was roasted evenly on all sides.

From the beginning of the present century, mainly in the United States, the waste heat from large power stations has been used to provide 'district heating' for houses and businesses in their neighbourhood. In recent years the principle has been extended and the technology of 'combined heat and power' (CHP) has been extensively developed in Scandinavia, Russia and the United States.

Combined heat and power installations are usually relatively small. They often raise steam by burning gases from organic materials and waste. Such installations cannot compete with large power stations if the object is the maximum generation of electricity to be supplied to an extensive network. But they can be markedly more efficient if the energy account covers both forms of energy – heat as well as electrical. Another method of generating heat is the heat pump. Carnot's argument clearly shows that a heat engine, driven backwards, can pump heat from a 'cold' body (e.g. a river) to a 'hot' body (e.g. a concert hall) with a thermal efficiency of well over 100 per cent. This does not violate the conservation of energy, as a simple hydraulic analogy will show. One tonne of water falling 100 metres and driving a water engine with efficiency of 60 per cent will provide enough power to raise 6 tonnes of water 10 metres. If we just consider the quantities of water/heat involved and ignore the fall in metres/degrees Celsius, then the efficiency is 600 per cent.

The late Professor Derek Price introduced the useful classification of big science/little science. We may extend the idea to the technological realm. Atomic energy and space rockets are, unquestionably, the products of big technology, both in themselves and in the research/development and manufacturing institutions they imply. On the other hand, a vast number of things used by ordinary individuals may be fairly described as belonging to little technology. Although advanced research may have been involved in their creation and development, no great social organizations have been involved; no Peenemunde, Los Alamos, Hanwell or Harwell.

The Transmission of Technology

Twenty years ago Dr E. F. Schumacher published a collection of essays under the title *Small Is beautiful*. They proved persuasive and helped to popularize the 'Green' movement. Schumacher's thesis was, essentially, a plea for smaller economic and technological units; suited, as he would say, to the scale of the individual rather than to that of the huge corporation. Now it is certainly true that there are rigidities in some of modern technology. A warplane is not so much as weapon as part of a 'weapons system', requiring sophisticated ancillary services before it can be effective (the Russian cannons that destroyed the Light Brigade in 1854 needed no elaborate infrastructure). A civilian airliner also depends on advanced technological services to an extent that far exceeds those that were required for nineteenth-century trains and steamships. As we saw, the latter, in their day, relied on relatively far more sophisticated services than were needed by the stage coaches and sailing ships they replaced. Nevertheless, it is doubtful whether Dr Schumacher's recommendations could receive more than limited application in an economically advanced country, given the expectations and wishes of most people.

Schumacher's argument was also applied to developing countries that were being increasingly left behind in the forward sweep of modern technology. Such countries, on gaining independence since 1945, had hastened to adopt the latest technology – a state airline, computers, a modern steelworks, a national television system – although the standards of living and of technological understanding of the great majority of the people were primitive. The results of this rush into the modern world were usually unsuccessful. As we saw – and as Svante Lindqvist clearly demonstrated – to be successful, immediate technology transfer requires a degree of compatibility between the donor and recipient cultures.

Schumacher's suggested alternative was 'intermediate', or small-scale, technology appropriate to the numbers and practical skills of the peoples concerned. This humane proposal requires forbearance on the part of the developing countries: a willingness to accept that the jam will have to be postponed beyond tomorrow. Intermediate technology, Dr Han Suyin tells me, is being successfully applied, but China is an old-established civilization with a highly creative people living in an ordered society.

Nothing is more remarkable than the way in which Japan, in little over a hundred years, has changed from being a feudal society, barely altered in social structure for many hundreds of years, into a modern country and a world leader in industry and technology. Clearly, the innate abilities of the Japanese people were suppressed by social institutions and further inhibited by geographical isolation. Much the same was true of China and the countries of south-east Asia. These communities began the take-off into technological and industrial growth by copying the products, even the dress and other outward trappings, of the west. This was the right policy; the English followed it in the seventeenth and eighteenth centuries with conspicuous success, just as the south Europeans had done, *vis à vis* the Arabs, in the earlier middle ages. Imitation is the gateway to technological advance. On the other hand, as we noted, compatibility is necessary for successful technology transfer. One possible clue to the success of the Japanese is their highly refined art (cf. Renaissance Europe) together with their long-established skills as metal workers and gardeners. For the rest we recall that modern technology began in the Middle East, travelled westward across Europe, reaching Britain last of all before crossing the Atlantic to the Americas. It reached Japan and the other countries of east Asia from both Europe and America. It has therefore migrated round the world. So far, experience suggests that the period of technological supremacy for any nation or empire is limited and that, after a time, leadership passes on to a successor.

Domestic Technology

Up to the last decades of the nineteenth century technology had had little impact on domestic arrangements. The growth of the iron industry in the late eighteenth century had provided homes with wood- or coal-burning cast iron stoves, grates and ranges together with basic cooking utensils. Otherwise things remained much the same as they had been in the days when Shakespeare wrote 'While greasy Joan doth keel the pot', an observation that prompts uncomfortable thoughts about domestic hygiene. The introduction of kerosine, or paraffin, after the founding of the mineral oil industry in the mid-nineteenth century, meant a new fuel for lighting as well as for cooking and heating, particularly in rural areas. The coal gas industry provided for lighting, cooking and heating in urban areas but the highly efficient gas mantle, invented by Auer, appeared too late to challenge the new electric light bulb successfully.

In middle-class and affluent homes generally, the domestic burden was carried by a huge army of servant girls. The existence of this cheap labour force discouraged innovation. Middle- , even lower middle-class, families, who could afford one or more servants, had no reason to demand domestic innovations and labour-saving devices. The servant girl got up early, raked out the dead ashes, cleaned the grates and fireplaces, blackleaded the stove and range, laid and lit the fires. Thereafter she would be expected to clean and dust the rooms, lay the tables, do the washing and certainly help with the cooking. As for the greater part, the working classes, they were usually unable to afford domestic innovations. Here was a 'Catch 22' situation. The impasse was broken, first, in the United States, where a shortage of domestic servants was first noticed by the early 1860s. No doubt greater social mobility and the rapid development of the North American continent accounted for this shortage. In any case changes were under way throughout the English-speaking world that exacerbated the 'servant problem'. The rise of office employment with the independence, enhanced status and higher wages it offered certainly contributed to the decline of the servant-girl class. This change was facilitated by the training courses in typing, shorthand, office procedures, book-keeping, etc. offered by the new technical colleges and night-schools. Over much the same period the rise of the department store, staffed by shop assistants, meant further job opportunities for working-class girls. Finally, experience in two World Wars taught that women could efficiently perform tasks that had previously been reserved for men. This marked the end of the traditional servant girl.

The disappearance of the servant girl was accompanied by a wide range of innovations in the individual home[1]. These were made possible by the rise of the electricity supply industry. The gas industry could – and still does – compete with electricity as regards space heating and cooking. It was not nearly so easy to compete with electricity when it came to the refrigerator and freezer. Although refrigeration had been practised on a small scale from the beginning of the nineteenth century, it was only when electricity was cheaply and freely available that the domestic refrigerator became commonplace. And no other form of energy could compete with electricity where small- or fractional-horsepower motors were required. Only electricity could drive domestic washing machines, spin driers, vacuum cleaners, hair driers, air conditioners and a host of minor gadgets about the home. Parenthetically, we note that when electric light replaced gas light, houses could be built with much lower ceilings. Nineteenth-century town houses had high ceilings so that the waste gas and fumes from the gas burners were raised well above the heads of the residents. With electric light this was no longer necessary, so that building costs were reduced and energy was saved. Electricity was, in Joule's memorable words, indeed 'a grand agent'.

With the new labour-saving devices the 'lady of the house' could quickly and easily do the jobs previously reserved for servants. Moreover, the new devices were made of sheet steel and not cast iron; they were in light colours, enamelled or in vitreous glaze, and not matt black. At the same time, dark little rooms such as pantries, sculleries, larders and wash houses became redundant.

The vacuum cleaner was rather more than a labour-saving cleaning machine. It was a positive aid to better health. A trained biologist, H. G. Wells was, as we saw, well aware that dirt, natural dirt, meant disease. Dirt was not only ugly and ungodly, it was increasingly under-stood, in the decades following Pasteur's work, to be dangerous. So too were parasites such as lice and fleas. The vacuum cleaner removed disease-breeding dirt brought into the house on people's shoes and boots; it also swept up fleas, lice and their eggs. The makers of vacuum cleaners were well aware of this and made it a strong selling point. New materials, such as plastics, that were non-absorbent and easily cleaned were contributed by the chemical industry, as were new and highly efficient disinfectants and detergents. In these ways technology harmonized domestic economy with the prescriptions of modern medicine.

As Wells had recognized, the rapid improvement in public transport technology, which began at the end of the nineteenth century, enabled town dwellers to settle further and further out into the countryside. The suburban train, electrified and with modern signalling and control systems, the motor bus, the street-car or tram-car and the automobile meant that increasing numbers of city workers could afford to live in the semi-country of suburbia, so that the arts of town and city life, so well developed in the eighteenth century, have been lost following the diaspora into the country. The new, greatly extended, cities required a new technology of food preparation, preservation and transport. Food could no longer be brought from the farm to the doorstep. Without modern food technology milk had to be boiled if it was to be kept for much more than 12 hours, meats had to be cooked and salted and eggs pickled. But the twentieth-century suburbanite would not accept the 'hard tack' that would have satisfied a seaman in the days of sail. Accordingly a highly efficient technology has developed for the preservation, packaging and distribution of food; without it the mega-cities of today would not be possible.

Most recently, technology has been applied to leisure activities. Such sports as mountaineering, sailing, ski-ing, scuba diving as well as traditional bat-and-ball games have all been subject to technological improvement. Whether these advances have added to the pleasure derived from such activities may be doubted and it is also doubtful if they have had any beneficial effects outside the individual sports in question. Sports technology does not seem to be a strategic technology. An exception was motor sport, for, among other advances, fuel injection was developed by German engineers for the racing cars of the 1930s and is now common in domestic automobiles.

The mega-city has its drawbacks, material as well as social. The contemporary practice of commuting means that millions of city workers spend hours each day packed into crowded trains or in traffic jams on roads. Whatever the pleasures and advantages of suburban life, commuting amounts to a shocking waste of human and physical energy. As Graham Wallas remarked, if Aristotle came back and stood on London Bridge during the rush-hour he would think the world had gone mad. What is true of London is true of all the major cities of the world. Cynics believe that many, possibly most, city workers spend their time sending bits of paper to each other. Whatever they really do, modern electronic technology has surely made much commuting unnecessary. Formal conferences and informal discussions can be held between people who are many miles apart. Tape recording together with disk storage of information with instant retrieval mean that enormous quantities of paper are no longer necessary. Other advances – photocopying and fax machines, for example – have simplified office procedures further still. It is now possible to imagine that, if security can be assured, the big offices in the middle of major cities will be dispersed to the homes of the employees, each of which can be equipped with efficient, but progressively cheaper electronic aids. On the other hand, it is well known that people work better and more happily in small groups than as isolated individuals. Put another way, this suggests that communal work fulfils a psychological need and that commuting, uncomfortable and wasteful of energy though it is, will be preferred to staying at home to work. This, however, is a speculation outside the realm of technology.

The nineteenth-century civil engineers made big cities possible and tolerable to live in by ensuring reliable water supply and efficient sewerage. Water treatment resulted in the elimination of harmful bacteria and an end to water-borne diseases such as cholera (although there have been occasional lapses as when typhoid has broken through). One thing the nineteenth-century engineers and scientists did not tackle was pollution. It is hard to see how they could have done, without bringing economic growth to a standstill. Those old enough to remember the filthy, choking London fogs of forty and more years ago – lovingly if not entirely realistically depicted in many Hollywood films – may wonder what all the present fuss is about. With electrification the fogs have long gone and, at the same time, the pollution of many rivers has been reduced. In fact pollution has changed; the automobile, not the steam engine, is now the guilty party and governments are taking steps to cut down their emission of lead and carbon dioxide. Vigilance, in the matter of pollution, is essential; exaggeration of the dangers can only be counter-productive.

There is nothing particularly modern about human exploitation of the environment. Two thousand and more years ago the lands of north-east Africa – Egypt and Libya – were the granaries of the Roman empire. Over many centuries excessive utilization turned them into deserts. Modern society cannot be blamed for this. We can go back even further. The Lake District of England was, two hundred years ago, greatly admired by poets and other persons of sensitivity. It represented Nature, unspoiled by industry, unaffected by the smoke, clangour and fumes of factories, forges, mines and mills. In fact, they deceived themselves. When human tribes began to settle in the British isles after the last ice age they started to cut down and burn off the native forests so that crops could be grown or animals pastured and this continued up to the late middle ages. The Lakeland mountains were duly cleared of trees and shrubs and since those very early days sheep farming has kept the mountains artificially bare of trees. The same process took place in the Highlands of Scotland; little remains of the ancient Caledonian forest. In every age, from that of stone axes to that of the tractor, the bulldozer and the combine harvester, human have sought to change the face of the land so that they might exploit it more successfully.

Modern peoples, therefore, have not been notably more careless of the natural environment than their ancestors, recent and remote. In fact, modern society may be said to be remarkably considerate of the environment. On the contrary, there is much concern about 'endangered species' and great efforts are made to save all species that are, or seem to be, in danger of extinction. It is easy to forget that from pre-Cambrian times onward 'endangered species' have been plentiful enough and extinction has been a common fate without any human intervention. The one species that it is reasonable to feel anxious about is the human race. The accelerating growth of world population brings with it appalling prospects of mass starvation and disease. Technology can offer little, apart from improved contraceptive methods; but these can only be effective in societies that accept them. It may be that the best technology with which to combat the population–starvation threat is intermediate technology.

The Origins of the Modern Computer

The history of the realization of 'Babbage's dream', as it has been called, is still unclear in important details and it may well be that we are still too close to the major advances in the development of the computer to draw very firm conclusions; nevertheless some valid inferences can be made and interesting conclusions drawn[2].

The two major movements that joined forces to produce the modern computer are the proliferation and improvement of office machinery during the present century and the advance of electrotechnics, particularly of electronics in the radar age. Over this period there have been three main motives for improving computers: commercial, military and scientific. The commercial motive needs no further discussion. The military motive breaks down into three, or perhaps four: the calculation of ballistic tables for artillery; the decryption of intercepted enemy messages in cypher; and the design of high-speed aircraft and atomic bombs. The scientific motives were twofold: the solution of specific scientific problems; and the desire, which presumably motivated Babbage when he contemplated the analytical engine, to build as advanced a machine as possible to push against the frontiers of knowledge.

It is clear that, with the exception of the last, the above motives, and their many subdivisions, would be best satisfied by specialized machines. Only the last would seem to open up the prospect of a universal, Babbage-type, computer. But the course of technological innovation rarely runs along easily predictable paths.

Konrad Zuse was led by purely intellectual interests to become a computer pioneer. Zuse, who began by working on his own, in the nineteenth-century tradition of the self-supporting devotee, started on his first computer in 1934. It was purely mechanical. His next computer, begun in 1936 and designated Z2, added electromagnetic operation, using relays of the sort developed for telephone exchanges. He rejected electronics on the grounds that such devices were still unreliable whereas relays had been developed over many years and were thoroughly proven; moreover, they were freely available and cheap. The construction of his third computer, which he began in 1939 and which relied entirely on relays, was interrupted when he was called up for military service. After a year he was released in 1941 to complete Z3 and to begin his fourth machine. Z3 was a program-controlled, universal computer of advanced design. From this time onward Zuse's work was supported by the German Aeronautical Institute (DVL) and his computers were used in the design of the V2 rocket as well as in aircraft development. Z4 remained in use for some eight years after the War. In all he was to build 21 computers. Zuse seems to have known little, if anything, of his predecessors and yet his ideas were remarkably clear and far-sighted. He incorporated binary arithmetic[3] from the beginning and made provision for a stored program.

Three years after Zuse began work Howard Aitken, of Harvard University, published a specification for a computing machine together with a survey of previous work in the field. He mentioned Babbage's work and pointed out that Babbage had been the first to propose a machine specifically for scientific purposes. There were, he remarked, important differences between a machine designed for business or accounting use and one intended for scientific work. The latter had, for example, to cope with negative as well as positive numbers and it had to deal with complex mathematical functions unlikely to appear in commercial work. The machine built by IBM (International Business Machines) staff to Aitken's specification was massive, being some fifty feet (sixteen metres) long and requiring a four-horsepower motor to drive it. For all Aitken's insight and learning, it was conservative in design, being largely mechanical and without conditional (branching) logic. Construction began in 1939 and was not completed until 1944. The machine was most probably used for ballistic calculations.

Apart from the old-established analogue computers the only other notable machines in the United States at this time were those designed by George Stibitz at the Bell Laboratories. The first, in 1938, was intended for the solution of problems in the design of electric power circuits. Although intended for a limited purpose, it was an efficient design. It has its place in history for it was the first computer to be operated remotely, over a telephone link. Subsequent Stibitz machines were used for ballistic calculations.

Alan Turing (1912–54) was a highly gifted British mathematician who was possibly a genius; we can never know, for his career was interrupted by war service[4] and, tragically, he committed suicide in 1954. In 1936, aged 24, he wrote a memorable paper 'On Computable Numbers, with an Application to the *Entscheidungsproblem*' (he defined 'computable' numbers as those that can be calculated, like π, the ratio of diameter to circumference of a circle, or rational fractions). The purpose of the paper was to prove that, contrary to the view of David Hilbert, there are insoluble problems in mathematics. He was able to show that there were numbers that could not be computed; a key element in his proof was the operation of what he conceived to be a universal computer. His idea of a universal computer was, however, abstract, with no specification of hardware.

The key to the development of the computer was the pulse technique, as used in radar, in conjunction with binary arithmetic. The presence or the absence of a pulse, alternatively a positive and a negative pulse, in a particular sequence could be taken as indicating 1 or 0 on the binary scale. As pulses are commonly a fraction of a microsecond long this means that great speeds are possible compared with mechanical and electromechanical methods. The information, expressed in binary code, has to be processed according to the instructions given to the machine, and the answer given in the same binary code.

On the day after the outbreak of the Second World War Turing joined the staff at Bletchley Park, a highly secret British intelligence establishment for the decryption of intercepted enemy messages. Particularly baffling were the radio messages cyphered by means of 'Enigma' machines. These machines were invented before the War and were fairly widely known. On depressing a key, corresponding to a letter of the alphabet, an electric current was sent through a contact on the rim of a wheel and then across the wheel to another contact, from whence it passed to a contact on a second wheel. In this way the signal passed randomly through a series of four or more wheels to emerge at the other end corresponding to an entirely different letter. On pressing another key the wheels rotated so that the message that emerged was hopelessly distorted. If the receiver had an identical set of wheels, decryption was simple and fast; if not, the cypher was, at first, believed to be unbreakable.

It was Turing's job to help break the Enigma code. Among his colleagues working on the same job was M. H. A. Newman, whose Cambridge lectures he had attended and who had helped him with his paper on computable numbers. It was Newman who saw that key aspects of the Enigma problem could be solved by electronic machines. The first machines, called 'Robinsons', were not reliable; they were, in effect, prototypes. The next machine, called 'Colossus' (1943), was much more successful. It incorporated 1500 vacuum tubes, or valves, and information was fed in by high-speed punched paper tapes. The next, Mark II Colossus (1944), had 2500 valves and set the pattern for eight more that were built before the end of the War. Max Newman was in charge of electronic design, Turing worked on the logic aspects of the machine, while actual construction – given top priority – was carried out by Post Office engineers. It has been claimed that the Colossus machines developed at Bletchley Park were the first electronic computers. Unfortunately, with that enthusiasm for secrecy characteristic of some British politicians, few details of the machines have been published and, after the War, all were destroyed.

Thirty years later, in 1975, a few photographs and some cryptic claims were made public. It was stated that it used binary arithmetic with electronic pulses, had electronic storage and conditional (branching) logic. Babbage had envisaged this last technique by which the computer is able to change the course of calculation according to the results at particular points. It would certainly be safe to conclude that the Colossus series were fast, special-purpose logical machines. One day, perhaps, parts of Colossus, Mark II, may be built and the nature and operation of the machine better understood (cf. the replica Newcomen engine). Almost certainly there are now too few of the original valves, or vacuum tubes, left to allow of anything like a complete restoration. If key elements could be reconstructed, though, the place of Colossus in the history of the computer might be more clearly established.

There is a point of general importance at issue here. As machines, or artefacts, become obsolete they are usually destroyed. They are old, out of date, superseded, not worth keeping. Slowly, as the years pass, they become more and more interesting; valuable, even, until, when there are few left, they become precious relics. If they are all lost, if nothing remains but a few fragments and some drawings, then we should try to re-create the object to find out how it was constructed and how it worked. This has been done in the cases of the Greek trireme, Viking and Saxon boats, the Newcomen engine and, most recently, the *Planet* locomotive. Much has been learned from these re-creations and there is no reason why the same should not apply to early computers.

Before the Second World War John V. Atanasoff, in Iowa State University, had started to build an electronic digital calculator with storage facility. This, like the later Colossi, was to use binary arithmetic, but the project was not completed. Of much greater significance was the work undertaken at the Moore School of Electrical Engineering of the University of Pennsylvania. The members of staff were involved in the calculation of ballistic tables for the US Army, using an analogue computer for the task. John Mauchly, one of those engaged on the work, knew of Atanasoff's ideas and thought of replacing the mechanical analogue computer with a digital electronic one. J. Presper Eckert, another of the team, was, in the meantime, working on the adaptation of parts of the analogue computer to electronics. A partnership formed between these two and from that partnership came the first electronic computer, the huge Electronic Numerical Integrator and Calculator (ENIAC) (1943–46). The machine incorporated a form of conditional (branching) logic.

ENIAC was, by virtue of its size, most impressive; it was also something of a dinosaur. It required more than 100 horsepower of electrical energy to operate some 18 000 vacuum tubes, or valves, and many thousands of relays and resistors. The excessive number of valves owed something to the fact that it used decimal (to base 10) and not binary arithmetic. Eckert and Mauchly had carefully considered the advantages of binary arithmetic but had good reasons for preferring decimal arithmetic. In compensation it was a thousand times faster than the mechanical and electromechanical computers of the time.

A serious shortcoming of the ENIAC, which prevented its being a true universal computer, was that setting up a fresh program required extensive switching and replugging that could take days. In effect the machine had to be rearranged for each new program. The solution that occurred to Eckert and Mauchly as well as to John von Neumann (1904–57), the distinguished mathematician at Princeton University, was to store the program in the memory so that the computer could, as it were, program itself as it proceeded. Von Neumann, who became associated with the Moore School in 1944, saw that a very large memory was therefore the desideratum for a universal computer. He set out clearly the principle of the stored program in a report at the end of June 1945. This was to be put into effect in the next Eckert and Mauchly machine, EDVAC – Electronic Discrete Variable Calculator – completed in 1951. As it happened the first computer to incorporate the von Neumann principles and the EDVAC method of storage was that built at Cambridge University by Dr (later Professor) M. V. Wilkes, who had attended lectures and discussions at the Moore School. This was EDSAC – the Electronic Delay Storage Automatic Computer – which can reasonably claim to be the first universal computer. It was working by 1949. A similar machine was the Automatic Computing Engine (ACE) at the National Physical Laboratory with which Turing was associated and that was working by 1950. ENIAC did the jobs that were required of it. In fact it did more: it was far faster than any purely mechanical or electromechanical computer; it pointed to the future, it showed people the way things could and would go. In this respect it was rather like Baird television, although considerably more successful.

The high speed of pulse techniques raised the problem of the most efficient and capacious storage system. Relays had been used and magnetic tape was available. These methods, however, were both slow and relatively bulky. An electronic device that could be used as a unit of a storage system was the Eccles–Jordan 'flip-flop', a combination of two electronic valves so connected that if one was conducting the other was 'biased off', or non-conducting. On the arrival of a pulse they would switch over, the conducting valve becoming non-conducting and vice versa; these alternations correspond to 0 and 1 on the binary system. The flip-flop was fast but far too many valves would be required to give adequate storage by von Neumann's standards.

The memory device proposed by Eckert, Mauchly and von Neumann and used in EDVAC, EDSAC, ACE and other early computers was the acoustic delay line. This was not a particularly novel invention; it had already been applied in radar. It depended on the fact that sound is propagated through, for example, air with a velocity of approximately 1000 feet (300 metres) per second. Accordingly, as everyone knows, there is an appreciable time lag before an echo returns to the originator of the sound. Now, if the echo could be restored to the original strength and quality and sent off again and again, as a rubber ball can be bounced repeatedly against a wall, then the original sound could be said to be stored. This was the simple principle that lay behind the acoustic delay line. Mercury was found to be the most suitable fluid: a short tube full of mercury a few feet long was fitted at each end with a quartz crystal. An electric pulse applied to the crystal would cause it to vibrate, thus sending an acoustic pulse along the mercury column. The pulse, on reaching the other crystal, caused it to vibrate, so generating a somewhat distorted electric pulse. This pulse could, however, trigger a small electronic device that would send an undistorted pulse back to the second quartz crystal. In this way an acoustic pulse could be bounced, to and fro, along the tube indefinitely. Such a delay line could accommodate a pattern of many pulses and absences (1 and 0), separated by time intervals of one microsecond. The mercury delay line was not as fast as the flip-flop – pulses could be recovered in milliseconds rather than microseconds – but that was acceptable in view of the enormous economy in space and cost, and it was a thousand times faster than any mechanical or electromechanical device.

After the end of the War in 1945 Max Newman became Professor of Pure Mathematics at Manchester University, where D. R. Hartree had built a large differential analyser before the War and where Jevons had built a 'mathematical engine' in 1866. Without delay Newman applied for a grant to build a computing laboratory[5]. He was successful in this and, before long, he was joined by Turing, who gave up his post at the National Physical Laboratory.

Shortly after Newman moved to Manchester, F. C. Williams, who had been on the staff at Manchester before the War, returned as Professor of Electrical Engineering. Williams had spent the War working on radar and he brought back with him the idea of cathode ray tube storage. The technique was simple enough although its realization was brilliant. A short pulse of negatively charged electrons, accelerated down a cathode ray tube and focused to a small spot on the screen, caused a sudden *positive* charge to appear on the screen. This was because the high-speed electrons knocked more electrons out of the screen than arrived. This sudden positive charge caused a pulse to appear on a curved metal plate placed just in front of the glass face of the cathode ray tube. A similar pulse can produce a similar spot charge on the screen a short distance from the first charge. In this way a whole row of point charges (and bright spots) can be 'written' across the screen and similar rows below the first until the screen is covered by row upon row of point charges and bright spots.

The charges on the screen take about a quarter or a fifth of a second – a long time in electronics terms – to leak away; so they can be refreshed or replenished by bringing the beam back to each one in turn and switching on a short pulse in each case. In this way a pattern of charges will be stored on the cathode ray tube face. The form, or strength, of the spot charges can be varied so that one form corresponds to binary 0 and the other to binary 1. The same beam that 'writes' the charges on the screen can be used to 'read' the pattern of 0s and 1s, the pulses being picked up on the metal screen, amplified and fed into the computer.

The Williams storage tube made possible the first entirely electronic computer; there was nothing mechanical about it, not even the simple mechanics of the acoustic delay line. A prototype computer using Williams tubes worked successfully in 1948; it is claimed that the first stored program was run on 21 June, but this was a modest effort to show that it could be done. Storage capacity could be increased easily to the size required by a universal computer with stored program; all that was needed were additional cathode ray tubes; a magnetic drum provided back-up storage. Storage was fast and efficient and (random) access was immediate. Turing explained one of its advantages very clearly when he likened such computers to books, on which the reader has almost immediate access to any desired page, while older computers were like the papyrus scrolls of antiquity, in which access meant that the scroll had to be unrolled from the very beginning. The book, as was pointed out above, played a key part in the first information revolution.

The Manchester Mark 1 computer, using Williams tubes, was working by April 1949, by which time the University had entered into an agreement with the Ferranti company to manufacture computers for the commercial market[6] (February 1951). Very shortly afterwards the catering firm of J. Lyons entered the computer business. Lyons had run a chain of restaurants and cafés in London and the major British provincial cities; they had also established a network for the distribution of pastries and confectionery all over the country. The smallest general store in the smallest village in England sells Lyons confectionery. Such an organization could readily make use of computers to organize and keep track of orders and invoices. J. Lyons was one of the first, if not the first, large commercial firm to go over to computers. In collaboration with the Cambridge team they installed a computer based on EDSAC and even started to manufacture computers. The LEO company (Lyons Electronic Office) designed and made EDSAC-type computers. These were, therefore, descendants of ENIAC but served very different purposes indeed from those for which ENIAC was designed.

Eckert and Mauchly, realizing the commercial possibilities of a universal computer, had left the Moore School at the end of the War to start their own company to manufacture and sell computers. Their first product was not a success but their second – UNIVAC or Universal Automatic Computer – was very successful. Unfortunately they found the costs of research and development, manufacturing and marketing computers extremely high and the company ran into financial difficulties. The Remington–Rand corporation took them over and marketed the UNIVAC successfully. The merging of small computer companies into bigger and bigger companies was a pattern that was to be repeated in the years to come in all countries. The two obstacles of research and development costs and marketing difficulties meant that only large companies with funds for 'R & D' and marketing experience and resources could develop and sell the large computer. For most businessmen electronics was an occult art and the idea of rearranging their businesses to accommodate one of the new computers was daunting. Only firms with extensive sales and service networks together with an established reputation could assuage such fears. Over the years the various pioneer firms formed combinations to cope with the expenses of the computer business. IBM, a direct descendant of the original Hollerith company, was one of these firms; others, formed by amalgamation, included Unisys and Honeywell. In Britain the residuary legatee of various amalgamations was International Computers Limited (ICL); in France, Machines Bull; and in Germany, the Siemens company.

By 1953 both the mercury delay line and the Williams tube were overtaken by a new and simpler method of storage. This was the magnetic ring system introduced by Jay W. Forrester of MIT. It was based on the simplest and oldest idea in the whole history of electricity and magnetism: the polarity of magnetism. North and south magnetic poles can be taken to correspond to 0 and 1 in the binary system. Small rings of 'ferrite', a magnetic material developed by Japanese and Dutch researchers, set in a matrix of wires proved the best arrangement. Once again, although the basic idea was simple, the development and systematic application required much further work. And yet the ferrite store was not to be the end of the story. Further changes were on the way.

The first was the application of the transistor to the computer in 1953. The transistor proved capable of systematic improvement and, with increased demand, its price fell steadily. Far smaller, reliable, using much less energy, dissipating far less heat and with an indefinite life span, the transistor soon displaced the vacuum tube, or valve. Computers could be much smaller; or, as had been the case with the high-pressure steam engine, could be made far more powerful and versatile for the same size.

The march of technological improvement continued. Another innovation was the introduction of magnetic tape, developed in Germany during the War, to replace magnetic drums for back-up storage. More important was the introduction of the integrated circuit in the 1960s. Silicon chips suitably treated made possible the integrated circuit in which the transistor and circuit element could be combined to act as a microprocessor. In this way a great range of computers are available; from very large and powerful ones, suitable for government departments and international corporations, down to small ones, suitable for domestic use.

The making of progressively smaller electronic devices – radio sets and the like – or miniaturization as it is called, began during the Second World War. In 1940 radar sets were about as big as garages; by 1945 many were the size of small suitcases. The changing requirements of war had brought this about. Initially, radar sets were designed for the detection of distant bomber fleets or battleships and, at closer range, for gunnery. The urgent needs of the war against U-boats and to improve the accuracy of bombing led to the development of radar sets that could be fitted in aircraft. The invention of the proximity fuse reduced the scale still further. To make this possible a series of tiny vacuum tubes, or valves, was developed. These, in turn, were replaced by transistors and integrated circuits. The electronic computer that had originally occupied a large hall came down in size so that one of equal power could be carried in a handbag. With this development an enormous civilian market opened up.

To satisfy this huge and diverse market, or series of markets, a large number of small firms have been created and many able and ambitious young people brought into what is now a world-wide industry. Although the great firms, whose names are household words, are an essential feature of computer technology, there remains ample room for small and specialized firms. For this reason we are surely justified in describing computer technology as little technology. This does not, of course, imply that the technology is simple, or slight, or unsophisticated.

It would be correct to say that since 1955 America has led the way in computer technology, although in recent years Japan has grown level. However, enough has been said to show that no one person, or even group of persons, invented the modern universal computer; nor was there a specific time and locality when and where the first one was completed. 'We made the invention and the foreigners stole it' is, as we have seen, an old complaint made by the peoples of many nations. It is a specious claim. In fact, very few major inventions can be unambiguously credited to one individual. If we base our judgement on the first notion, the first inkling, then the Greeks and the Chinese can claim to have invented most things, while the abacus, remote ancestor of the computer, belongs to pre-history. Such an assessment would overlook the practical and conceptual difficulties involved in any significant invention as well as the economic and social problems. If we give the credit to those responsible for the first machine to be marketed successfully, then we do injustice to those who, in the past, had made important contributions on the way. In the case of the universal computer it may be said that it evolved as the result of a number of detailed additions, some major, many minor, that led eventually to its completion. And it may be that over this time only a few, perhaps no more than two or three, had a clear idea of the ultimate goal. It has been observed that only three great men had a clear idea of a universal computer: Babbage, Zuse and von Neumann. No doubt this is a fair assessment, but the contributions of Atanasoff, Aitken, Shannon, Stibitz, Eckert, Mauchly, Goldstine, Turing, Williams, Forrester and others must not be overlooked. Many, perhaps most, of these men were 'outsiders'. Babbage was a mathematician and a wealthy devotee; Zuse began as an amateur; von Neumann and Turing were mathematicians who looked on computers as means to (mathematical) ends; and Williams was an electrical and electronics engineer who, from about 1955 onwards, lost interest in computers and turned his attention to other things.

A question of interest to historians is this: how important a role did Babbage play in the invention of the universal computer? The answer must be, regretfully, not a very important one. Had Babbage never lived, it would not have been necessary to have invented him as far as the computer is concerned. Much the same may be said of Cayley and the airplane; and, no doubt, of many other pioneers 'born before their time'. The slow evolution of the analogue computer followed, as we saw, a line through the planimeter, Morin's *compteur*, James Thomson's tidal predictor, the speculations of William Thomson and Clerk Maxwell and the gunnery predictors of the 1900s, to the mechanical differential analysers of Vannevar Bush, D. R. Hartree and others in the 1920s and 1930s. This line owed nothing to Babbage; nor did the development of pulse techniques in radar. As we saw, Eckert and Mauchly began by applying electronics to an analogue computer, intended for the computation of gunnery tables; an enterprise that was to lead to the construction of ENIAC. And ENIAC was the main ancestor of all American electronic computers.

Two questions that are unresolved are the origins of Colossus and any possible communication between its designers and computer pioneers in the United States. We have little information about any awareness of Babbage's work on the parts of Newman and Turing, although it is said that Turing knew something of Babbage's ideas while he was at Bletchley Park. War-time secrecy notwithstanding, much more positive evidence is required before significant credit can be given to Babbage. It seems unlikely therefore that Babbage directly influenced the designers of Colossus. As regards the second question, while it is true that American experts worked at Bletchley, they would have been concerned with cryptanalysis, not with gunnery and missile problems. Professor Brian Randell has stated that such was the secrecy concerning Colossus that it is most unlikely that any knowledge of it could have reached the Moore School. Professor Arthur W. Burks, who worked on ENIAC, confirmed that they knew nothing of either Zuse's work or of Colossus. This confirms Norbert Wiener's complaint about the effects of strict secrecy on war-time science. Two research organizations belonging to two different military authorities would, after much effort and at great cost, solve the same problem while each was unaware that it was duplicating the efforts of the other. It is therefore unlikely that there was any significant interchange of ideas – in either direction – between computer pioneers in the United States and the Bletchley team.

A final question arising from a consideration of ENIAC, Colossus and similar machines is this: was the computer the offspring of war? The answer must surely be that it was, but only to a limited extent. Turing's paper in which he conceived of a universal computer was published in 1936. Since June 1945 when von Neumann wrote his seminal analysis, the computer has advanced out of all recognition. We are therefore left with the conclusion that certain war-time advances, particularly in radar technology, helped to expedite the invention of the universal computer as did certain war-time military requirements. For the rest it may be assumed that the universal computer would have been invented had there been no war, but quite probably it would have taken longer.

From one point of view there can be no question that the computer represents one of the great strategic technologies; arguably in its applications and scope exceeding all others, with the possible exception of metal working, which we noted as the first great strategic technology. The computer is now universal. With the possible exception of closed religious orders, the computer is found in all aspects of individual and communal life. Like the television set and the vacuum cleaner, it is a common item in the home; in the form of computer games it offers the frivolous technology of the modern world. Apart from commerce, finance and government, it is used in the design of bridges, air traffic control, the issue of airline tickets, the control of road traffic, meteorological forecasting, and the control of manufacturing and assembly plants, refineries and processing plants. It finds an application in a multitude of other human activities. It is used in musical composition and in literary scholarship; in the form of the word processor it is placed between the keyboard and the type ribbon, thus computerizing the typewriter.

Babbage himself had prophesied that 'As soon as an Analytical Engine exists, it will necessarily guide the future course of science.' Bold words indeed for 1864; did he, perhaps, realize that the universal computer would have an enormous impact on technology as well as on public, social and intellectual affairs generally?

The universal computer is a product of the second half of the twentieth century. It is the most distinctive technological creation of that period. Only Babbage and his associates in the previous century had any understanding of its possibilities. Atomic energy and space travel are also unique products of the present century, but their impact on the individual has been indirect. That is one reason why they have been classified as big technology, while the computer, by virtue of individual numbers and size, belongs to little technology.

It is fitting, too, that as the Anti-Kythera mechanism, perhaps the first example of a mathematical technology, began our story so the computer ends it. And furthermore: just as the electromechanical television system of Baird was replaced by the more efficient and flexible electronic system, so computing machines have moved from the mechanical to the mechanical–electronic to the electronic, first with vacuum tubes or valves and then with transistors and integrated circuits. Consistent with these changes the electronic engineer, the computer engineer and the programmer appeared during the last seventy years. Like the chemical engineer, they represent the new technology of engineering without wheels. Much has been written recently of a 'post-industrial society'. If, by this, is meant a society without industry then such a society, if it came about, would exist at the level of primitive humanity, with all the discomforts, suffering and diseases that would be entailed. There will, foreseeably, always be demand for aircraft, automobiles (or their successors), trains, ships, machine tools; in short the mainstream articles of mechanical engineering. But the rise of the computer, while it cannot replace these items, is an instance of the rise of a radically new technology: the engineering of intelligence. Babbage, in his day, had some insight into this (see chapter 16). Just how far it will go cannot possibly be predicted: there is a particularly good example (see chapter 19) to warn the historian against predictions, however soundly based they may seem. It may be that we have already reached the limit; it is more likely that we are on or are approaching the steepest part of the ascent and that the limits of electronic technology are not in sight.

19

Notes Towards
a Philosophy of Technology

'The abilities of the individual are a debt to the
common stock of public happiness and accom-
modation.'   JOHN SMEATON

Summary

The word 'technology' was coined in the seventeenth
century. Since then technology has been, as Bacon
understood so clearly, the instrument of economic and
political power of the wealthiest nations of the world.
Before 1600 technological leadership, again as Bacon
realized, lay with southern Germany and northern Italy;
between 1700 and 1900 it was the nations of western
Europe together with the United States that were the main
exponents of technology. From 1900 onwards and more
particularly since 1945, Japan and the nations of south-
east Asia have begun to take the lead, as a cursory glance
in the garage or round the living rooms and kitchen of a
modern home will confirm.

The dictionary definitions of technology as 'the scientific
study of industrial processes' or 'the application of science
to industry' are unsatisfactory. They reflect the
uncertainties of the historians who have studied the
contributions of science and technology to industrial
progress. Does the word 'science' mean mathematics,
chemistry, physics and the like or does it imply something
broader? The frequent identification of applied chemistry

and applied physics with technology and industrial progress lies behind the first interpretation. The second is less specific and awaits further study.

It is instructive to begin with Aristotle's ideas. The founder of the science of biology, he insisted that four questions had to be answered if a satisfactory scientific explanation was to be given: What is it? What is it made of? Who or what made it? And, lastly, what is its purpose? The answers to these questions are usually called the four causes (formal and material, efficient and final). The last question implies that science must always take account of purpose, that the aim of science is basically teleological. It may well be that Aristotle drew to some extent on his experience and understanding of the technics of his time when he wrote that if ships and houses grew by nature and were not made by men they would be much as they are now. Purpose and adaptation to purpose are essential in the worlds of technics and of living things. Since the seventeenth century, however, science, or at least physical science, has turned its back on teleology. Technology cannot do so. It was surely no accident that W. J. M. Rankine, a most sophisticated engineer, was a classical scholar, and his Aristotelianism was apparent in the dualism he posed of actual and potential energy. The dualism was obscured when Thomson and Tait substituted the word 'kinetic' for 'actual' to suit the arrangement of their book. And Rankine's distinctive positivist philosophy indicates another affinity with Aristotle. For a marked feature of technology has been the necessary positivism of many of its practitioners.

The emphasis on purpose marks a distinction between science and technology. Many major advances – Dalton's atomic theory for example – have been made by scientists whose original objectives were quite different from their ultimate achievements. At the beginning they could not have foreseen the outcome of their researches. This makes the planning of science and the social control of science (whether with benevolent intention or otherwise) sterile. The technologist, the inventor, on the other hand, almost always works towards a foreseen objective. It would be difficult to imagine how an engineer setting out to design a new type of bridge or to build a revolutionary new heat engine could end up producing a new type of ship, or freezer.

And yet science and technology, as I have indicated, are closely related; so much so that some of the boundaries between them are obscure to the point of non-existence. It would be natural to suppose, therefore, that the philosophy of science – a recognized field of enquiry since the last century – would throw light on the philosophy of technology. This, surprisingly, is not the case.

The philosophy of science today is still dominated by two schools: one associated with Sir Karl Popper and the other with Professor Thomas S. Kuhn.[1] Popper rejects Bacon's inductive method, accepts in its place the hypothetico-deductive system and insists that valid scientific theories are, in principle, falsifiable. It is the possibility of falsification that distinguishes scientific statements from those of politicians, Marxists, psychoanalysts and many philosophers, particularly those of the neo-Hegelian school. It is difficult to see how the Popperian philosophy of science can apply to technology where the question of falsifiability does not arise and where the criterion of validity is pragmatic: does it work?

Popper's view of science is academic; that is, he considers only 'pure' science, and thus excludes technology. Exponents of Popper's philosophy might hold that technology, determined by the market, does not come within the scope of philosophical scrutiny, but that would be unsatisfactory. Since technology is a basic human activity of enormous consequence, it must fall within the range of philosophical attention; unless, that is, philosophy is content with a subordinate, limited role in the effort to analyse and understand human actions, knowledge and belief.

Kuhn's philosophy derived from his work as a historian of science. Kuhn recognizes two processes at work in the advance of science. There is what he terms normal science: the routine processes and methods of everyday science. And there is revolutionary science, when the steady advance of normal science results in a crisis necessitating the substitution of a new 'paradigm', or set of governing ideas and laws, in place of the old paradigm. A prime example of a scientific revolution was the new chemistry established by Antoine Lavoisier in the eighteenth century. Critics of Kuhn's philosophy have denied the dualism of normal and revolutionary science, either asserting, on Popperian grounds, that all valid science is revolutionary or maintaining that it is unclear whether Kuhn's dualism is a recognition of the fact that there are, in science, periods of steady progress alternating with dramatic changes or whether it is a fundamental theory of scientific progress. Kuhn has modified his ideas in the light of these and other criticisms. Unquestionably, however, he has directed attention to an important characteristic of the progress of science.

The Procedures of Technology

The distinguished historian of technology, S. C. Gilfillan[2], pointed out many years ago that there have been two complementary processes in the development of technology: evolutionary improvement and revolutionary invention. These processes cannot be defined with great precision, although evolutionary improvement is a fact of common experience: automobiles, for example, improve year by year following the addition of new components brought in from ancillary industries, due to the research carried out in the manufacturing firms themselves and to the application of Smeatonian principles. Revolutionary inventions fall into two categories: those that are made in a pre-existing technology, as, for example, the Watt engine fitted into and eventually transformed steam engine technology; and those that create a new technology where one had not existed before. Gutenberg's printing press, Marconi's 'wireless telegraphy' and the universal computer are examples in the second category.

Gilfillan's concepts of evolutionary improvement and revolutionary invention resemble Kuhn's normal and revolutionary science. In fact, the latter could be taken as a bold extrapolation of Gilfillan's idea to the philosophy of science. Furthermore, there is a feature of scientific and technological advance that is common to both but it has not, so far, been commented on. As Gilfillan pointed out, revolutionary advances in technology are usually brought about by outsiders, by men from outside the relevant technology. It is often the brash newcomer who has the revolutionary idea. Among them we place James Watt (an instrument maker), Sadi Carnot (a half-pay army officer) and Nicolaus Otto (a travelling salesman). For the obvious reason, there can only be 'outsiders', like Gutenberg, Marconi and Eli Whitney, in the case of inventions in the second category.

Outsiders sometimes play a similar, key role in 'pure' science. In chemistry, for example, the atomic theory was the result not of 'normal' chemistry but of the speculations and researches of a meteorologist. Thermodynamics was created substantially by engineers and not by 'normal' scientific researchers in heat. The philosophies of Popper and Kuhn do not explicitly recognize this characteristic of science. But Kuhn can more readily make provision for it; as can the alternative 'ecological' approach of Toulmin. Apart from Aristotelianism, we conclude that a philosophical approach to technology is likely to be more fruitful by these routes than by that proposed by Popper.

Gilfillan, together with Messrs Jewkes, Sawers and Stillerman[3], refuted the once popular legend that inventions are sometimes made by accident. This, however, requires qualification. The psychology of invention suggests that certain events – accidents – can act as catalysts, setting off trains of thought that lead to inventions. We have Triewald's account of the accident that led to a major improvement of the Newcomen engine (chapter 5) and the Parker brothers' description of the accident that led them to improve their turbine (chapter 12). Hargreaves' curiosity was said to have been aroused by seeing a spinning wheel lying on its side and still revolving. This, the story goes, led to his invention of the spinning jenny. It is, however, impossible to imagine how Hargreaves could have set out – purposed – to invent, say, an improved butter churn only to find that he had invented the spinning jenny!

Things are somewhat different in the case of chemical inventions, where the link with science is close. Popular history has it that the explosive power of black powder (gunpowder) was discovered by accident. It could hardly have been otherwise. There was no precedent or analogy to lead anyone to imagine, and therefore to seek, an *explosive substance* (at that time unimaginable). Bacon's argument (chapter 4) makes this clear. Far more recently we have the well authenticated case of the accidental discovery/invention of the first aniline dyestuff by W. H. Perkin in 1856. We conclude that some inventions, particularly in the chemical and biological fields, can be made by accident.

At the heart of technology lies the ability to recognize a human need, or desire (actual or potential) and then to devise a means – an invention or a new design – to satisfy it economically. Having done so, the model or prototype has, usually, to be scaled up and adapted to become a marketable item. The process of turning the full-scale product into something that satisfies market requirements of safety, cost/profit effectiveness and customer acceptance is a difficult one. Boulton and Watt's well documented efforts to develop and sell the steam engine are a classic example of this process. Emerson's 'better mousetrap' may have involved great originality; what he failed to realize, or failed to point out, was that the mousetrap had to be sold: the last item in the programme. The public have to be told about the new invention and, having been told, to be persuaded that it is worth having. This is not usually easy; old habits, old loyalties have to be disposed of, fears have to be allayed, the protective conservatism and inertia of most people have to be overcome.

The great majority of inventions are, and always have been, what I have called empirical. A good example is the agricultural swipe[4], a machine for cutting coarse vegetation on rough ground. Such inventions have been made by arranging familiar components or materials in a novel way and without resort to abstract or scientific principles; just as Bacon supposed the printing press to have been invented. The extent to which all such inventions could have been immediately understood by Archimedes – Bacon's touchstone – may be doubted. It is a mistake to endow the leading figures of the past with our thoughts, our ideas, our experiences.

On the other hand, it is certainly true that some of the most fruitful inventions were derived from scientific discoveries. Von Guericke's speculations led to the first steam engine, although he could not have envisaged such an invention. Moll and Henry's work turned the electromagnet into a device that made the electric motor possible; something that was not originally contemplated by Sturgeon (still less by Arago, who showed how an electric current could magnetize a steel needle). Hertz did not foresee the use of 'Hertzian waves' for communication. In none of these cases could the invention have been inferred or deduced from the discovery. In every case a separate and additional act of invention, based on the new experience represented by the discovery, was necessary.

It is difficult to imagine that the discoverer of black powder foresaw the gun.

Conversely, scientific advances have been derived from technology. Familiar instances here are the mechanization of time by medieval clockmakers which preceded the Scientific Revolution of the seventeenth century, the rise of thermodynamics and the concepts of energy. In these cases the additional act of invention, of scientific imagination, was required. Science is based on human experience and human experience is enlarged and deepened by technological advance. To suppose that science could have been created by individuals working in predetermined and limited areas of experience, exclusive of the great areas of technological knowledge, is to suppose that Nature has been arranged in accordance with the presumptions and convenience of academic administrators and examiners.

Invention is not necessarily limited by the extent of scientific knowledge. The medieval cathedral builders designed their astonishing creations without the benefit of established theories of structures and strength of materials. They relied on accumulated practical knowledge and their own intuition. John Smeaton when he was designing and building the Eddystone lighthouse could not wait for science to provide him with answers to the very difficult problems involved; while Marconi succeeded in transmitting radio signals across the Atlantic when science would have ruled it impossible; a feat that, as we saw, led to new scientific knowledge. Randall and Boot invented the cavity magnetron while unaware of much of the published work on magnetrons. Semmelweis and Snow made their valid contributions to medicine although the germ theory that would have validated their procedures was not yet formulated. It comes to this: it would be ridiculous to suppose that invention has to wait humbly, cap in hand, for science to open the door before it can proceed. Technology is purposive and it tends, as Rankine implied, to be positivist. The criterion is simply, does it work?

Professor Billington has emphasized the sharp distinction between machines and structures. The former have limited lives; they wear out and are duly replaced by new, and probably better, machines. Structures are designed to be permanent (the medieval cathedrals can be taken as exemplary). Theory, he claims convincingly, cannot prescribe the form for the structural engineer. The intuition – the artistry – of the structural engineer determines the form, after which theory can be invoked, if required, to settle the details (e.g. Smeaton's design for the Eddystone lighthouse, based on the form of a branch springing from a tree). Science and theory, in structural engineering, are, therefore, good servants, but can never be understanding masters. To some extent, at least, this is generally true in other branches of technology.

Medieval cathedrals are in regular use today; Tudor manor houses and seventeenth-century thatched cottages are greatly prized by their owner–occupiers; but the mechanical artefacts of those times, where they survive, are in museums. Of modern mechanical artefacts, we can take the engine of the Me 109 as the standard representative of the in-line liquid-cooled aero engine developed in the first half of the twentieth century. Thereafter such aero engines could only be improved increasingly slowly until they reached a stage beyond which further advance would have been immensely difficult. During this period of maturity the rival form of engine, the jet, was being developed. This was simpler, more reliable, more powerful and with far greater potential. Much the same process was apparent in the case of the reciprocating steam locomotive, which reached near-perfection early in the century at about the same time that its successor, the electric locomotive, was under development. The beautiful clipper ship of the mid-nineteenth century was yet another example. It may, of course, be that such aesthetic approbation is merely the consequence of the fact that, in each case, these machines were the last of the line.

The supersession of these machines did not represent complete loss: far from it. Apart from the enhanced manufacturing skills, the gradual clarification of the major parameters in the working of heat engines amounted to the creation of a system of formal knowledge, analogous to a system of scientific knowledge[5]. Important stages were the Carnot cycle, the Rankine cycle, the entropy–temperature and the Sankey diagrams. This relatively slow advance was in sharp contrast to the rapid development of electronic engineering in the present century. Here, there were few delays; the principles were quickly mastered and an elaborate system of knowledge created (this point was acknowledged by Polanyi).

These systems of knowledge have been co-ordinated and formalized by academic and professional bodies over the past two hundred years; textbooks have been written, professors appointed, journals founded, examinations instituted. The civil (i.e. originally non-military) engineers were the first to achieve institutional and professional status. They were followed by the mechanical engineers, the telegraph (later, the electrical) engineers and the chemical engineers. In the present century other distinct technological groups have achieved institutional and professional status. This has made it more difficult for technologists and scientists – and certainly for the general public – to gain anything like a comprehensive view of technological developments. Too much, however, should not be made of this. Only with the wisdom of hindsight can we say that there was a time when an educated individual could understand all contemporary science or technology. If we already know what is to be important in the future so that we can ignore the rest, then a wide understanding is easy enough.

In an attempt to distinguish between science and

technology Michael Polanyi insisted that technology depended entirely on the market. As an amusing example, he pointed out that an apparatus designed to convert champagne into bath water, no matter how ingenious, no matter how elegant, would be condemned by the market and thrown on the scrap heap. The implication was that the market was everything, the technological knowledge represented by the apparatus was of no interest. The interpretation is, however, incomplete; the argument partial. If it were as ingenious and as elegant as he postulated, then we might suppose that some, at least, of its principles would find other applications; that fruitful developments could flow from it. The principle of the suspension wheel, designed for water-wheels that long ago failed the market test, and for Cayley's airplane that never flew, has found continuing useful applications in bicycles and perambulators. Watt's steam engine is as dead as the dodo, but all modern heat engines have descended from it. In the same way, Copernicus' astronomy is no less dead, but the Newtonian revolution was built on the reformed version of it due to Kepler.

Implicit in this argument is a question of great interest. No one would dispute that the take-up of inventions and technological innovations is determined by economic considerations – the market, in short – and social acceptability. The question now is, whether or to what extent it is money that motivates the individual inventor and technologist. Are they primarily concerned with the money their innovation will earn or save, either for themselves or for others (i.e. the employer or society in general)? There are, at present, no answers to these questions. All that can be said is that it would be unwise to assume that all or even most innovators are motivated by purely mercenary hopes. Personal satisfaction and the hope of social distinction are, no doubt, factors in the lives of many technologists as well as in the lives of many 'pure' scientists.

In Defence of Heroes

The words hero/heroine are commonly applied to those who risk their lives on behalf of others. They are therefore appropriate for soldiers, policemen, doctors and nurses. *Per contra*, they would be inappropriate applied to academics, accountants, grocers, secretaries, stock-brokers, etc., who do not, in the normal courses of their careers, take significant physical risks. During the nineteenth century, however, the word hero began to be applied to certain engineers who had risen to the top of the profession. A leading popularizer of the notion of the hero-engineer was Samuel Smiles, in his day an excellent historian of technology whose works are still worth reading. There was, of course, nothing new in the practice of eulogizing those publicly admitted to be leaders. Bacon and Newton were two such men (as the *Encyclopédie* asserted). Smiles had an additional motive, another justification besides technical achievement. He wanted to muster evidence to support the doctrines of *laissez faire* and utilitarian radicalism. Smiles set out his version of this belief in his *Self-help*, a work whose title neatly summarizes his thesis in just two words.

Smiles found, in the lives of the engineers who had carried through Britain's Industrial Revolution, plenty of examples to support the doctrine of self-help. Men like Watt and the Stephensons had overcome formidable physical obstacles and often strong human opposition to carry out their work. From the essence, the common factor of these studies, a triumphant vindication of the doctrine of self-help can be inferred; such men, Smiles asserted, had often risen in the world from humble beginnings with no material advantages and little education beyond the elementary.

With the decline of utilitarian radicalism and *laissez faire* as acceptable social and economic doctrines has gone a sharp fall in Smiles' reputation as a social philosopher and (regrettably) as a historian. The hero-engineer is no longer acceptable. Historians of technology know that there is a long prehistory of relatively minor innovations tending towards and underpinning each and every allegedly major innovation. From this it could be inferred that the hero-engineer is a Victorian myth, that no inventions were ever truly revolutionary and that the whole course of technological history has been evolutionary. Now there is some truth in this position; there are also the possibilities of errors.

A well-established industry has its own technology, or its own technological discipline. There will be recognized hierarchies of authority supported by systems of apprenticeship, training and promotion. The industry will have a corresponding professional institution and journal. In outline the same principle can be traced back to the medieval guilds. Such a system does not inhibit, in fact it may well encourage, evolutionary improvement but it may – and for familiar reasons – resist radical innovation for which there will be few or no precedents and which, as we saw, tends to be brought about by individuals from outside the industry or technology.

The objection, then, is that a strictly evolutionary history of technology would obscure this, surely important, feature of the processes of invention and innovation. Furthermore, with a revolutionary invention a new language and a new vocabulary have often to be created, and 'new men' appear – much to the bewilderment and often the disapproval of the older generation. How many stage-coach drivers, for example, could have understood the terms used by locomotive engineers in 1831? Older people are baffled by the language of computing; their children take to it easily.

A second objection is *ad hominem*. The historian in his or her study can easily describe the early history of the steam engine as an evolutionary process, beginning with imaginative speculators such as Branca, de Caus, the Earl of Worcester and going on to include von Guericke, Hautefeuille, Huygens, Papin and then Savery, before Newcomen rounds off the story with his successful engine of 1712. However, our experiences in building and operating an exact replica of the 1712 engine (at one third scale it stands five metres high) has convinced us of the original genius of Newcomen, a real hero-engineer. Problems, not mentioned in any of the literature, were met and overcome; the true functions of the key components were fully understood and their relationship to the operation of the engine appreciated. More recently, Mr. Michael Bailey and his team have begun experiments on the full-scale, working replica of the locomotive *Planet* that they have built for the Manchester Museum of Science and Industry. Much has already been learned that is not to be found in the written records. It follows that any levelling down of technological achievement, and with it of the hero-engineer, must tend to obscure key features of the process and make difficult any objective evaluation of individual cases. A final lesson is that *there is great scope for practical experimentation to supplement (and correct) history based solely on documentary evidence.* In addition to these specific objections, there is the general point that, the course of political history being determined, at least in part, by statesmen, kings, conquerors and prelates, the reader will expect to find analogous figures in the course of the history of technology. That expectation should be met. A history without notable figures, without major episodes and in which all is ascribed to social action, would be an unsatisfactory and, in the last resort, a sterile affair.

An interesting evolutionary interpretation of the history of technology has recently been published by George Basalla[6], who begins with a discussion of Ortega y Gasset's theory that technology evolves not in response to the necessities of life but to its superfluities. Human beings, according to Ortega's theory, did not need even the simplest of tools or facilities; for ages they lived without cooking, agriculture, stone axes, before choosing to have these things. There seem, however, to be one or two difficulties with this position. In the first place how could peoples choose not to have, say, stone axes before such things existed? Choice, where there is no alternative, is no choice at all. And, secondly, one of the defining characteristics of human beings is that they developed technics. While we may not be completely satisfied with Babbage's definition of '*man* . . . as a *tool-making animal*', we can hardly deny that every known culture, past or present, has been characterised by the *manufacture* and use of tools. As regards the earliest known cultures, we know a little of their technics, rather less of their art and of their religious beliefs, nothing at all of their literature (if any) or folk lore, their music (if any) or their systems of law.

It is true that isolation and/or political decision (equals choice) by an authoritarian government can prevent the spread of technology, and political dogmas have denied people what, by any other standard, was a need. So need is a matter of choice too? In the last resort I can choose not to live. It would seem that Ortega pushed the concept of collective choice to the limit at which 'need' was virtually eliminated. A further difficulty confronts us if we take this position. How can we account for the early history of technology and for the differing responses of various cultures? In Daniel Defoe's optimistic novel, *Robinson Crusoe*, the castaway creates for himself a reasonable lifestyle; in William Golding's novel, *The Lord of the Flies*, a party of schoolboys, marooned on an island with food to sustain life, rapidly descend to savagery. Whether they chose to do so may be debated. What seems to be clear is that the available evidence makes Golding's story more plausible than Defoe's. Tasmanian man, cut off from mainland Australia about 12 000 years ago when Tasmania became an island, proceeded to forget some of the technics – such as fishing – of his mainland compatriots. Was this by choice? Or is there a fundamental question here: what is the minimum size below which a simple community cannot maintain, still less advance, its technics?

Sixty years ago Herbert Butterfield published a notable critique of the Whig interpretation of history, which is to say the study of the past with reference to the present, more particularly with regard to questions of moral judgement. The historian of technology faces an analogous problem though in a less challenging form. The history of the railroad, for example, can certainly be presented, without distortion, as an evolutionary process. But it would be absurd to suppose that the German miners who invented the railroad could foresee its eventual development, and it would be worse than absurd to criticize them for failing to realize the possibilities of its wider application. In the same way the history of the steam engine has usually been presented in evolutionary terms. However, as R. L. Hills has pointed out[7], the engine was, for over a hundred years, thought of as a *pressure* engine, analogous to the water pressure engines of the time. Watt, possibly, Carnot, certainly, thought of it as a heat engine. Only after the establishment of thermodynamics could it be thought of as a heat energy engine. The historian of technology must always remember that within the broad evolutionary framework there are always a number of separate histories. In the case of the steam engine, there are the histories of the pressure (atmospheric and steam) engine, of the heat engine and of the heat engine whose performance is dictated by the laws of thermodynamics. The understanding, motives and expectations of the inventors and engineers in these different phases may be expected to differ, perhaps widely. This granted, an evolutionary history of fire/atmospheric/steam/heat (including gas turbine, or jet) engines, although it might do some violence to the ideas and goals of different epochs, should reveal how the efforts of engineers and scientists were directed to increasing command over energy resources and how these efforts led, among other things, to a great extension of scientific knowledge.

Factors in the Progress of Technology

The basic requirements for the progress of technology are easily summarized. The beliefs and psychology of a people should be such that they are receptive to new ideas and inventions from whatever quarter they may come. This implies that the dominant philosophy or religion should be, at least, permissive in this respect; and so should physical factors, such as geography. In addition, a minimum degree of individual freedom is essential; this means freedom for all to travel, to learn, to change jobs, to experiment and to invent. It is obvious that there must be economic and other social incentives to invention and that there be a supply of skilled assistants, or technicians. The latter requirement implies a need for suitable systems of training and education. And experience over the last three hundred years shows that systematic experimental and development methods, such as that pioneered by Smeaton, together with close association with science, are characteristic of modern technology.

The Newcomen engine and its distant modern descendant, the jet engine, are both heat engines. Although their modes of operation are different, the underlying principle – the conversion of heat energy into mechanical energy – is the same in both cases. The modern jet engine, however, depends on materials that have to stand temperatures and stresses far beyond those involved in the working of the Newcomen engine. Furthermore, the components of the jet engine have to be machined to an accuracy that was quite unattainable only a few score years ago. Generalizing, it can be said that the introduction of revolutionary inventions depends on the availability of materials – in particular new metals – and tools that can meet unprecedented requirements. Invention and innovation in stone-age times must have been a slow and difficult business since the only materials – and tools – available were stone, bone and timber.

We have noticed how, from early bronze and iron working and the inventions of the alphabet and arithmetic up to the computer, certain technologies and their associated industries have assumed strategic importance, stimulating progress in other technologies and industries. The computer was in large measure the offspring of a prior technology, electronics, just as the railroad was the offspring of mining technology. The question then arises, can any other universal or common elements be found in the history of technology, in the same way that the atomic theory provided chemistry with its most important governing idea and energy did the same for physics? Liebig gave a useful answer to the question. Civilization, he wrote, is the economy of power. Many would object to this apparent over-simplification. Art, music, poetry, philosophy, the rule of law are vital components of civilization. Yet without economy of power, or rather energy, the extent to which these other activities can flourish must be limited and life – nasty, brutish and short – reduced to a struggle to survive. The first person to fit a sail, or something to act like a sail, to a canoe, coracle, or raft, achieved a significant economy in power. The invention of the wheel was another major advance in the economy of power. Running through the history of inventions therefore is the linking thread of the economy of power, or rather energy. With the knowledge we now possess, we can see that the history of agriculture – one of the basic technologies – is one of a sustained drive for increased economy of energy. This is why packaged foods carry information as to the number of joules to be credited to the contents; and why advertisements stress the energy contents of foods and drinks they promote.

Doubts creep in at this point. Can we justify a 'triumphalist' history of technology that follows the golden threads of success through the years and that largely ignores the many failures on either side? Who is to say that our way has been the only, or the best, or even a desirable way? What of the frivolous technologies of the present age? What about Bacon's aphorism about obeying Nature? The only answer is that I have been concerned with what happened, not with what might or should have happened. It may be that matters will take a different course in the future, but that is for later generations to attend to. For the present, however, I recall that Jevons' dire warning had no perceptible effect on subsequent generations although oil consumption over recent decades increased far more rapidly than coal consumption in Jevons' worst possible forecast. Optimism and complacency usually manage to override the direst warnings from the most authoritative prophets.

Historians of technology have shown very little interest in the question of the possible role of religion in the progress of technology although, *a priori*, the relations between religious beliefs and practices and the advance of technology would seem to be close. Few historians, of science as of technology, have shown much interest in possible relationships with law, a notable exception here being Hans Kelsen. This is surprising, for Roman law is the other great inheritance that helped form the distinctive character of medieval Europe, where modern science, technology and the various arts began. At this point we should recall Maine's dictum that the development of law will be inhibited by too close an association with religion. Once law is emancipated from the charismatic rules of religion, it can become a rational system of thought and therefore consistent with the development of science and technology. It seems that Christian Europe maintained – at least for many centuries – a fruitful balance of religion, philosophy, law and early technics; all relatively independent of each other but mutually supportive and all therefore free to develop.

War, it has often been said, has stimulated technology and invention. However, as was pointed out, the Napoleonic wars had little or no direct effect on the strategic technologies of steam power and mechanical transport, on the development of structures and the textile industry. And we wonder what inventions and innovations might have been made had there been no major wars this century. The question cannot be answered; but it would be unsound to assume that continued peace would have meant fewer inventions. A war economy is a command economy that is marked by extreme technological secrecy and the conscription of technological manpower (both are hindrances to innovation; cf. the 'Colossus' affair (chapter 18) and the argument about slavery (chapter 2)). It is agreed that a peace-time, market economy is more efficient than a command economy, so we can suppose that without major wars in this century the world would have seen more, although possibly somewhat different, technological innovations.

The thesis is more persuasive when it is put like this: the prospect of war stimulates certain inventions without imposing a stifling command economy. A wealthy nation at peace can afford to allow some R & D resources to be allocated to military projects that may prove successful; the same nation at war may well lack the time and the resources – and deny the intellectual freedom – to do so. The individual factor should not be overlooked. A number of scientists, technologists and inventors were killed in action or on war service in the First World War. Many potential scientists or technologists must have been killed before their talents had had a chance to flourish. Finally, we should remember the warning that Joule, a deeply conservative man, gave in the address he wrote, but was never to deliver, as President of the British Association in 1884. Foreseeing that wars would become more and more destructive, he added that, 'by applying itself to an improper object science may eventually fall by its own hand. In reference to this subject we must also deplore the prostitution of science for the aggrandisement of individuals and nations, the result being that the weaker is destroyed and the stronger race is established on its ruins.'

The complexity that characterizes major technological advances and the frequent dependence of invention on the intervention of outsiders imply that predicting the future course of technology with any degree of accuracy is practically impossible. Imaginative writers such as Jules Verne and H. G. Wells made some surprisingly good forecasts, but almost always in general terms. The numerous specious solutions to the energy problem proposed since 1800, all fairly widely accepted by competent engineers, indicate how strong desire for an innovation can encourage the belief that it has been achieved. Few have examined instances of technological progress with as much care and learning as Gilfillan. He felt sufficiently confident to predict, in 1927, a great future for the, now largely forgotten, rotor ship. Of course, methods in the social sciences have improved greatly since Gilfillan wrote, but the complexities of technology have increased also.

It was widely believed, for some time after the Second World War, that German rocket engineers and technicians had been taken to Russia in order to develop missile and space technology in that country. At the same time von Braun and other German engineers went to the United States to help with the American space programme. The implication was that neither Russia nor even America could have developed successful rockets without the help of the men 'in the secret'. This was atavistic thinking, a return to the belief that the heretical scientists – alchemists, astrologers, magicians and philosophers – could command mysteries beyond the powers of ordinary folk. In fact it is evident that the United States could have designed and built successful rockets without the assistance of von Braun and his associates. As for Russia, we now know that no German engineers went to that country after 1945. Besides K. E. Tsiolkovsky, the nineteenth-century Russian prophet of space technology, Russia had, in Sergei Pavlovich Korolev, an inspired designer who had been working on rockets even before the Second World War; the Russian rocket and space programme was developed under his guidance. The diffusionist theory, according to which inventions must have one source from which they spread to different communities, or nations, does not apply in the case of the space rocket. Much the same argument holds for the early history of nuclear power. In this case there were some celebrated trials in which people were accused of betraying secrets to the potential enemy. Heavy sentences were imposed on those found guilty although, as Professor Shils pointed out at the time[8], it is difficult to believe that anything of importance can have been involved. Nature's secrets are open to all who investigate them.

We should recall that two cultures, or civilizations, or nations, at roughly the same technical level, may be expected to produce, more or less simultaneously, the same inventions quite independently, the nature of the inventions being determined by the general level of technical resources and public demand. A diffusionist theory that sought to trace each and every invention to one particular source is unacceptable.

There has, in recent years, been much discussion of the roles of women in various professions. We are reminded that women have made major contributions to literature, art and science. The perplexing problem is, why have there been, at least until recently, so very few notable women technologists? In fact female technologists of any distinction are hard to find (this is true of all nations)[9]. Women, equally with men, are users of technology. Unless one accepts a male conspiracy theory the only plausible explanation is that up to now the influential members of society – women as much as men – have not regarded technology as a suitable career for women and that families have followed this lead. If this prejudice can be broken down we may expect a radical acceleration in technology, associated perhaps with a perceptible change in direction. Particularly relevant to this problem is Professor Ruth Cowan's observation[10] that industrialisation tended to reduce the work load of the husband rather than that of the wife.

It seems likely that, following the precedent of Byron's daughter, who was Babbage's research associate, and the recent example of Rear Admiral Grace Hopper, USN (1906–92)[11], computer technology will be the first major technology in which women should play a full part. As a new technology it is without the established male hierarchies, assumptions and prejudices of older technologies, and it is reasonable to assume that the only significant hindrances will come from parental prejudice. This is a further reason for describing computer technology as a unique strategic technology. In any case it is probable that the first people or nation to break down the widespread prejudice (if that is what it is) against women in technology will gain a marked advantage over its rivals.

The Public Perception of Technology

Technology is often blamed for the evils of the modern world. Some recent critics have blamed science – and this may be assumed to include technology – for the alleged lack of spirituality of the present age, for the decline in religious belief, for present-day materialism. The criticisms are ill-founded. Those who yearn for more spirituality might consider living in one of the fundamentalist nations or communities of the world where they would find spirituality in abundance. As for religious belief, the most damaging attacks have come not from scientists, who by and large are not polemicists, but from numerous social critics, philosophers, novelists, strident nationalists and totalitarian politicians. Materialism might be deplorable but is surely no more than the result of public demand for an ever higher standard of living. The critics are, whether they realize it or not, proposing a state of Gothic ignorance as the ideal. They might reflect that had it not been for science and technology they would, in all probability, not be here to make their criticisms. You cannot enjoy the benefits of modern medicine and public health – of which the most determined of critics would surely approve – without accepting the rest of science and technology. Medicine and public health are dependent on other technologies which, at first sight, appear to have little to do with them.

It may be a tired cliché that technology abolished distance and time but there was a considerable element of truth in it. By 1914 travel was more comfortable, faster, safer and easier than it had ever been before. Pre-1914 Baedekers assured their readers that passports were no longer necessary for travel to the civilized countries of Europe; America and the British commonwealth were, with regrettable exceptions, open to all-comers. We are now, after the post-1914 disasters, trying to return to that condition. And if we consider recent atrocities that would justify the century now ending being called the century of barbarism, or, according to Dr Jonathan Miller, the unforgivable century, we shall find that they had little enough to do with technology. They sprang from, their efficient causes were, extreme nationalism and perverse political philosophy. Yet we do not hear calls for a ban on the teaching of political theory, or a moratorium on the writing of nationalist histories, particularly those that glorify military victories. Beneath much of the modern dread and distrust of science and technology lie atavistic fears associated with the taboo on the natural. No such fears attach to the outpourings of political windbags or the scribblings of chauvinistic historians.

On the other hand, it seems at first sight remarkable that the changes brought about by technology since the eighteenth century have hardly been noticed by contemporary poets, dramatists and novelists. For their part, most poets, dramatists and novelists wrote for leisured middle-class people, the majority of whom lived far from the actual scenes of industry and whose education had included little science and no technology. Some, like William Wordsworth, (initially) scorned the railways; in the next generation they took them for granted; in the generation after that they scorned the automobile; in generations to come they will scorn whatever replaces the automobile.

The impact of technology is clearly apparent in the realm of architecture. In their time the great cotton and woollen mills of the north of England – unique, unprecedented buildings – were much admired by informed critics such as the distinguished Prussian architect C. F. Schinkel. People came from all over Britain and Europe to see them. As W. Cooke Taylor remarked, a Manchester industrialist would rather you admired his mill than his mansion. It would be wrong to dismiss this as philistinism: these mills were (the survivors still are) fine buildings. And it is nonsensical to describe them as 'dark satanic mills'. Cotton or wool cannot be carded, spun, woven in the dark; there must be abundant light. In any case the new iron and then steel technology carried structural engineering far beyond the mills. Iron structures made public buildings such as theatres and opera houses safer from fire and also far better from the audience's point of view. Steeply raked galleries with many rows of seats were possible with iron and, later, steel framing.

New materials[12] opened up possibilities for architects that were not immediately exploited. The use of iron and glass in the structure of railroad stations was not thought of as art at the time but has come to be recognized as such in the present century. Concrete, reinforced and prestressed, has made possible slender, elegant and bold structures such as road bridges and roofs for large exhibition halls, factories or airports.

The representation of new technology by artists raises questions of perception and understanding. It is clear, from paintings and drawings of the first locomotives, on the Liverpool and Manchester Railway, for example, that the artists had no understanding of the machines and, accordingly, depicted them in barest outline with little more subtlety than a child would have done. Ever since the Renaissance, artists had studied human and animal anatomy in order to draw and paint figures that were anatomically correct. Confronted by an entirely new machine, artists had no analogous guidelines. They did not understand – how could they? – the relationship between the different components; they could not differentiate between the essential and the accidental and unimportant features, and they had little understanding of the forces that determined the action of the machine. They could, of course, cheat by putting a human figure or a wreath of smoke in front of items they could not interpret. This defect of perception characterizes pictures of prototypical machines from the first printing presses onwards. For the rest, pictures of static objects such as mills and bridges presented no problems to contemporary artists. So, for example, eighteenth-century artists had no difficulty in painting Arkwright's mill or Coalbrookdale furnaces.

Technology in the World

It is natural to wonder if there are different national styles in the pursuit of technology[13]. Do, for example, Germans excel in highly scientific innovations while Americans are conspicuous for practical ingenuity? No doubt there is a degree of truth in such generalizations, just as one can assert, with due caution, the validity of national character.

Yet differing national styles do not alter the fact that technology, like science, is international. As with science, the value of such differences as exist is that they allow for diverse approaches. A technology, or a science, that was uniformly the same the world over would be sterile, or at least not nearly as productive as it could and should be.

Simple, basic, or empirical invention has characterized the human race since its very beginning. Two thousand and more years ago, in the countries of the eastern Mediterranean, the foundations of natural science were laid and there were indications of a more systematic technics, or technology. Following what I think was a profound change in human ambitions and understanding in medieval Europe, there was an acceleration in the rate of invention, and by the Renaissance there was evidence of technology, or systematic technics, a corpus of accepted technical knowledge as represented by such books as Agricola's *De re metallica* and Biringuccio's *Pirotechnia*. Renaissance technology, in the forms of mining, metal working, civil engineering and architecture, flourished in Germany and Italy; while the advanced technology of navigation flourished on the other side of Europe, in Portugal and Spain. It was the latter technology that made possible the discovery of the world, with all the revolutionary consequences that flowed from it. By the eighteenth century the technological centre of gravity had moved to Holland, France, Scandinavia and Britain. In that century the scope and nature of technology became more clearly defined and the nature and procedures of technologists understood, due to the patent system and the establishment of regular journals. It would be true to say that since the seventeenth century technology and science have advanced together. At the beginning of the nineteenth century *science technologique* was acknowledged in France, largely through the work of the *polytechniciens*. And, by the end of that century, the social institutions of technology had evolved in a similar way to those of natural science.

We are accustomed to divide scientists into two distinct groups: experimentalists, who work in laboratories, and theoreticians, who work with pencil and paper (and aided, now, by computer). Technologists form a more diverse, perhaps richer, community. The membership ranges from those inventors who know, or use, no scientific or technological theory, to those who work on the furthest frontiers of theory, as for example in computer technology or atomic energy. Broadly speaking, the ancient field of technics was enriched by the rise of science during and after the seventeenth century. This, coupled with the very diverse fields that technology embraces, has led to the wide diversity of technologists in the modern world.

Simple, empirical technics is the earliest form of little technology; now, with the infusion of sophisticated technology and science, it includes home computers and the other complex products we have mentioned. Big technics begins with the civilizations: the Great Wall of China, the Pyramids of Egypt, Roman aqueducts and roads. Big technics and big technology entail decisions at the highest level of state, the employment of experts, a massive labour force and large-scale investment. When the relationship between science and technology was strengthened by the institution, in about 1870 in Germany, of the industrial research laboratory, big technology made possible the exploitation of atomic energy and the development of space travel, both of which require massive social and political involvement as well as extensive scientific research.

During the present century the process has been extended to an extent that the ordinary, educated person finds it difficult to distinguish between science and technology. Asked to say what they consider the most important scientific advances since 1945 non-technical people are quite likely to reply: the jet engine, the solution of the genetic code, colour television, the computer and space travel. In expressing this opinion they are not making a mistake. As Wittgenstein remarked, what is a cow is a matter of public opinion. It is the public at large that determines the meaning of words. We conclude that, the role of purpose notwithstanding, in the last resort technology and science are aspects of the same thing. They constitute the inseparable procedures by which we attempt to understand and to control the natural world for the benefit and, ultimately, for the survival of humanity. In this way the history of technology may, sooner or later, come to be accepted as a main branch of history that also incorporates the history of science as well as the study of humanity's oldest artefacts. One thing is certain, a history of science that ignores the history of technology must be unsatisfactory.

Here, perhaps, are the ultimate problems of the philosophy of the history of technology. And their further consideration will surely require reference to the mature views of Aristotle and of Bacon.

NOTES

Chapter 1

A. G. Drachmann, *The Mechanical Technology of Greek and Roman Antiquity* (Copenhagen: Munksgaard, 1963).

R. J. Forbes, *Studies in Ancient Technology*, vols 7–9 (Leyden: E. J. Brill, 1963, 1964).

Vitruvius, *On Architecture* (Cambridge, Mass.: Harvard University Press; and London: Heinemann, 1962), 2 vols in Loeb Library.

1 J. C. Maxwell, *Theory of Heat* (London: 1872) 75.

2 D. F. S. Scott, *Wilhelm von Humboldt and the Idea of a University* (Durham: University of Durham, 1960); C. E. McClelland, *State, Society and University in Germany, 1700–1914* (Cambridge: CUP, 1980).

3 The moral responsibilities of modern scientists and technologists have been discussed by, for example, J. R. Ravetz, *Scientific Knowledge and its Social Problems* (Oxford: OUP, 1971).

4 H. Kelsen, *Society and Nature: A Sociological Inquiry* (London: Routledge & Kegan Paul, 1946).

5 O. Neugebauer, *The Exact Sciences in Antiquity* (Princeton, NJ: Princeton University Press, 1952; Harper Torch Books, 1962).

6 The concept of a strategic technology was introduced by A. P. Usher.

7 A. N. Whitehead, *Science and the Modern World* (Cambridge: CUP, 1926; reprint 1953).

Chapter 2

A. C. Crombie, *Augustine to Galileo, the History of Science AD 400–1650* (London: Mercury Books, 1952; paperback edition 1962).

Lynn White, *Medieval Technology and Social Change* (Oxford: OUP, 1962).

A. G. Keller, *A Theatre of Machines* (London: Chapman & Hall, 1964).

1 G. E. R. Lloyd, *Matter, Reason and Experience* (Cambridge: CUP, 1979).

2 D. J. de S. Price, *Gears from the Greeks, the AntiKythera Mechanism, a Calendrical Computer from c. 80 BC* (New York: Neale Watson Science History Publications, 1975).

3 H. S. Maine, *Ancient Law* (Oxford: OUP, 1939).

4 B. Gille (ed.), *Histoire des Techniques* (Paris: Editions Gallimard, vol. 1 1978).

5 Cf. Geoffrey Chaucer, 'The Reeve's Tale'.

6 It is regrettable that there is still no study of Arab science and technology comparable to Dr Needham's monumental work. This is the more surprising in view of the great wealth of certain Arab nations.

7 The expulsion of the Jews from England in 1290 may have contributed to Britain's relative technical backwardness in the later middle ages. The expulsion was formally rescinded by Oliver Cromwell in the middle of the seventeenth century.

8 R. L. Hills, *Paper Making in Britain 1488–1988* (London: The Athlone Press, 1988). This book has an extensive bibliography.

9 Carlo Cipolla, *Clocks and Culture, 1300–1700* (London: Collins, 1967); E. L. Edwardes, *Weight Driven Clocks of the Middle Ages and Renaissance* (Altrincham: John Sherratt, 1965); D. S. Landes, *Revolution in Time: Clocks and the Making of the Modern World* (Cambridge, Mass: Harvard University Press, 1983).

10 The functioning of gears may seem obvious. But not to the tyro and the uninformed; most people will recall bewilderment when first confronted with the concept of 'mechanical advantage' in school physics.

11 R. Lenoble, 'La Pensée Scientifique', in *Histoire de la Science*, ed. Maurice Daumas (Paris: Editions Gallimard, 1957).

CHAPTER 3

J. L. E. Dreyer, *A History of Astronomy from Thales to Kepler* (New York: Dover Books, 1953).

A. Pannekoek, *A History of Astronomy* (London: Allen & Unwin, 1961).

G. R. Elton, *Reformation Europe, 1517–1559* (London: Collins, 1963).

1 V. Scholderer, *Johann Gutenberg, the Inventor of Printing* (London: British Museum, 1963). This short work has an excellent bibliography.
2 R. L. Hills, *op. cit.*
3 A. G. Keller, *op. cit.*
4 A. C. Crombie, *op. cit.*
5 D. W. Waters, *The Art of Navigation in Elizabethan and Early Stuart Times* (London: Hollis & Carter, 1958); E. G. R. Taylor, *The Mathematical Practitioners of Tudor and Stuart England* (Cambridge: CUP, 1969).
6 I am grateful to Professor Rattansi for this quotation.
7 T. S. Kuhn, *The Copernican Revolution* (Cambridge, Mass.: Harvard University Press, 1957); J. R. Ravetz, *Astronomy and Cosmology in the Achievement of Copernicus* (Poland, 1965); A. Armitage, *Copernicus, the Founder of Modern Astronomy* (New York & London: T. Yoseloff, 1957).
8 The Austrian unit of currency, the 'thaler', was named after Joachimsthal. The word 'dollar' was derived from thaler.
9 R. G. Collingwood, *The Idea of Nature* (Oxford: OUP, 1944).

CHAPTER 4

C. V. Wedgwood, *The Thirty Years War* (London: Jonathan Cape, 1938, 1964).
G. N. Clark, *The Seventeenth Century* (Oxford: OUP, 1950).
A. R. Hall, *The Scientific Revolution* (London and New York: Longmans, 1954, 1962; Boston: Beacon Press, 1956).
C. Hill, *Intellectual Origins of the English Revolution* (Oxford: OUP, 1965).
C. Webster, *The Great Instauration* (London: Duckworth, 1975).

1 P. Rossi, *Francis Bacon: from Magic to Science*, trans. Sacha Rabinovitch (Chicago: University of Chicago Press, 1968; London: Routledge & Kegan Paul, 1968). See also Macaulay's essay, 'Lord Bacon', in *The Edinburgh Review* (July 1837).
2 The last six words of the quotation suggest that Bacon's thoughts were still influenced by the taboo on the natural.
3 Gilbert was favoured by, and received a pension from, Queen Elizabeth. Bacon, who to his chagrin was shunned by Elizabeth, may therefore have had a grudge against Gilbert.
4 Liebig was among the first of the moderns to denounce Bacon's

inductive method. However, a late friend and colleague, Dr Wilfred Farrar, pointed out that Liebig's criticism was not entirely disinterested. The Baconian method had been used to refute some of his work on agricultural chemistry.

5 Galileo Galilei, *Two New Sciences*, transl. H. Crew and A. de Salvio (London: Macmillan, 1914; New York: Dover Books, n.d.); Galileo Galilei, *On Motion and On Mechanics*, transl. and with notes by I. E. Drabkin and Stillman Drake (Madison: University of Wisconsin Press, 1960); see also C. Golino *Galileo Reappraised* (Berkeley: University of California Press, 1966) and P. Redondi, *Galileo, Heretic* (Princeton, NJ: Princeton University Press, 1987; London: Allen Lane, 1988).

6 F. Reuleaux, *The Kinematics of Machinery* (London, 1876; New York: Dover Books, 1963).

7 R. Descartes, *Discours de la Methode* (1637), chapter 5. Descartes wrote in French, avoiding the Latin of the academics. R. Descartes,

Discourse on Method, transl. Arthur Wollaston (London: Penguin Books, 1960).

8 Through the efforts of Professors Cohen, A. R. and M. B. Hall, Turnbull and Whiteside, Newton's very extensive works and correspondence have now been published in authoritative editions (Cambridge: CUP).

9 Margaret 'Espinasse, *Robert Hooke* (London: Heinemann, 1957). It is regrettable that there is no more modern biography of this remarkable man.

10 O. von Guericke, *Experimenta Nova Magdeburgica de Vacuo Spatio* (reprint, Aalen: Otto Zeller, 1962).

11 Father Lana Terzi reasoned, correctly, that large spheres, exhausted of air, must rise up, being lighter than the surrounding air. Unfortunately a material strong enough to resist the pressure of the surrounding air and sufficiently light in weight has yet to be found so the idea, in theory correct, proved quite impracticable.

CHAPTER 5

J. T. Desaguliers, *A Course of Experimental Philosophy*, 2 vols (London, 1744).

S. A. Bedini. *Thinkers and Tinkers: Early American Men of Science* (New York: Charles Scribner's Sons, 1975).

1 C. MacLeod, *Inventing the Industrial Revolution, The English Patent System, 1660–1800* (Cambridge: CUP, 1988).

2 H. M. Soloman, *Public Welfare, Science and Propaganda in Seventeenth Century France* (Princeton, NJ: Princeton University Press, 1972).

3 Lord Ernle, *English Farming, Past and Present* (London: Longmans Green & Co, 5th edn, 1936).

4 A. Young, *Letters Concerning the Present State of the French Nation* (London, 1769); *Travels in France During the Years 1787, 1788 and 1789*, ed. Constantia Maxwell (Cambridge: CUP, 1950).

5 C. K. Hyde, *Technological Change and the British Iron Industry, 1700–1870* (Princeton NJ: Princeton University Press, 1977); R. A. Mott, 'Abraham Darby (I and II) and the Coal-Iron Industry', *Transactions of the Newcomen Society*, **31** (1957): pp 49–93; T. S. Ashton, *Iron and Steel in the Industrial Revolution* (Manchester: Manchester University Press, 1924).

6 Jean Petot, *Histoire de l'Administration des Ponts et Chaussées, 1599–1815* (Paris: Libraire Marcel Reviere, 1959).

7 J. Heyman, *Coulomb's Memoir on Statics: an Essay on the History of Civil Engineering* (Cambridge: CUP, 1972); S. B. Hamilton, 'The French Civil Engineers of the Eighteenth Century', *Transactions of the Newcomen Society*, **22** (1941–2): 149.

8 A. Parent, 'Sur la Plus Grande Perfection Possible des Machines', *Histoire de l'Académie Royale des Sciences* (1704), pp. 116, 323. Parent's theory and conclusions were accepted by the authorities of the time and, in the cases of a few obscurantist British writers, up to the beginning of the nineteenth century, although by that time they had long been superseded.

9 Von Guericke, *op. cit.*

10 Thomas Savery, *The Miner's Friend, Or An Engine to Raise Water by Fire* (London, 1702, reprint, 1827), pp. 1–53; J. S. P. Buckland, 'Thomas Savery: His Steam Engine Workshops', *Transactions of the Newcomen Society*, **56** (1984–5): 1–20.

11 A. Stowers, 'Thomas Newcomen's First Steam-Engine 250 Years Ago and the Initial Development of Steam Power', *Transactions of the Newcomen Society*, **34** (1961–2): 133–49; J. S. Allen, 'The 1712 and Other Newcomen Engines of the Earl of Dudley', *Transactions of the Newcomen Society*, **37**

(1964–5): 57–84; see also
J. S. Allen in *Transactions of
the Newcomen Society*, **42**
(1968–70); **41** (1968–9); **43**
(1970–71); **45** (1972–3); **50**
(1978–9) and **51**
(1979–80); Marie B.
Rowlands, 'Stonier Parrott
and the Newcomen
Engine', *Transactions of the
Newcomen Society*, **41**
(1968–9): 49–67; A.
Smith, 'Steam and the
City: the Committee of
Proprietors of the
Invention for Raising

Water by Fire,
1715–1735', *Transactions of
the Newcomen Society*, **49**
(1977–8): 5–20; R. L.
Hills, 'A One-Third Scale
Working Model of the
Newcomen Engine of
1712', *Transactions of the
Newcomen Society*, **44**
(1971–2): 63–78.
12 K. Baedeker, *Handbook for
Paris* (Leipzig: K. Baedeker,
1904), p. 356. Baedeker is
unquestionably an
impartial authority.

CHAPTER 6

L. T. C. Rolt and J. S. Allen, *The Steam Engine of Thomas
Newcomen* (Hartington: Moorland Publishing, 1977).

1 J. R. Harris, 'The
Employment of Steam
Power in the Eighteenth
Century', *History*, **52** (June
1967); J. Kanefsky and J.
Robey, 'Steam-Engines in
Eighteenth Century Britain:
a Quantitative Assessment',
Technology and Culture, **21**,
No. 2, (1980): 161–86. See
also the papers by R. A.
Mott, A. W. A. White and
J. N. Rhodes in *Transactions
of the Newcomen Society*; G.
J. Hollister-Short, 'The
Introduction of the
Newcomen Engine into
Europe', *Transactions of the
Newcomen Society*, **48**
(1976–79): 11–24.
2 M. Teich, 'The Diffusion
of Steam-, Water-, and Air-

Power to and from
Slovakia during the 18th
Century and the Problem
of the Industrial
Revolution' in *L'Acquisition
des Techniques par les Pays
Non-Initiateurs* (Paris:
Editions du CNRS, 1973);
D. M. Farrar, *The Royal
Hungarian Mining Academy,
Schemnitz, Some Aspects of
Technical Education in the
Eighteenth Century*
(Manchester: unpublished
M.Sc. thesis, UMIST,
1971); D. S. L. Cardwell,
'Power Technologies and
the Advance of Science,
1700–1825', *Technology and
Culture*, **6**, No. 2 (1965):
188–207; T. S. Reynolds.
'Scientific Influences on

Technology: The Case of the Overshot Water-Wheel, 1752–1754', *Technology and Culture*, **20**, No 2 (1979): 270–295.

3 S. Lindqvist, *Technology on Trial: the Introduction of Steam Power Technology into Sweden* (Stockholm: Almqvist & Wiksell, 1984).

4 The Newcomen engine, although often called an atmospheric engine, was essentially a steam engine. Power came from steam pressure; the atmosphere acted only as a spring, pushed back by the steam and then released. The ultimate source of power was heat energy from the burning fuel.

5 A. W. Skempton (ed.), *John Smeaton, F.R.S.* (London: Thomas Telford, 1981); J. Smeaton, *A Catalogue of Engineering Drawings, 1741–1792*, Extra Publication No. 5 (London: Newcomen Society, 1950); J. Farey, *A Treatise on the Steam Engine* (London, 1827); D. S. L. Cardwell,

Technology, Science and History (London: Heinemann, 1972); as *Turning Points in Western Technology* (New York: Science History Publications, 1972).

6 R. L. Hills, *Power in the Industrial Revolution* (Manchester: Manchester University Press, 1970); R. S. Fitton, *The Arkwrights, Spinners of Fortune* (Manchester: Manchester University Press, 1989).

7 A. and N. L. Clow, *The Chemical Revolution* (London: Batchworth Press, 1952).

8 H. Malet, *Bridgewater: the Canal Duke, 1736–1803* (Manchester: Manchester University Press, 1977); C. T. G. Boucher, *James Brindley, Engineer* (Norwich, 1968); A. Rees, *Cyclopaedia* (1819) art. 'Canals'.

9 D. Hudson and K. W. Luckhurst, *The Royal Society of Arts, 1754–1954* (London: John Murray, 1954); R. E. Schofield, *The Lunar Society of Birmingham* (Oxford: OUP, 1963).

CHAPTER 7

H. L. Beales, *The Industrial Revolution, 1750–1850* (reprint, New York: A. M. Kelley, 1967).

Phyllis Deane, *The First Industrial Revolution* 2nd edn, (Cambridge: CUP, 1979).

D. S. Landes, *The Unbound Prometheus* (Cambridge: CUP, 1969).

P. Mathias, *The Transformation of England* (London: Methuen, 1979).

1 J. le R. d'Alembert, *Discours Preliminaires de l'Encyclopédie* (Paris, 1929).
2 I. B. Cohen, *Franklin and Newton* (Philadelphia: American Philosophical Society, 1956).
3 A. Raistrick, *Dynasty of Iron Founders: the Darbys and Coalbrookedale* (London & New York: Longmans Green, 1953).
4 E. Robinson and D. McKie, *Partners in Science: Letters of James Watt and Joseph Black* (London: Constable, 1970).
5 J. P. Muirhead, *The Life of James Watt* (London, 1859); *The Origin and Progress of the Mechanical Inventions of James Watt*, ed. J. P. Muirhead (3 vols, London, 1854); H. W. Dickinson and Rhys Jenkins, *James Watt and the Steam-Engine* (London, 1927); H. W. Dickinson, *James Watt: Craftsman and Engineer* (Cambridge: CUP, 1935) and *Matthew Boulton* (Cambridge: CUP, 1936); Jennifer Tann, *The Selected Papers of Boulton and Watt*, vol. 1: *the Engine Partnership, 1775–1825* (Cambridge, Mass: MIT Press, 1981); D. S. L. Cardwell, *Steam Power in the Eighteenth Century* (London: Sheed & Ward, 1963); D. S. L. Cardwell, *From Watt to Clausius*, 2nd ed (Ames: Iowa State University Press, 1989). Dr

R. L. Hills is preparing a new biography of Watt.
6 R. L. Hills, *Power in the Industrial Revolution, op. cit.*
7 E. Roll, *An Early Experiment in Industrial Organisation, Being a History of the Firm of Boulton and Watt, 1775–1805* (London: Longmans Green, 1930); J. Lord, *Capital and Steam Power, 1750–1800*, 2nd edn with an introduction by W. H. Chaloner (London: Frank Cass, 1966); E. Robinson and A. E. Musson, *James Watt and the Steam Revolution* (London: Adams and Dart, 1969).
8 J. Farey, *op. cit.*
9 A. J. Pacey, *The Maze of Ingenuity* (London: Allen Lane, 1974); R. S. Fitton and A. P. Wadsworth, *The Strutts and the Arkwrights, 1758–1830* (Manchester: Manchester University Press, 1973); see also R. S. Fitzgerald, 'The Development of the Cast Iron Frame in Textile Mills to 1850', *Industrial Archaeology Review*, **10**, No. 2 (Spring 1988). 127–45.
10 R. S. Fitton and A. P. Wadsworth, *op. cit.*
11 E. L. Kemp, 'Thomas Paine and his "Pontifical" Matters', *Transactions of the Newcomen Society*, **49** (1977–8): 21–43.
12 C. K. Hyde, *op. cit.*; C. W. Roberts, *A Legacy from Victorian Enterprise*

(Gloucester: Allen Sutton, 1983); A. Birch, *The Economic History of the British Iron and Steel Industry 1784–1879* (London: Frank Cass, 1967).

CHAPTER 8

1 J. R. Partington, *A History of Chemistry*, vol 3 (London: Macmillan, 1962).
2 J. Petot, *op. cit.*, F. B. Artz, *A History of Technical Education in France, 1500–1850* (Society for the History of Technology, Chicago; and Cambridge, Mass.: MIT, Press, 1965).
3 D. M. Farrar, *op. cit.*
4 The point about state involvement is made by Lindqvist, *op. cit.*
5 J. McClellan, *Science Reorganized: Scientific Societies in the Eighteenth Century* (New York: Columbia University Press, 1985).
6 C. Aspin and S. D. Chapman, *James Hargreaves and the Spinning Jenny* (Helmshore: Local History Society, 1964).

13 J. G. A. Rhodin, 'Christofer Polhammar, Ennobled Polhem', *Transactions of the Newcomen Society*, **7** (1926–7): 17–22; S. Lindqvist. *op. cit.*

7 It is said that when Jacquard was asked to explain his invention to a committee, the chairman, who was Napoleon himself, began by asking if M. Jacquard claimed to be able to do what the Almighty Himself could not do: tie a knot in a taut string.
8 J. G. Smith, *The Origins and Development of the Heavy Chemical Industry in France* (Oxford: OUP, 1979).
9 A. and N. L. Clow, *op. cit.*
10 C. C. Gillispie, *The Montgolfier Brothers and the Invention of Aviation* (Princeton, NJ: Princeton University Press, 1983).
11 J. Farey, *op. cit.*
12 Hamilton, *op. cit.*; Heyman, *op. cit.*

CHAPTER 9

D. B. Barton, *The Cornish Beam Engine* (Exeter: Cornwall Books, 1989).
S. Smiles, *Lives of the Engineers* (London, 1874–79).
La Science Contemporaine, 1, Le XIX Siècle, ed. R. Taton (Paris: Presses Universitaires de France, 1961).

1 P. J. Booker, *A History of Engineering Drawing* (London: Chatto and

Windus, 1963).
2 For French science and technology during this

period see M. Crosland, *The Society of Arcueil* (London: Heinemann, 1967) and *Gay Lussac, Scientist and Bourgeois* (Cambridge: CUP, 1978); Robert Fox, *The Caloric Theory of Gases, from Lavoisier to Regnault* (Oxford: OUP, 1971); W. A. Smeaton, *Fourcroy, Chemist and Revolutionary* (London: privately published, 1962). These books have extensive bibliographies.

3 J. L. Pritchard, *Sir George Cayley, Inventor of the Aeroplane* (London: Max Parrish, 1961).

4 F. Trevithick, *Life of Richard Trevithick* (London, 1872). See also the modern biographies by H. W. Dickinson and A. F. Titley and by L. T. C. Rolt.

5 W. Fairbairn, *The Life of Sir William Fairbairn*, with an introduction by A. E. Musson (Newton Abbot: David & Charles, 1970); T. S. Reynolds, *Stronger than a Hundred Men: A History of the Vertical Water-Wheel* (Baltimore: Johns Hopkins University Press, 1983). See also article on water-wheels in Rees, *Cyclopaedia*.

6 L. T. C. Rolt, *Tools for the Job* (London: Batsford, 1965), published in the U.S. as *A Short History of Machine Tools*; R. S. Woodbury, *History of the Lathe to 1850* (1961), *History of the Gear-Cutting Machine* (1958), *History of the Milling Machine* (1959) (all Cambridge, Mass.: MIT Press).

7 G. S. Catterall, *Life and Work of Richard Roberts* (Manchester: unpublished M.Sc. thesis, UMIST 1975).

8 D. J. Jeremy, 'Technological Diffusion – the Case of the Differential Gear', *Industrial Archaeology Review*, **5**, No. 3 (Autumn 1981): 217–27.

9 F. Greenaway, *John Dalton and the Atom* (London: Heinemann, 1966); D. S. L. Cardwell, (ed.) *John Dalton and the Progress of Science* (Manchester: Manchester University Press, 1968).

10 R. A. Paxton, 'Menai Bridge (1818-1826) and its Influence on Suspension Bridge Development', *Transactions of the Newcomen Society*, **49** (1977–8): 87–110; D. Billington, *op. cit.*

CHAPTER 10

Among the many books on the history of railroads may be mentioned Hamilton Ellis, *British Railway History: an Outline from the Accession of William IV to the Nationalization of the Railways*, 2 vols (London: Allen & Unwin, 1954, 1959); A. W. Bruce, *The*

524 · Notes

Steam Locomotive in America (New York: W. W. Norton, 1952); F. J. G. Haut, *The History of the Electric Locomotive* (London: Allen & Unwin, 1969); authoritative and with extensive references: M. C. Duffy, *Strategic Technology: American Innovation and the Electric Railway*, 2 vols (Manchester: unpublished Ph.D. thesis, UMIST, 1989); R. S. Fitzgerald, *Liverpool Road Station, Manchester, an Architectural and Historical Survey* (Manchester: Manchester University Press and the Royal Commission on Historical Monuments, 1980).

1 J. Payen, *La Machine Locomotive en France, des Origines au Milieu du XIXe Siècle* (Paris: CNRS and Presses Universitaires de Lyon, 1986).

2 M. P. Crosland and C. W. Smith, 'The Transmission of Physics from France to Britain, 1800–1840', *Historical Studies in the Physical Sciences (1978)*, pp. 1–61; see also Fox, *The Caloric Theory of Gases, op cit.* and D. S. L. Cardwell, *James Joule* (Manchester: Manchester University Press, 1989), pp. 157–8 and *passim*.

3 D. S. L. Cardwell, *From Watt to Clausius, op. cit.*, pp. 196–238. Sadi Carnot, *Réflexions sur la Puissance Motrice du Feu*, edition critique par Robert Fox (Paris: J. Vrin, 1978) and English edition (Manchester: Manchester University Press, 1986).

4 By the end of the eighteenth century it had been shown that compressing air, or a gas, heats it up; expanding it

cools it down; cf. the puzzling phenomenon associated with Hoell's 'Heronic' engine.

5 E. Daub, 'The Regenerator Principle in the Stirling and Ericsson Hot Air Engines', *British Journal for the History of Science*, **7** (1974): 259–77. Professor Fox states that there is no evidence to show that Carnot knew about Stirling's engine (private communication).

6 W. Thomson, *Proceedings, Glasgow Philosophical Society*, **2** (1844–48): 169.

7 F. M. G. de Pambour, *A Practical Treatise on Locomotive Engines Upon Railways* (London, 1836).

8 E. W. Constant, 'Scientific Theory and Technological Testability, Dynamometers and Water Turbines in the 19th Century', *Transactions of the Newcomen Society*, **24**, No. 2 (1983): 183–98.

9 G. Hodkinson, *William Sturgeon, 1783–1850, His Life and Work to 1840* (Manchester: unpublished M.Sc. thesis, UMIST,

1979); see also J. P. Joule, 'A Short Account of the Life and Writings of the Late Mr William Sturgeon', *Manchester Memoirs*, **14** (1857): pp 53–83.

10 Among the many books about Faraday, L. Pearce Williams, *Michael Faraday* (London: Chapman & Hall, 1965) and G. N. Cantor, *Michael Faraday, Sandemanian and Scientist* (Basingstoke: Macmillan 1991) are commended. Faraday's *Experimental Researches* have been republished (New York: Dover Books, 1965).

Faraday's bicentenary in 1991 gave rise to a number of extraordinary claims made on his behalf in the British national press, by the BBC and by the Institution of Electrical Engineers. It was repeatedly stated that he invented the electric motor, which was quite untrue, that he invented the dynamo, which was partly true, and that all modern electrical technology can be traced back to him, which was palpable nonsense. Here it is enough to note that Faraday was a great man and that he was not an inventor; he was an experimental philosopher.

11 A. Ritter von Urbanitzky, *Electricity in the Service of Man*, transl. R. Wormell (London, 1890).

CHAPTER 11

1 W. V. Farrar in M. Crosland (ed.), *The Emergence of Science in Western Europe* (London: Macmillan, 1975); see also the article, 'Liebig' in the *Dictionary of Scientific Biography*; C. E. McClelland, *State, Society and University in Germany, 1700–1914* (Cambridge: CUP, 1980).

2 A. Pugsley, *The Works of Isambard Kingdom Brunel* (Cambridge: CUP, 1976).

3 N. Rosenberg and W. G. Vincenti, *The Britannia Tubular Bridge, the Generation and Diffusion of Technological Knowledge* (Boston, Mass.: MIT Press, 1978); D. Smith, 'Structural Model Testing and the Design of British Railway Bridges in the 19th Century', *Transactions of the Newcomen Society*, **48** (1976–77), 77–90.

4 G. Borrow, *Wild Wales* (London, 1862).

5 Pugsley, *op. cit.*; G. S. Emmerson, *The Greatest Iron Ship, S.S. 'Great Eastern'* (Newton Abbot: David and Charles, n.d.).

6 A. T. Crichton, 'William

and Robert Edmund Froude and the Evolution of the Ship Model Experimental Tank', *Transactions of the Newcomen Society*, **61** (1989–90): 33–49; see also T. Wright, in *Annals of Science*, (forthcoming).

7 D. A. Steineman and S. R. Watson, *Bridges and their Builders* (New York: Dover Books, 1957).

8 M. R. Smith, 'John H.Hall, Simeon North and the Milling Machine, the Nature of Innovation among Ante Bellum Arms Makers', *Technology and Culture*, **14**, No. 4 (Oct. 1973): 573–91; D. Hake, 'The Rise of the American System of Manufactures, 1800–1870', *Technology and Culture*, **21** (1980): 67–70.

9 L. J. Jones, 'John Ridley and the South Australian "Stripper" ', *History of Technology*, **5** (1980): 55–101.

10 Rev. Prof. H. Moseley, E. Hodgkinson and J. Enys, 'Report on the Construction of a Constant Indicator for Steam-Engines', *British Association Report, 1841* (Plymouth); and *BA Reports, 1842*.

11 The distance travelled on the disc by the little wheel in unit time is the product of the angular velocity of the disc multiplied by the distance between the centre of the disc and the little wheel.

CHAPTER 12

N. Rosenberg, *Perspectives on Technology* (Cambridge: CUP, 1976).

1 D. S. L. Cardwell, 'Science and Technology: the Work of James Prescott Joule', *Technology and Culture*, **17**, No. 4 (1976): 674–87. The power of a steam-engine is limited by the rate at which the boiler can supply steam; the power of a hydraulic engine is set by the quantity of water flowing per minute. But the attractive force exerted by a moving electromagnet was believed to be quite independent of its speed.

2 *The Catalogue of the International Exhibition of 1851* (London).

3 Society of Arts, 2 vols, *Lectures on the Results of the Great Exhibition* (London, 1852).

4 *The American System of Manufactures* (The Report of the Committee on the Machinery of the United States, 1955, and the special reports of George Wallis and Joseph

Whitworth, 1854, edited with an introduction by Nathan Rosenberg for the University Press) (Edinburgh: Edinburgh University Press, 1969); D. Hake, 'The Rise of the American System of Manufactures', *Technology and Culture*, **21**, No. 3 (1980): 67–70.

5 W. K. V. Gale, 'The Bessemer Steelmaking Process', *Transactions of the Newcomen Society*, **46** (1972–3): 17–26.

6 E. T. Layton, 'Scientific Technology: the Hydraulic Turbine and the Origins of American Industrial Research, 1845–1900', *Technology and Culture*, **20**, No. 1 (1979): 64–89.

7 The familiar rotary lawn sprinkler is a simple example of a reaction turbine. It would be far more efficient as a source of power if the energy that throws the water over the lawn were used to drive the spindle, coupled to a load, round. But then it would be useless as a lawn sprinkler!

8 M. R. Fox, *Dye-makers of Great Britain, 1856–1976* (Manchester: Imperial Chemical Industries, 1987); L. F. Haber, *The Rise of the Chemical Industry in the Nineteenth Century* (Oxford: OUP, 1958).

9 Holmes' original paper, together with those of Harvey, Lister, Simpson etc., is printed in C. N. B. Camac, *Classics of Medicine and Surgery* (New York: Dover Books, 1959).

CHAPTER 13

C. W. Smith and M. N. Wise, *Energy and Empire: A Biographical Study of Lord Kelvin* (Cambridge: CUP, 1989).

S. G. Brush, *The Kind of Motion We Call Heat: A History of the Kinetic Theory of Gases in the Nineteenth Century*, 2 vols (Amsterdam: North Holland, 1976).

1 D. S. L. Cardwell, *James Joule, op. cit.*

2 K. Hutchison, 'W. J. M. Rankine and the Rise of Thermodynamics', *British Journal for the History of Science*, **14**, No. 46 (1981): 1–26. D. F. Channell, 'The Harmony of Theory and Practice: the Engineering Science of W. J. M. Rankine', *Technology and Culture*, **23**, No. 1 (1982): 39–52.

3 See *Proceedings of the Institution of Shipbuilders and Engineers in Scotland*, vol. 1 (1857–8), onwards; R. L. Hills, *Power from Steam, a History of the Stationary*

Steam-Engine (Cambridge: CUP, 1989).

4 D. S. L. Cardwell, *James Joule, op. cit.*, pp. 45, 219 and *passim*. Joule's discovery of the origin of the current from a magneto led him, among other things, to suggest that a sufficiently powerful magneto, driven by a steam engine, could be used to effect the electrical resistance welding of metals.

5 A. C. Lynch, 'History of the Electrical Units and Early Standards', *Institution of Electrical Engineers Proceedings*, vol. 132, Pt A, No. 8 (1985): 564–73.

6 *The Scientific Papers of J. Clerk Maxwell*, ed. W. D. Niven (Cambridge: CUP, 1890).

7 W. Johnson, 'Admiral John A. B. Dahlgren, (1803–1870), His Life, Times and Technical Work in the US Naval Ordnance', *International Journal of Impact Engineering*, 8, No. 4 (1989): 255–387. See also C. S. Peterson, *Admiral John A. Dahlgren, Father of US Naval*

Ordnance (New York, 1935).

8 The first sea-going armoured warship was France's *La Gloire*; she was closely followed by Britain's HMS *Warrior* (now restored and preserved) of 1861. Even before the *Monitor* had been laid down a British shipyard had built an ironclad warship for the Danish navy. The designer, Captain Cowper Coles, fitted the ship with two revolving turrets, each with two guns (this was to be the standard for the next forty years). The drawback of the sea-going turret ship was that with heavy turrets the ship rolled excessively in a seaway, unless the turret deck was low down, in which case she would be a very wet ship. The problem became tragically clear when, in 1870, the turret ship HMS *Captain* capsized in a gale in the Bay of Biscay. All the crew and the designer, Captain Coles, were lost.

CHAPTER 14

R. A. Buchanan, *The Engineers: A History of the Engineering Profession in Britain, 1750–1914* (London: Kingsley, 1989).

K. Gispen, *New Profession, Old Order: Engineers and German Society, 1815–1914* (Cambridge: CUP, 1989).

Thomas P. Hughes, *Networks of Power: Electrification in Western Society, 1880–1930* (Baltimore: Johns Hopkins University Press, 1983).

1 *Inventor and Entrepreneur; Recollections of Werner von Siemens*, 2nd edn (London: Lund Humphries, 1966).

2 Of the other Rankine books, the *Manual of the Steam-Engine* (1859) went through 17 editions to 1908, the *Manual of Civil Engineering* (1862) through 17 editions to 1911 and the *Manual of Machinery and Millwork* (1869) through 7 editions to 1893.

3 M. R. Fox, *op. cit.* and L. F. Haber, *op. cit.*

4 For accounts of J. B. Dancer and micro-photography, see W. Browning, 'John Benjamin Dancer, F.R.A.S., 1812–1887', *Manchester Memoirs*, **107** (1964–5), pp 115–42; H. Milligan, 'New Light on J. B. Dancer', *Manchester Memoirs*, **115** (1972–3): 80–8.

5 C. Babbage, *The Exposition of 1851* (London, 1851).

6 K. J. Barlow, *A History of Gas Engines. 1791–1900*, 2 vols (Manchester: unpublished Ph.D. Thesis, UMIST, 1979). Dr Barlow is now writing a definitive history of the gas engine.

7 A. Beau de Rochas, *Nouvelles Recherches* (Paris, 1862).

8 William Thomson had discusssed this possibility but had concluded that the best solution was to raise the temperature and pressure of the steam. He advocated superheating the steam.

9 R. L. Hills, *Power from Steam, op. cit.*

10 Eda F. Kranakis, 'The French Connection: Giffard's Injector and the Nature of Heat', *Technology and Culture*, **23**, No. 1 (1982): 3–38.

11 P. Dunsheath, *A History of Electrical Engineering* (London: Faber and Faber, 1962; New York: Pitmans, 1962); W. A. Atherton, *From Compass to Computer* (San Francisco: San Francisco Press, 1984).

12 Professor Hughes points out that Joule's discovery of the convertibility of electrical into other forms of energy, widely accepted by the early 1880s, was a fruitful influence on Edison. Thomas P. Hughes, *American Genesis: A Century of Invention and Technological Enthusiasm, 1870–1970* (New York: Viking, 1989).

13 Osborne Reynolds, 'The Transmission of Energy', the Cantor Lectures, *Journal of the Society of Arts*, **32** (1883): 973–9, 985–9, 995–9.

14 Is it possible that the great enthusiasm for electric power in the early years of the Soviet empire ('socialism plus electricity equals communism') owed something to Engels'

prophecy?

15 R. Diesel, *Theory and Construction of the Rational Heat Motor*, transl. Bryan Donkin (London, 1894).

16 The papers in *The Minutes of Proceedings of the Institution of Civil Engineers*, 1888 to 1901, contain full details of the discussions concerning the Rankine cycle, the entropy – temperature diagram and the Sankey diagram.

17 At the end of the century there was anxiety about an alleged decline of British engineering leadership: 'The danger which, as many thought, now threatened English engineering, lay in the more thorough education and superior mathematical knowledge of so many foreign engineers.' Comment by Mark Robinson in P. W. Willans, 'Steam Engine Trials', *Minutes and Proceedings of the Institution of Civil Engineers*, **114** (1893): 2.

18 Energy, we noted, is conserved and cannot be renewed. In *all* natural and man-made processes some energy becomes less available, or degraded. Entropy is, in (Sir) James Swinburne's words, the measure of this accrued waste.

CHAPTER 15

1 In the first part of the nineteenth century Manchester was the leading centre for the doctrines of *laissez faire* (cf. the 'Manchester school'); at the end of the century Birmingham pioneered municipal enterprise. See C. Gill and A. Briggs, *A History of Birmingham*, 2 vols (Oxford: OUP, 1952).

2 The 'eugenics' movement at the turn of the century came perilously close to advocating the race doctrines later put into practice by the Nazis. Francis Galton, a leading exponent, was described by the late Professor Medawar as a 'spiritual fascist' (*Times Literary Supplement*, 24 January 1975).

3 Wells' ideas and prophecies are reported in contemporary issues of *The Fortnightly Review* and other journals.

4 J. B. Rae, *The American Automobile Industry* (Boston, Mass.: Twayne Publishers, 1984).

5 H. Hertz, *Electric Waves*, transl. E. E. Jones (London and New York, 1893). Hertz ranks as one of the greatest experimentalists of the nineteenth century.

6 Reported in *Nature*, **64** (May–Oct., 1901): 636.
7 Among important papers dealing with the early history and prehistory of Hertzian telegraphy are J. O. Marsh and R. G. Roberts, 'David Edward Hughes, Inventor and Scientist', *Proceedings of the Institution of Electrical Engineers*, **126** No. 9, (1979): 929–35; D. W. Jordan 'The Adoption of Self Induction by Telegraphy, 1886–1889', *Annals of Science*, **39** (1982): 433–61; D. W. Jordan, 'D. E. Hughes, Self-Induction and Skin Effect', *Centaurus*, **26** (1982): 123–53; B. J. Hunt, ' "Practice vs. Theory", the British Electrical Debate, 1888–91', *Isis*, **74** (1983): 341–55; C. Susskind, *Popov and the Beginnings of Radiotelegraphy* (San Francisco Press, 1962); P. Rowlands, *Oliver Lodge* (Liverpool: Liverpool University Press, 1990). All these works have extensive references and bibliographies.
8 *British Association Report* (1874), Belfast.
9 W. A. Atherton, *op. cit.*
10 C. H. Gibbs-Smith, *Aviation, an Historical Survey* (London: HMSO, 1970).
11 A. Pollen, *The Great Gunnery Scandal: The Mystery of Jutland* (London: Collins, 1980); A. J. Marder, *From the Dreadnought to Scapa Flow, the Royal Navy in the Fisher Era, 1904–1919* (Oxford: OUP,1965); J. Thomson, 'On an Integrating Machine Having a New Kinematic Principle', *Proceedings of the Royal Society*, **24** (1876): 262–65.

CHAPTER 16

1 G. Breit and M. A. Tuve, letter to *Nature*, **116** (1925): 357, (following communication to *Terr. Mag.*, **70** (1925): 15–16); E. V. Appleton and M. A. F. Barrett, *Nature*, **115** (1925): 333–4; Breit and Tuve, *Physics Review*, **28** (1926): 554; E. V. Appleton and G. Builder, *Nature*, **127** (1931): 970.
2 S. S. Swords, *Technical History of the Beginnings of Radar* (London: Peter Peregrinus, 1986); R. A. Watson Watt, *Three Steps to Victory* (London: Odhams Press, 1958).
3 M. J. Lazarus, 'Electromagnetic Radiation: Megahertz to Gigahertz, a Tribute to Heinrich Rudolf Hertz and John Turton

Randall', *Procedings of the Institution of Electrical Engineers*, **133**, No. 2, Pt A (1986): 109–18.

4 The Prime Ministers of Australia and New Zealand had to spend two months at sea in order to attend Empire conferences in London. Airships, it was pointed out, could cut the travel time by at least half. The R-101 crashed in bad weather on 5 October 1930, on its first flight to India; 48 of the 54 people on board were killed. The sister ship, R-100, was scrapped and no more airships were built in Britain.

KLM, (Royal Dutch Airlines), was founded in 1920; Imperial Airways on 1 April 1924; DLH at the end of 1925. European airlines flew more than twice as many route miles . as airlines in the United States.

In 1925 German domestic airlines operated 39 national and inter-national routes; Imperial Airways operated six, with one pioneer route between Egypt and Iraq. German airlines, which were heavily subsidized, carried five times as many passengers as Imperial Airways. See *Memorandum of the Secretary of State for Air,* Imperial Conference 1926 (London: HMSO, 1926) and Air Ministry, *Report of Directorate of Civil Aviation* (London: HMSO, 1926).

5 J. B. Rae, *Climb to Greatness: The American Aircraft Industry, 1920–1960* (Cambridge, Mass.: MIT Press, 1968).

6 E. Morgan and E. Shacklady, *Spitfire; The History* (London: Key Publishing, 1990).

7 F. Whittle. *Jet: The Story of a Pioneer* (London: Muller, 1953).

8 P. and E. Morrison, *Charles Babbage, Selected Writings* (New York: Dover Books, 1961).

CHAPTER 17

Daniel J. Kevles, *The Physicists: The History of a Scientific Community in Modern America* (New York: Alfred A. Knopf, 1978).

1 H. G. Wells' novel was written at a time when it was plausible to believe that there were canals on Mars and that they were the product of a superior civilization.

2 Major-General W. Dornberger, *V 2,* transl. J. Clough and G. Halliday

(London: Hurst & Blackett, 1954).

3 Frederick L. Ordway III and W. von Braun, *A History of Rocketry and Space Travel* (New York, Thomas Crowell & Co, 1967); N. L. Johnson, *A Handbook of Soviet Lunar and Planetary Exploration* (San Diego: San Diego University, 1970).

4 Isaac Newton, *Mathematical Principles of Natural Philosophy*, transl. Andrew Motte (London: 1627).

5 Many substances that cannot be ignited in a normal atmosphere burn readily in pure oxygen. After this tragedy the oxygen atmosphere in capsules was modified by the addition of inert helium.

6 M. Born, *Atomic Physics*, 3rd edn (London and Glasgow: Blackie, 1944), p. 195.

7 H. D. Smyth, *A General Account of the Development of Methods of Using Atomic Energy for Military Purposes Under the Auspices of the United States Government, 1940–1945* (Washington DC: Government Printing Office, 1945). This is the official and authoritative history of the first two atomic bombs.

8 For a comprehensive account, from the German point of view, of the development of the pre-nuclear submarine, see E. Rossler *The U-Boat: The Evolution and Technical History of German Submarines*, transl. H. Frenberg (London and Melbourne: Arms and Armour Press, 1981).

9 K. Jay, *Calder Hall, The Story of Britain's First Atomic Power Station* (London: Methuen & Co., 1956).

10 W. S. Jevons, *The Coal Question, op. cit.*

CHAPTER 18

E. F. Schumacher, *Small is Beautiful* (London: Blond & Briggs, 1973).

1 Martha M. Truscott (ed.), *Dynamos and Virgins Revisited: Women and Technological Change in History* (Metuchen, NJ: Scarecrow Press, 1979); Anne Rowe, *The Development of Domestic Appliances, 1920–1935* (Manchester: unpublished M.Sc. thesis, UMIST, 1985); Ruth S. Cowan, *op. cit.*, Chapter 19.

2 There are many books, papers, etc., on the history and development of computers. For an early history of the electronic

computer, see B. V. Bowden (ed.), *Faster than Thought* (London: Pitman, 1953). More recent authoritative works are B. Randell, (ed.), *The Origins of Digital Computers* (New York: Springer Verlag, 1975); N. Metropolis, J. Howlett and G-C. Rota (eds), *A History of Computing in the Twentieth Century* (New York: Academic Press, 1980). In addition to 37 authoritative articles, this work also contains Professor Randell's supplementary bibliography. Reference should also be made to the publications of the Charles Babbage Institute, University of Minnesota, Minneapolis, MN 55455, USA. A detailed biography of A. M. Turing has been written by A. Hodges, *The Enigma of Intelligence* (London: Burnett Books, 1983; London: Unwin Paperbacks, 1985). For a general survey see: M. J. Mitchell, *The Evolution of the Electronic Computer* (Manchester: M.Sc. thesis, UMIST, 1976).

3 With binary arithmetic only two digits are used, 0 and 1, so that the decimal sequence 0, 1, 2, 3, 4, 5, . . . becomes 0, 1, 10, 11, 100, 101, 110 . . . The great advantage was that the 'on' and 'off' of an electric or electronic switch, or relay, could correspond to 1 and 0 in binary arithmetic.

4 British manpower policy was far more sensible than that of Hitlerite Germany. Britain had taken urgent and comprehensive steps to conserve scientific manpower. The lesson of 1914–18 had been well learned; scientists were not to be drafted into, or even allowed to volunteer for the armed forces. The Royal Society, on behalf of the government, had drawn up a full Scientific and Technical Register for the whole nation. It included undergraduate students as well as qualified scientists and technologists. Alan Turing would certainly have been on the Register. Scientists and technologists on it but of military age could, of course, be accepted for service in technical branches as well as in certain combat branches such as the Fleet Air Arm, which had top priority and in which a high rate of loss was expected.

5 S. H. Lavington, *A History of Manchester Computers* (Manchester: NCC Publications, 1975).

6 The late Lord (B. V.) Bowden, who once worked for the Ferranti

company, claimed to have been the first computer salesman in the world. He recounted that he had made this claim to uniqueness to a fellow-

passenger on a trip across the Atlantic. The latter, however, insisted that his job was even more unusual: *he* sold lighthouses.

CHAPTER 19

1 T. S. Kuhn, *The Structure of Scientific Revolutions* (Chicago: University of Chicago Press, 1964, 1970); Sir K. R. Popper, *The Logic of Scientific Discovery* (London: Hutchinson, 1959; New York: Basic Books, 1959); I. Lakatos and A. Musgrave, *Criticism and the Growth of Knowledge* (Cambridge: CUP, 1970) contains useful discussions of the various works of Kuhn, Popper and others; S. Toulmin, *Human Understanding* (Oxford: OUP, 1972). See also M. Polanyi, *Personal Knowledge* (London: Routledge & Kegan Paul, 1959; Chicago: University of Chicago Press, 1958). Popper has stressed that science students should be encouraged to be critical. If, by that, he means that they should not accept concepts and laws until they fully understand them, the advice is surely sound. But if he means that they should always be critical as a matter of

principle then the advice could lead to the stultification of science. The whole secret is knowing just when, how and where to be critical.

2 S. C. Gilfillan, *The Sociology of Invention* (Chicago: 1935; Cambridge. Mass.: MIT Press, 1972) and *Inventing the Ship* (Chicago: 1935).

3 J. Jewkes, D. Sawers and R. Stillerman, *The Sources of Invention* (London: Macmillan, 1969).

4 The author has personal knowledge of the circumstances and history of the invention of the swipe (by a retired stockbroker).

5 On this important point see Professor E. Layton's two papers, 'Mirror Image Twins: the Communities of Science and Technology in 19th Century America' *Technology and Culture*, **12** (1971): 562–80 and, 'Technology as Knowledge', *Technology and Culture*, **15**, No. 1 (1974): 31–41.

6 G. Basalla, *The Evolution of Technology* (Cambridge:

CUP, 1988).

7 R. L. Hills, *Transactions of the Newcomen Society* (1971–2): 63–77.

8 E. Shils, *The Torment of Secrecy* (Glencoe, Ill.: The Free Press, 1956).

9 A Madam Errani invented the first carburettor (1869); Patent No. 3174.

10 Ruth S. Cowan, *More Work for Mother; The Ironies of Household Technology from the Open Hearth to the Microwave* (New York: Basic Books, 1983).

11 It is said that Grace Hopper had read Ada Augusta Lovelace's papers on Babbage's engines.

12 N. Davey, *A History of Building Materials* (London: Phoenix House, 1961); Patricia Cusak, 'François Hennebique: The Specialist Organisation and the Success of Ferro-Concrete. 1892–1909', *Transactions of the Newcomen Society*, **56** (1984–5): 71–86; Billington, *op. cit.*

13 R. Fox and G. Weisz, *The Organisation of Science and Technology in France, 1808–1914* (Paris: Maisons des Sciences de l'Homme; and Cambridge: CUP, 1980); E. P. Hennock,

'Technological Education in England, 1850–1926: the Uses of a German Model', *History of Education*, **19**, No. 4 (1990): 299–331; W. Krohn, E. Layton and P. Weingart, *The Dynamics of Science and Technology: Social Values, Technical Norms and Social Criteria in the Development of Knowledge* (Dordrecht: Reidel, 1978); D. S. L. Cardwell, *The Organisation of Science in England*, 2nd edn (London: Heinemann, 1981); T. Shinn, 'The Genesis of French Industrial Research, 1880–1940', *Social Science Information*, **19**, No. 3 (1980): 607–40 and 'From "Corps" to "Profession": the Emergence and Definition of Industrial Engineering in modern France', in Fox and Weisz *op. cit.*; R. S. Turner, 'The Growth of Professorial Research in Prussia, 1818 to 1848 – Causes and Context', *Historical Studies in Physical Science*, **3** (1971): 137–82; P. Hoch, 'The Rise of Physics Laboratories in the Electrical Industry', *Physical Technology*, **16** (1985): 177–83.

SHORT GENERAL BIBLIOGRAPHY

R. A. Buchanan, *History and Industrial Civilisation* (London: Macmillan, 1979).

A. F. Burstall, *A History of Mechanical Engineering* (London: Faber & Faber, 1963).

M. Daumas, *Histoire Générale des Techniques*, 3 vols (Paris: Presses Universitaires de France, 1962). English translation by Eileen B. Hennessy, *History of Technology and Inventions: Progress Through the Ages*, vol. 1, *Origins of Technological Civilisation to 1458*, vol. 2, *1458–1725*, vol. 3, *1725–1860* (New York: Crown, 1979–80).

E. S. Ferguson, *Bibliography of the History of Technology* (Cambridge, Mass.: MIT Press, 1968).

R. J. Forbes and E. J. Dijksterhuis, *A History of Science and Technology*, 2 vols (London: Penguin Books, 1973).

B. Gille, (ed.), *Histoire des Techniques*, 2 vols (Paris: Editions Gallimard, 1978). Unfortunately the English translation (1986) leaves much to be desired.

F. Klemm, *A History of Western Technology* (London: Allen & Unwin, 1959).

M. Kranzberg and W. H. Davenport (eds), *Technology and Culture: An Anthology* (New York: Schocker Books, 1972).

M. Kranzberg and C. W. Pursell (eds), *Technology in Western Civilisation*, 2 vols (New York: Oxford University Press, 1967).

W. H. McNeill, *The Pursuit of Power: Technology, Armed Force and Society since AD 1000* (Oxford: Basil Blackwell, 1983).

J. Mokyr, *The Lever of Riches, Technological Creativity and Economic Progress* (Oxford: OUP, 1990).

J. Needham, *Science and Civilisation in China* (Cambridge: CUP, 1954 onwards).

J. R. Partington, *A History of Chemistry*, 4 vols (London: Macmillan, vol. 1 1960, vol. 2 1961, vol. 3 1962, vol 4. 1964).

C. Singer, E. J. Holmyard, A. R. Hall and T. I. Williams, *A History of Technology*, 7 vols (London: OUP, 1954–78).
T. K. Derry and T. I Williams, *A Short History of Technology* (London: OUP, 1960).

H. Straub, *A History of Civil Engineering* (London: L. Hill, 1960).

A. P. Usher, *A History of Mechanical Inventions* (Cambridge, Mass.: Harvard University Press, 1954).

A. Wolf, *A History of Science, Technology and Philosophy in the Sixteenth and Seventeenth Centuries*, and *A History of Science, Technology and Philosophy in the Eighteenth Century* (both London: Allen & Unwin, 1935, 1962).

Encyclopaedia Britannica (from 1754).

A Rees, *Cyclopaedia*, (1819).

Technology and Culture.

Transactions of the Newcomen Society.

The Open University, Milton Keynes, MK7 6AA, UK, has published excellent monographs on economic history, the history of technology and the history of science.

INDEX

560 · *Index*